Astronavigation

K.A. Zischka

Astronavigation

A Method for Determining Exact Position by the Stars

Springer

K.A. Zischka
Sparrows Point, MD
USA

ISBN 978-3-319-47993-4   ISBN 978-3-319-47994-1 (eBook)
DOI 10.1007/978-3-319-47994-1

Library of Congress Control Number: 2016955314

© Springer International Publishing AG 2018
This work is subject to copyright. All rights are reserved by the Publisher, whether the whole or part of the material is concerned, specifically the rights of translation, reprinting, reuse of illustrations, recitation, broadcasting, reproduction on microfilms or in any other physical way, and transmission or information storage and retrieval, electronic adaptation, computer software, or by similar or dissimilar methodology now known or hereafter developed.
The use of general descriptive names, registered names, trademarks, service marks, etc. in this publication does not imply, even in the absence of a specific statement, that such names are exempt from the relevant protective laws and regulations and therefore free for general use.
The publisher, the authors and the editors are safe to assume that the advice and information in this book are believed to be true and accurate at the date of publication. Neither the publisher nor the authors or the editors give a warranty, express or implied, with respect to the material contained herein or for any errors or omissions that may have been made.

Printed on acid-free paper

This Springer imprint is published by Springer Nature
The registered company is Springer International Publishing AG
The registered company address is: Gewerbestrasse 11, 6330 Cham, Switzerland

Preface

As a mathematician, I conceived the idea to write about the scientific part of navigation long before GPS became available to the public. From the outset, it was meant to be invariant with regard to the fast-changing technology available to navigators. It was also meant to be a manual that would make the navigator less dependent on the availability of other ephemerides. However, I did not want to turn the clock back by ignoring the state of developments in the fields of calculator and PC technology. Even the most casual observer must admit that we have achieved a level in our general education that renders anyone who does not know how to operate a personal computer illiterate. Today, the average student aged twelve or above should be able to handle the algebraic symbolic language as employed in all advances calculators without even understanding the underlying mathematics of the formulae used in navigation. I also wanted to draw a clear dividing line between the art and science of navigation (based on clear definitions, the important concept of stating the underlying assumptions, and the rigorous applications thereof) and the laws of physics as they apply to navigation.

I also wanted to show that the approximate methods used in celestial navigation, which are based on the original methods of Capt. Sumner and St. Hilair, are merely special cases of a general mathematical method that consists in approximating the two transcendental equations—the fundamental equations of navigation—by two linear equations.

A similar statement can be made with regard to the Lunar/Distance method for finding the approximation to the time at sea. This method is also an approximation to the problem of solving the Fundamental Equation with respect to the parameter of time. Therefore, it is also no longer necessary to look for the elusive error analysis of the Line of Position (LOP) method. Not knowing the error or the realistic upper bound for it (as a function of distance and azimuth from the true position) has always been a shortcoming for this method.

From a mathematical point of view, the distinction between Celestial Navigation and Astro Navigation is very clear. Celestial Navigation is an approximate method requiring an estimated position. Astro Navigation is an exact method not requiring an estimated position. Furthermore, in Celestial Navigation it is necessary to select the Celestial Objects very carefully to, one, avoid improper spacing and, two, choosing them too close to the zenith of the observer. In addition, the initial guess or Dead Reckoning Position (DRP or DR) needs to be sufficiently close to the true position. Therefore, in Celestial Navigation, all of the azimuths of the assumed triangles have to be sufficiently close to the true azimuths.

On the other hand, this does not apply to Astro Navigation (AN). There are very few cases where the methods of AN fail to work, as, for instance, in cases where the observed bodies are too close together or in such pathological cases where the given parameters are erroneous. (Those cases are referred to in future chapters as "Ill-Conditioned".)

Basically, this book is about anything pertaining to navigation that can be quantitatively expressed and ultimately computed. This book also does not rely on the use of any special calculator or PC. Nor does it depend on any other algorithms than the generally accepted mathematical ones. In theory, this amounts to saying that in the event where no calculator is available, a navigator can solve his positioning problem by using the provided formulae and a copy of the old, standby Logarithmic Tables.

I wanted to provide the reader with several independent methods for determining a position and other relevant data. I have also provided more than one method to give the navigator several alternatives for doing so. This also applies to choosing a suitable formula for calculating refraction and dip, which are considered by some to be the most limiting factors to accuracy in Celestial Navigation.

In addition to the above objectives, I have addressed several problems unique to navigation such as the problem of navigating without a sextant or even a clock. These problems have been addressed analytically and not just for emergency purposes but also to examine their underlying principles. All of the final formulae presented here are governed by a self-imposed rule for simplicity and comprehensibility and can be evaluated on any scientific calculator without the user even understanding their mathematical derivations. I would also like to stress that this book is not an exercise in spherical trigonometry (although it employs some basic spherical trigonometry equations sparingly).

In the first part of this book the Earth is assumed to be a sphere. Only in the section that deals with parallaxes is the Earth assumed to be a Spheroid of Revolution. The first chapter deals with the concept of Terrestrial Navigation and provides the rigorous formulae needed for Rhumb Line and Great Circle navigation. The chapter also covers the basics of the underlying mathematical projections on which navigational charts are based. In particular, it provides error analysis for approximating Mercator Plotting Sheets by the Non-Mercator Plotting Sheets that are sold commercially or are self-made. Furthermore, it also provides an analytic estimate for approximating a small area of the surface of the Earth by a plane Euclidian surface.

For centuries navigators have been using the Line of Position (LOP) method to avoid numerous complex mathematics and trigonometric calculations. In this book, I use the "Orientation Schematic" that enables a navigator to apply the exact formulae for determining his or her position at sea or in the air. In this context, the number of required trigonometric calculations in the Exact Method does not exceed the number of the same calculations required by the LOP method.

In addition to the derivations of the exact formulae, I have included other independent methods and, at least, one iterative method for determining position. In addition to the exact methods, I have covered approximate methods and the least square method of error analysis as applied to the non-exact formulae of navigation. However, it should be noted that this method becomes meaningless in cases where the data used is sufficiently erroneous.

In a separate and independent section of the first part of this book, I have devoted space to the error analysis of navigational data, specifically to the analysis of random errors, thereby providing navigators with a method for assessing their own proficiency with respect to measuring sextant altitudes simultaneously with the corresponding watch-time.

I have also attempted to dispel the notion that the Lunar Distance method is the only practical method for a sailor to determine time and therefore longitude at sea. I have provided the reader with an approximate iterative method for calculating time that is based solely on altitude and azimuth observations.

If the reader uses the first part of this book alone, he will have to depend fully on the availability of a current Nautical Almanac (NA). However, the second part of this book provides the reader with an option that replaces the NA.

Very little knowledge of astronomy is required to understand and use the first part of this book as a navigation manual and to use the provided formulae, the reader merely has to be familiar the trigonometric functions. Spherical trigonometric formulae have been kept to a bare minimum.

However, the second part of this book employs the basics of Positional Astronomy. It covers all the relevant topics as related to navigation and develops the formulae for precession, nutation, equation of the center, the equation of the equinox, the equation of time, the equation of the vertical, parallaxes, aberration and proper motion.

As I said, the main objective of this book is to provide the navigator with a comprehensive set of formulae that solely involve polynomials and trigonometric series that can be evaluated on any scientific calculator. With the help of those formulae, some additional data on star positions, and data on the perturbation of the Kepler motion of the Moon and planets, the navigator will no longer have to depend on the current NA. In cases where additional data is not available, he or she can still determine their position by relying on the data for the Sun and selective stars (provided herein). However, in the later cases, they will not be able to use the Moon and the planets for navigation.

Since the development of suitable formulae for an ephemeris was based on the motion of the Earth, Moon, and several planets about the Sun, or, more precisely, about their common center of masses, it was necessary to treat the Earth as a giant gyro that moves under the influence of the gravitational forces of the aforementioned bodies and describes a nearly elliptical orbit about the Sun. Therefore, in terms of Astro-Dynamics, we are dealing with a multi-body problem of a rigid body that can only be solved by means of employing approximations.

In terms of the underlying mathematics, we require the application of orthonormal transformations to the rigid coordinate system of the gyro Earth that, in turn, rotates about an inert system of coordinates that move along the orbit Sun–Earth. The problem of actually solving the equations that govern the motion of gyro Earth under the influence of said gravitational forces can be done only by employing approximations. Therefore, all formulae and data pertaining to the motion of the moon and the planets are approximations.

Of course, astronomers have been able to determine and predict the position of heavenly bodies without the use of explicit astro-mechanics a long time before Newton developed the concept of modern mechanics. However, the development of viable ephemerides that depend primarily on observations constitutes a tremendous task that requires years of effort on the part of many astronomers.

Sparrows Point, USA   K.A. Zischka

Contents

Part I Analytical Approach to Navigation

1 Terrestrial Navigation... 3
   1.1 On the Design of Conformal-Mercator and Non-conformal
         Charts and Plotting Sheets 3
   1.2 Rhumb-Line or Loxodrome Navigation 7
   1.3 Approximations of Loxodromes by Straight Lines
         on the Plotting Sheet 11
   1.4 Applications and Numerical Examples 14
   1.5 Gnomonic or Great-Circle Navigation 20
   1.6 Numerical Examples and More Chart Projections............. 24

2 Astro-navigation 29
   2.1 Lines of Position, Position Fix, Navigational Triangle
         and Fix by Computation.................................. 29
   2.2 Celestial Sphere, Equatorial and Horizon System
         of Coordinates, Navigational Triangle and the Ecliptic
         Coordinate System 34
   2.3 Conclusions and Numerical Examples 42
   2.4 The Use of the Exact Equations for Finding the Position at Sea
         or Air by Employing Two or More Altitude Measurements
         Together with the Corresponding Measurements of Time....... 44
   2.5 Conclusions and Numerical Examples 59
   2.6 An Exact Method Based on Cartesian Coordinates
         and Vector Representations............................... 63
   2.7 Numerical Examples and Conclusions 73
   2.8 On Approximate Solutions for Finding the Position at Sea
         or Air by Employing Two or More Altitude Observations 77
   2.9 An Approximate Method Based on Matrices and the Least
         Square Approximation 91
   2.10 Sumner's Line of Assumed Position Method as Scientific
         Method.. 94

	2.11	Numerical Example and Logarithmic Algorithm.	97
	2.12	How an Approximate Position at Sea or Air Can Be Found if an Approximate Value for the Azimuth or the Parallactic Angle Is Known in Addition to One Altitude.	103
	2.13	On the Effect of a Change in Time on the Altitude and Azimuth.	110
	2.14	How to Determine Latitude at Sea or Air Without the Use of a Clock	112
	2.15	On Calculating the Interval Between Meridian Passage and Maximum Altitude and Finding Approximate Longitude and Latitude of a Moving Vessel, and Longitude by Equal Altitudes.	116
	2.16	To Find Latitude by Observing Polaris When Exact UTC and Longitude or an Approximation Is Available.	125
	2.17	The Most Probable Position When Only One LOP and DRP Are Known.	128
	2.18	How to Calculate the Time of Rising and Setting of Celestial Objects and How to Use the Measured Time of These Phenomena to Find Longitude	133
	2.19	On the Identification of Stars and Planets.	139
	2.20	How to Navigate Without a Sextant.	147
	2.21	On Finding Time and Longitude at Sea, the Equation of Computed Time (ECT), and Being Completely Lost	149
3	**Methods for Reducing Measured Altitude to Apparent Altitude**. . . .	173
	3.1	Navigational Refraction that Includes Astronomical Refraction for Low Altitude Observations	173
	3.2	The Dip of the Horizon as a Function of Temperature and Pressure.	185
	3.3	Planetary Parallax and Semi-diameter of the Sun and Moon	191
	3.4	Time and Timekeeping.	196
	3.5	On the Minimization Procedure for the Random Errors in Determining Altitude and Time	200
4	**Some of the Instruments and Mathematics Used by the Navigator**	209
	4.1	Some of the Formulae and Mathematics Used by the Navigator.	209
	4.2	Some of the Instruments Used by the Navigator.	230

Part II Formulae and Algorithms of Positional Astronomy

5	**Elements of Astronomy as Used in Navigation**	241
	5.1	Some Basic Concepts Describing the Motion of the Earth Around the Sun	241

5.2	An Approximation to the Time of Transit of Aries at Greenwich and the Greenwich Hour Angle GHA of ♈	244
5.3	The Right Ascension of RA of the Mean Sun, Mean Longitude, Mean Anomaly, Longitude of Perigee, Longitude of Epoch and Kepler's Equation	245
5.4	The Equation of the Center, Equation of Time and True Longitude of the Sun	248
5.5	Numerical Examples and Other Concepts of Time	250
5.6	An Approximate Method for Finding the Eccentricity, the Longitude of the Perigee and the Epoch	253
5.7	Some Improved Formulae for the Equation of Time and Center	257

6 Qualitative Description: The Relevant Astronomical Phenomena ... 259

6.1	On the Change of the Elements of the Orbit with Time	259
6.2	The Concept of the Julian Date (JD) and Time Expressed by Julian Centuries (T)	260
6.3	The Elements of Our Orbit as a Function of the Time T Expressed by Polynomials	265
6.4	Qualitative Aspects of Precession and Nutation	266
6.5	The Concept of Proper Motion for Stars	268
6.6	Aberration	269
6.7	Annual Stellar Parallax, Definitions of Mean, True and Apparent Place of a Celestial Object	271

7 Quantitative Treatise of Those Phenomena ... 275

7.1	Effects of Precession on the RA and the Approximate Method of Declination	275
7.2	Rotational Transformations and Rigorous Formulae for Precession	277
7.3	Approximate Formulae for the RA Θ and Declination δ as the Result of Two Rotations Only	280
7.4	Effects of Nutation on the RA and Declination	282
7.5	Effects of Proper Motion on the RA and Declination δ	285
7.6	Effects of Aberration on the RA and Declination δ	287
7.7	Effects of Annual Parallax on the RA and Declination δ	290
7.8	Calculating the Apparent RA and Declination δ, and the Equation of the Equinox	291

8 Ephemerides ... 295

8.1	Low Accuracy Ephemeris for the Sun, a Numerical Example	295
8.2	Intermediate Accuracy Ephemeris for the Sun	297
8.3	Low Accuracy Ephemeris for the Stars	300
8.4	Intermediate Accuracy Ephemeris for the Stars	303

8.5	Compressed Low Accuracy Ephemeris for the Sun and Stars for the Years 2014±	306
8.6	The Earth Viewed as a Gyro	308

Appendix A: Condensed Catalogue for the 57 Navigational Stars and Polaris ... 315

Appendix B: Greek Alphabet ... 317

Appendix C: Star Charts ... 319

References ... 321

Index ... 325

About the Author

K.A. Zischka was born in 1933 in Germany. After graduating from a Teacher College in Leipzig, he went into teaching and studied Chemistry by correspondence at the Technische Hochschule Dresden. He then returned to studying full time Mathematics and Physics at the Friedrich Schiller University in Jena where, after four years, he advanced to the level of Vordiplom-pre Master's degree. Subsequently, he was admitted to the Johann Wolfgang Goethe University in Frankfurt where he graduated with a Diploma in Mathematics and Theoretical Physics—Master's Degree. He then worked at the Research Center of the AEG in Frankfurt and continued studying towards a Dr. of Science Degree at the Technische Hochschule in Darmstad.

After graduating with distinction, he became Special Lecturer and then Assistant Professor at the University of Saskatchewan in Saskatoon, Canada. Later, because of his interest in sailing, he accepted an appointment at the University of Windsor, Ontario where he was elevated to the rank of Full Professor.

During his stay at the U. of W., he taught Applied Mathematics and Mathematical Physics. While on sabbaticals, he held a Visiting Appointment as Professor at the University in Karlsruhe, Germany and also had an appointment at the D.I. of F.R. in Bombay (Mumbai), India. Later, he worked for 2 years at a university in West Africa.

While living in Canada, he became a sailor, a bush-pilot, canoeist and something of an adventurer. (He kayaked solo down the Abititi River into the Hudson Bay.) He is also an avid fisherman, hunter, and survivalist and has owned several sailboats and float planes.

For the past 20 years, he has also been traveling and living in various parts of Central America and studying Anthropology, Theoretical Medicine, and Architecture. He recently acquired a 41-foot Sea Wolf Ketch in which he plans to retire to Mexico.

Introduction

> *The diversity of the phenomena of nature is so great, and the treasures hidden in the heavens so rich, precisely in order that the human mind shall never be lacking in fresh nourishment.*
>
> Johannes Kepler (German Astronomer, 1571–1630) [33]

This book is based on the premise that the Earth, as viewed by a navigator, can be approximated by a sphere. A more accurate description would result if one adopts the concept of the Earth as a Geoid. However, for astronomical and navigation purposes it suffices to consider the planet to be a Spheroid of Revolution with its minor axis coincident with the North/South axis of the Celestial Sphere.

For the navigator, treating the Earth as a sphere has two great advantages: the first is that it conforms to the concept of the Celestial Sphere; and the second is that the corresponding geometry is fairly simple. (Spherical trigonometry is easier than Elliptical Geometry.)

The first part of this book treats the Earth as a sphere with the single exception of the derivation of formula for the parallax of the Moon (Sect. 3.3), where we have to take the oblateness of the Earth into consideration to distinguish between the astronomical and geocentric altitudes.

The role which accurate charts play in navigation cannot be overemphasized. Without accurate charts, the celestial coordinates become meaningless. Therefore, it is imperative that the navigator has a basic knowledge of Cartography (and, if possible, Geodesy). The making of accurate charts is based on mathematical projections. Because of that, the most relevant projections used in making nautical and aeronautical charts, as well as star finders, are discussed in various sections of this book.

In addition to having accurate charts at hand, the navigator should also be familiar with the accuracy of the plotting sheets. Plotting sheets that employ a longitude scale that is constantly proportional to the latitude scale for a wide range of latitudes (Non-Mercator plotting sheets) introduce an additional error in the plotting techniques that, by themselves, are based on the false premise that the azimuth of the assumed or Dead Reckoning position (DR) is always sufficiently close to the true azimuth. An analysis of Non-Mercator plotting sheets is provided in Sect. 1.3.

Because of the ramifications that the use of Non-Mercator plotting sheets have on the accuracy of the plotting techniques, the first part of this book also addresses the problems of flattening the Earth locally. For the purpose of navigation, elliptic or spherical geometry can be approximated by Euclidean geometry in the small. Practically speaking, this shows the size of an area on the surface of the Earth that can be approximated by an equal area on the tangent plane and depicted on a sheet of paper if drawn according to a suitable scale. In this case, great circles become straight lines and the sum of the angles of a corresponding spherical triangle equate to a hundred and eighty degrees.

The book also includes an elementary exercise for the interested reader to show clearly that despite knowing the geometry in the small, one cannot tell for sure what the geometry in the large—the true shape—might be.

Historically, it has been a long time since Bernard Riemann (1826–1866) conceived of spaces different from the Euclidean ones that result in geometries that, in the small, can be approximated by what we actually perceive. Another distinguished mathematician, Nickolai Lobachevsky (1792–1856) actually proved that in the infinitesimally small neighborhood of a point on his hyperbolic plane the trigonometric formulae of Euclidean geometry were also valid. Much later, the great physicist Albert Einstein (1879–1955) tried to convince us with his General Theory of Relativity that our real space must have a curvature, i.e., that it cannot be a Euclidean space.

In the 1830s, Lobachevsky considered the possibility of replacing Euclidean geometry with his own astronomical space but failed to convince the experts because of the absence of supporting experimental data since the vast distances required in his space were of the order of $10^7 \times a$ (where 'a' denotes the length of the semi-major axis of our solar system). The failure of providing experimental proof, of course, does not disprove his theory and is very similar in nature to the failure of proving that the Earth is a sphere by measuring the angles of a triangle on the surface of the planet and showing that the sum of the three angles is different from 180°. This experiment had been carried out rigorously by the distinguished mathematician K.F. Gauss (1777–1855).

As you begin to read this book, you might ask yourself: Why astronavigation and not Celestial Navigation? Are not they the same thing? The reasons for the distinction between the two are not based on etymological criteria but on astronomical and mathematical ones.

Astronavigation is an independent navigation system based on the observation of stars, the Sun (which is also a star), the planets and the Moon. It distinguishes itself from the concept of Celestial Navigation, as still practiced by the majority of navigators, by its independence on other navigational systems such as Dead Reckoning and intelligent guessing. Celestial Navigation as practiced by navigators around the world is merely a method that improves on an initial approximation to the true position of the ship or aircraft. Hence it is always assumed that the navigator more or less knows his position.

Celestial Navigation has been used all around the world for centuries. This book does not question its usefulness as an aid to navigation provided the user is familiar with its limitations. So far, no one has come up with an adequate error analysis. However, if the navigator already has a sufficiently close approximation to his true position, Celestial Navigation will provide an improved position provided the navigator follow certain well-established procedures for the selection of the heavenly bodies employed.

Currently, the availability of sophisticated programmable calculators has given Celestial Navigation a new image. For with such calculators, it is immaterial how often the user has to apply certain subroutines in their program. It is feasible to repeat the process of obtaining a "fix" using the LOP method by merely substituting the previously used DR or assumed position with the so obtained resulting fix and by pushing one more button, thereby obtaining a successive fix. Once the difference between the two successive results becomes negligible, the CN navigator has found his true position—perhaps so, but not necessarily true. In the end, we are back to experimental mathematics and nothing said above about Celestial Navigation has really changed. In Sect. 2.3, this book presents an iterative method that covers the method mentioned above. The criteria used by the proponents of said method merely constitute a necessary condition for the convergence, but not a sufficient one.

In astronavigation the determinations for a "fix" is obtained independently of the DR or assumed position. It requires only one more additional parameter in order to eliminate the ambiguity inherent with the exact solution. This parameter can be readily obtained by one more astronomical observation as, for instance, by noting the approximate azimuth of two celestial objects or by using an approximate latitude.

The observer obtains his position by using exact spherical trigonometric formulae. Any resulting errors due to errors in measuring altitudes, time, and parameters like refraction, SD and HP, can be evaluated analytically since the underlying functions are analytical ones. Whenever numerical methods are being employed their error bounds can be determined.

From a mathematical point of view, this book is based on the question of the solvability of a system of two transcendental equations, also known as Fundamental Equations of Navigation (FEN), since each of these two equations represents a circle of equal altitude on the celestial sphere with the celestial object in its center—solvability not only with regards to latitude and longitude, but also with regards to the variable of "time" which enters those equations through the "time dependency" of all the other parameters. This book also analyzes the problems of the existence, non-uniqueness, and ill-conditioning of solutions of FEN and also provides approximate methods for solving it.

A famous physicist once said, "Look at the equation. It tells you everything you want to know." This book shows that methods of celestial navigation turn out to be special cases of the mathematical methods of linearization and iteration. In this context, astronavigation can provide navigators with their approximate position anywhere on the globe without prior knowledge of other approximations. It can also provide the navigator with the approximate time of observation by using the Equation of Computed Time.

My motives for writing the second part of this book derive from a long search for a comprehensible, concise and compact manual on Positional Astronomy which included the necessary ephemerides for the Sun and stars which were normally at a navigator's disposal. By compact, I mean, something that a navigator could carry in the large cargo pocket of his or her pants. By comprehensive, I mean, something that would give the navigator the relevant equations at a glance and also enable him to review the underlying theory should he or she care to do so.

When this idea of designing a manual first came to me, GPS was not available. Navigators had to rely on bulky tables and almanacs. Today, a wide variety of electronic devices have freed navigators from the need for such bulky books and tables. However, as anyone who has spent any time at sea knows, salt water and electronics are not compatible. At one point or another, some or all of a person's electronic devises can and do fail at sea. Part II of this book covers such an emergency and should make the reader fairly independent of the software as used in celestial navigation.

But aside from having something to cover an emergency, those of us who are seriously interested in all aspects of navigation like to have something handy to review, if not even to learn some of the basics of positional astronomy as part of the knowledge necessary to understand ephemerides. Accordingly, the objective of the second part of this book is twofold: namely to provide the reader with a concise and comprehensible manual on positional astronomy as it applies to astronavigation; and to furnish him with the concise algorithms for finding the position of the sun and the 57 navigational stars at any given instant.

All the algorithms used in the second part of this book can be executed on any simple, inexpensive scientific calculator, thus freeing a navigator from being tied to a programmable calculator or a PC.

The formulae for the algorithms are either exact ones or approximations to the exact ones. Therefore, the resulting algorithms vary with respect to their degree of accuracy leading to the development of three types of ephemerides.

Type 1: Low precision ephemerides for the sun and navigational stars should yield numerical results $\pm 1'$.

Type 2: Intermediate precision ephemerides should provide results with an accuracy of $\pm 20''$.

Type 3: Compressed low precision ephemerides should come fairly close to the low precision ephemerides.

All the algorithms provided in this book can be executed without knowing the underlying theory and therefore, this manual can be compressed to a few pages that will easily fit into a shirt pocket.

Appendix A includes all the data needed for the ephemerides for the 57 navigational stars. It includes RA, δ, μ_α, μ_δ, and Π. Therefore, the manual holder is not dependent on the availability of other publications.

For the reader who is also interested in Astro-Dynamics, a brief section on the earth as a gyro has been included showing where some of the limitations of a strictly theoretical approach enter the quantitative analysis of the theory of orbits of rigid and non-rigid bodies.

I should like to disclaim this as a "complete" treatise on the discipline of astronavigation. It is, however, more complete than anything I am aware of or have found on the subject of Celestial Navigation to date.

Part I
Analytical Approach to Navigation

In which the navigator will find an analytic approach to navigation elevated to the level of a scientific discipline, eliminating the systematic error prevailing in celestial navigation with a multitude of practical algorithms for calculating the position of a craft at sea or in the air.

Chapter 1
Terrestrial Navigation

1.1 On the Design of Conformal-Mercator and Non-conformal Charts and Plotting Sheets

In order to navigate on a Rhumb-Line, i.e., a curve on the globe that intersects all the meridians at the same angle C, it is desirable to construct a map/chart that provides an accurate one to one correspondence between the coordinates of latitude and longitude on the position on the globe and the corresponding coordinates on the map. Mathematically, we are seeking the formulae which enable us to compute the coordinates on the map if the coordinates on the globe are given and conversely if the coordinates on the map are given to compute the coordinates on the globe. Our objective, therefore, consists in finding a mapping system that is conformal, i.e., that map an infinitesimal triangle on the globe on to a similar triangle on the chart preserving the angles and proportionality of the sides. In addition, we will also require that all curves that intersect the meridians at a constant angle get mapped on to straight lines on the corresponding chart.

It is obvious then that all meridians and circles of equal latitude get mapped as straight lines that intersect each other at right angles, i.e., that are perpendicular. Moreover, it is also obvious that such mappings do not preserve such particular features as distances and areas which will appear distorted on the chart. If we choose the image of meridians and circles of constant latitude as our grid on the chart that meet the above requirements, we obtain charts that are generally referred to as Mercator-Charts, after its inventor. It should also be noted that not every chart that employs a longitude/latitude grid with uneven spacing is necessarily a Mercator-Chart. In those cases, a straight line on such a chart or plotting sheet does not represent a Rhumb-Line on the globe. (Unfortunately, some plotting sheets that are not conformal are misleadingly called Mercator-Charts.)

In this chapter, we will examine the mathematical relationship between true Mercator-Charts (or true Mercator plotting sheets) and Non-Mercator plotting sheets thereby providing the navigator with expressions for the errors introduce by replacing a point on the Mercator plotting sheet with the corresponding point on the Non-Mercator plotting sheet thus allowing the navigator to compute the curve to be followed on the globe if following a straight line on the plotting sheet and conversely.

In this section, we must first interpret geometrically what has been said above and then express it quantitatively, i.e., mathematically. Geometrically speaking, Mercator conformal charts are based on the projection of the surface of the globe on to a cylinder of radius equal to the mean radius of the Earth and being tangent to the globe at a great circle. The center of the projection hereby is always the center of the Earth. The resulting charts are maps on the plane surface of the unrolled cylinder and are referred to as Oblique Mercator Projections or charts. At this point, we shall only consider the most important example in detail as used in navigation, namely the case where the tangent circle is the equator. Accordingly, the resulting Mercator-Charts are referred to as the Standard Mercator-Chart.

First, let us consider the common feature of these projections, namely the concept of conformal mappings.

Figure 1.1.1a depicts a small-infinitesimal rectangular triangle about a point $A \sim (\lambda_A, \varphi_A)$ on the globe with one side ΔX along the circle of constant latitude φ_A and one side ΔY along the adjacent meridian and perpendicular to ΔX. The third side connecting these two sides we denote by ΔD and the angle between this side and ΔX by $90° - C°$. This infinitesimal triangle gets mapped on the plane infinitesimal triangle with sides ΔX, ΔY, and ΔD preserving all the angles of the triangle on the globe (Fig. 1.1.1b).

Of course, for clarity, the actual size of the "infinitesimal" triangles is highly exaggerated in the above diagrams. At this point in our discussion, we need to introduce the scale factor n for the actual drawing on paper. We choose a scale factor n, meaning that one millimeter, 10^{-3} m, on the chart corresponds to n Nautical Miles (NM) on the globe. As customary in navigation, we measure

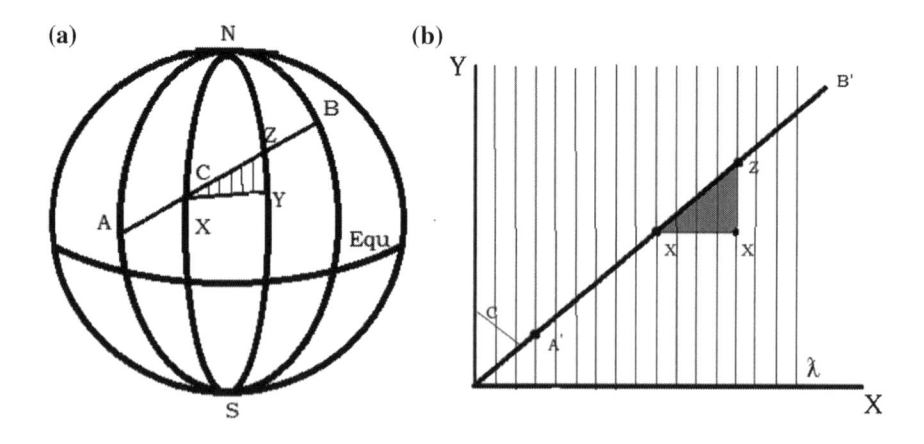

Fig. 1.1.1

1.1 On the Design of Conformal-Mercator ...

distances on the globe in terms of angular minutes of degree whereby one minute of circular measure subtends the distance of one nautical mile on the surface of the Earth.

It should be clear by now that the conformity of the mappings being used already assures the invariance of angles in those projections. However, the requirement that the Rhumb-Lines or Loxodromes get mapped as straight lines on the chart has not yet been imposed on the more general conformal mappings.

The actual mathematical requirements for the conformal mappings depicted in Fig. 1.1.1a, b are fairly simple and can be stated as follows:

1. a. $\Delta X = \alpha \Delta X = \cos \varphi \cdot \Delta \lambda$

    b. $\Delta Y = \alpha \Delta Y = \Delta \varphi$

where "α" denotes the so called conformal scaling factor (not to be confused with the constant scaling factor n). In general, α can be a function of λ or φ or both. Depending on the choice of α, we obtain one or the other type of conformal mappings.

In particular, if we choose the equator as the tangent circle we must require that $\Delta X = n^{-1} \cdot \Delta \lambda$, resulting in:

2. $\alpha = n \cdot \cos \varphi$

This choice for the conformal scaling factor also assures that any Rhumb-Line on the globe gets mapped as a straight line on the corresponding chart.

Choice #2, corresponding to the true Mercator-Chart, results in the mapping:

3. a. $X - X_A = n^{-1} \cdot (\lambda - \lambda_A)$

    b. $Y - Y_A = n^{-1} \cdot 1/\sin 1' \cdot \int_{\varphi A}^{\varphi B} \sec \varphi \cdot d\varphi$.

As the integral in (3b) can be evaluated by means of elementary calculus, it is appropriate to define a Rhumb-Line function as:

4. $R(\varphi) = 1/\sin 1' \cdot \int_0^{\varphi} \sec \varphi \, d\varphi = 1/\sin 1' \cdot \ln\left[\tan(\frac{\pi}{4} + \frac{\varphi}{2})\right] = 1/\sin 1' \cdot \ln\left[\tan(45° + \frac{\varphi}{2})\right]$

Again by using calculus, we can find the inverse $R^{-1}(z)$ of the Rhumb-Line function as:

5. $R^{-1}(z) = 2\left[\tan^{-1}(e^{\sin 1' \cdot z}) - \frac{\pi}{4}\right] = \varphi_r$ (in radian)

By employing the definition of the Rhumb-Line function, the formulae in (3) may be rewritten as follows:

6. a. $X - X_A = n^{-1} \cdot (\lambda - \lambda_A)$

    b. $Y - Y_A = n^{-1}[R(\varphi) - R(\varphi_A)]$

This mapping so defined not only maps curves, as for instance Loxodromes on to a flat sheet of paper (chart), but also maps entire regions of the globe on to flat surfaces. The calculations involved are simple and can even be solved by using logarithmic tables if necessary.

Formulae (6) enables us by using straight forward calculations to find the coordinate on the Mercator-Chart if the coordinates for λ and φ on the globe are given. (The assumption here is that λ_A, φ_A, and n are prescribed.) In the cases where x and y on the chart are given, you can find the coordinates on the globe by employing the inverse to $R(\varphi)$, formula (5), to obtain:

7. a. $\lambda - \lambda_A = n(X - X_A)$

   b. $\varphi = R^{-1}[n \cdot (Y - Y_A) + R(\varphi_A)]$

At this point, we have mathematically established the desired mapping (Mercator-Chart) as a means of depicting an area of the globe by a section of a flat map so that directions are preserved and the Rhumb-Lines on the globe are mapped on to straight lines on the chart. We have also shown that this is not possible by employing geometric concepts only. You have to use analytical ones.

Later we will devise approximations for simplifying the underlying calculations which will lead to simplified mappings with constant mapping factors and simplified geometric interpretations. In order to design a linear mapping system that approaches sufficiently close to the Mercator-Chart, we are going to have to take another look at formula (3) The right hand side of that equation consists of an integral which we may express for any fixed value of φ, denoted by φ_B, by means of the intermediate theorem of integral calculus.

By choosing the value of $\varphi = \varphi_B$ we have identified this value with the latitude of our endpoint $B \sim (\lambda_B \, \varphi_B)$—see Fig. 1.1.1a, b.

The application of the intermediate theorem of integral calculus yields:

$$Y_B - Y_A = n^{-1} \cdot 1/\sin 1' \cdot \int_{\varphi_A}^{\varphi_B} \sec \varphi \cdot d\varphi = n^{-1} \cdot \sec \underline{\varphi} \cdot (\varphi_B - \varphi_A); \quad \varphi_A \leq \underline{\varphi} \leq \varphi_B$$

The value $\underline{\varphi}$ is referred to as the mean-value of the above integral. From the definition of the above mean-value, we conclude that $\underline{\varphi}$ is given by:

8. $\sec \underline{\varphi} = \dfrac{R(\varphi_B) - R(\varphi_A)}{\varphi_B - \varphi_A} = \dfrac{1/\sin 1' \cdot \int_{\varphi_A}^{\varphi_B} \sec \varphi \, d\varphi}{\varphi_B - \varphi_A}$

Up to this point, everything has been mathematically exact. But this merely leaves us with the formulae for the images at the endpoints. However, formula (3) suggests that by applying the mean-value theorem, we may use the linear approximation:

9. a. $\tilde{X} - X_A = n^{-1} \cdot (\lambda - \lambda_A)$

   b. $\tilde{Y} - Y_A = n^{-1} \cdot \sec \psi \cdot (\varphi - \varphi_A); \quad \varphi_A \leq \psi \leq \varphi_B$

instead. (The value ψ is yet to be determined according to additional requirements.)

1.1 On the Design of Conformal-Mercator ...

Similarly we may also employ the linear mapping defined by:

10. a. $\tilde{X} - X_A = n^{-1} \cdot \cos\psi \cdot (\lambda - \lambda_A)$

    b. $\tilde{Y} - Y_A = n^{-1} \cdot (\varphi - \varphi_A)$

We have already seen what happens if we choose $\psi = \overline{\varphi}$, the mean-value. In this case the two end points are mapped exactly where the corresponding points on the Mercator-Chart are located. However, for any other values of λ and φ the images do not coincide with the images on the Mercator-Chart. Of course, the use of such linear approximations can only be justified if we can find bounds or estimates for the errors which we may have introduced in our calculations for finding the true values on the Mercator-Charts. (Sect. 1.3 addresses this problem in greater detail when we deal with the approximations of Loxodromes.)

In the next section, we will derive the equation of the Loxodromes and show how to navigate along one by computation.

1.2 Rhumb-Line or Loxodrome Navigation

Since the Rhumb-Line, or any other curve on the globe, constitutes a subset of our general mapping domain, we have to specify in formulae the subset of values of λ and φ, (Sect. 1.1 #1, the equation of the curve on the globe) in order to find the corresponding image on the Mercator-Chart or plotting sheet. This amounts to finding the representation of $\varphi = \varphi(\lambda)$ of the Loxodrome on the globe or, in other words, being able to assign for each value of the longitude λ the corresponding value of φ of the latitude.

Recalling that the Loxodrome is mathematically defined by the requirement that the angle C this curve forms with each meridian remains constant, we can deduce from Sect. 1.1 #1 that:

1. $\dfrac{dY}{dX} = \dfrac{dy}{dx} = \dfrac{d\varphi}{\cos\varphi \, d\lambda} = \cot c$

Integrating the last expression we find that: $\dfrac{1}{\sin 1'} \cdot \int_{\varphi_A}^{\varphi_B} \sec\varphi \, d\varphi = \cot C \cdot (\lambda - \lambda_A)$

and then by employing the definition of the Rhumb-Line function (Sect. 1.1 #4) we obtain the "Equation" of the Rhumb-Line or Loxodrome as:

2. $\mathbb{R}(\varphi) - \mathbb{R}(\varphi_A) = \cot C \cdot (\lambda - \lambda_A)$.

Although Eq. 2 is all we require for the mapping itself, it is still necessary to find the distance traveled on the Loxodrome between points $A \sim (\lambda_A \varphi_A)$ and $B \sim (\lambda_B \varphi_B)$ on the globe and/or the Mercator-Chart. Furthermore, the additional formulae provided herein will make it possible to find the coordinates of $B \sim (\lambda_B \varphi_B)$ if the course and distance $\mathbb{D}$ are given without employing the inverse $\mathbb{R}(\varphi)$ in (2).

With regard to the distances, it is necessary to distinguish between the actual distance $\mathbb{D}$ on the globe measured in Nautical Miles (or minutes of a degree) and the distance D between the image points of the Mercator-Chart. The distance $\mathbb{D}$ on the globe between A and B along the Loxodrome of angle-C can easily be found by referring again to Fig. 1.1.1a of Sect. 1.1. We can readily deduce that in the original triangle, the relation $\Delta Y / \Delta \mathbb{D} = \cos C$ holds and hence $\Delta \mathbb{D} = \sec C \cdot \cos \Delta Y = \sec C \cdot \Delta \varphi$

Integration of this relation yields the distance $\mathbb{D}$ as:

3. $\varphi_B - \varphi_A = \mathbb{D} \cdot \cos C$ (Note that the distance so defined is expressed in minutes of a degree = 1 NM.)

On the other hand, if we look at the corresponding triangle ΔX, ΔY, ΔD, and C on the Mercator-Chart (Fig. 1.1.1b, Sect. 1.1), we find that: $\dfrac{\Delta Y}{\Delta D} = \cos C$ holds, and therefore

$\Delta D = \sec C \cdot \Delta Y = n^{-1} \cdot \sec C \cdot \sec \varphi \cdot \Delta \varphi$

Integration yields: $D = n^{-1} \cdot \sec C \cdot [\mathbb{R}(\varphi_B) - \mathbb{R}(\varphi_A)]$ or

4. $\varphi_B - \varphi_A = n \cdot D \cdot \cos C \cdot \cos \underline{\varphi}$, where $\underline{\varphi}$ is defined by Sect. 1.1 #8.

It follows then from (3) and (4) that:

5. $\mathbb{D} = n \cdot D \cdot \cos \underline{\varphi}$.

At this point in our investigations it is advantageous to introduce another distance used in terrestrial navigation, namely the distance strictly traveled east or west when following a Loxodrome. This distance is usually referred to as the "DEPARTURE", denoted by "Dep." Mathematically speaking, departure is defined by the use of relation (1a). Then by integration of this relation we find that:

$X_B - X_A = \dfrac{1}{\sin 1'} \cdot \int_{\lambda_A}^{\lambda_B} \cos \varphi \cdot d\lambda = \text{Dep.}$

We can reduce this expression even further. From (1) we can deduce that:

$d\varphi = \cot C \cdot \cos \varphi \cdot d\lambda$. Therefore:

$\varphi_B - \varphi_A = \cot C \cdot \dfrac{1}{\sin 1'} \cdot \int_{\lambda_A}^{\lambda_B} \cos \varphi \cdot d\lambda = \cot C \cdot \text{Dep}$, or

6. $\text{Dep.} = \tan C \cdot (\varphi_B - \varphi_A)$.

We can further simplify this equation if we apply the mean-value theorem to the integral that defines the departure. Specifically, we have:

1.2 Rhumb-Line or Loxodrome Navigation

7. $\text{Dep} = \frac{1}{\sin 1'} \cdot \int_{\lambda_A}^{\lambda_B} \cos\varphi \cdot d\lambda = \cos\bar{\bar{\varphi}} \cdot (\lambda_B - \lambda_A)$, where "$\bar{\bar{\varphi}}$" denotes another mean-value. However, this new mean-value turns out to be our original mean-value $\underline{\varphi}$, defined by (Sect. 1.1 #8.). For, on the Loxodrome, according to (2) we have:

$\cot C \cdot (\lambda_B - \lambda_A) = \mathbb{R}(\lambda_B) - \mathbb{R}(\lambda_A) = \sec\underline{\varphi} \, (\varphi_B - \varphi_A)$, and according to (6): $\cos\bar{\bar{\varphi}} \cdot (\lambda_B - \lambda_A) = \tan C \cdot (\varphi_B - \varphi_A)$. Eliminating $\lambda_B - \lambda_A$, we deduce that: $\sec\underline{\varphi} = \sec\bar{\bar{\varphi}}$ and hence:

8. $\bar{\bar{\varphi}} = \underline{\varphi}$, as was to be shown.

Employing (8) in relation to (7) we find the simplified expression:

9. $\text{Dep} = \cos\underline{\varphi} \cdot (\lambda_B - \lambda_A)$.

And finally by substituting (8) into (6) we obtain:

10. $\text{Dep} = \mathbb{D} \cdot \sin C$.

Of course we will not require all ten expressions for navigation. Indeed, we will only need five expressions for exact navigation and as little as three for the approximate Rhumb-Line navigation.

The necessary algorithm for Loxodrome navigation can be stated as follows:

11. a. $\text{Dep} = \mathbb{D} \cdot \sin C$.
    b. $\varphi_B - \varphi_A = \mathbb{D} \cdot \cos C$.
    c. $\text{Dep} = \cos\underline{\varphi} \cdot (\lambda_B - \lambda_A)$
    d. $\cos\underline{\varphi} = (\varphi_B - \varphi_A)/(\mathbb{R}(\varphi_B) - \mathbb{R}(\varphi_A))$
    e. $\mathbb{R}(\varphi) = \frac{1}{\sin 1'} \cdot \ln[\tan(45 + \frac{\varphi}{2})]$ in min.°

Next we find the corresponding values X and Y on the Mercator-Chart. According to (6) we have:

$X - X_A = n^{-1} \cdot (\lambda - \lambda_A)$
$Y - Y_A = n^{-1} \cdot [\mathbb{R}(\varphi) - \mathbb{R}(\varphi_A)] = n^{-1} \cdot \cot C(\lambda - \lambda_A)$.

The last expression is given by (2) Therefore we deduce that:

12. $Y - Y_A = \cot C \cdot (X - X_A)$ the equation of the straight line on the Mercator-Chart.

Again we deduce that $D = n^{-1} \cdot \mathbb{D} \cdot \sec\underline{\varphi}$ (which has already been shown—see #5.

Up to this point in our analysis, all the mathematical expressions are exact. The Rhumb-Line navigation defined by (11) above, is exact. However, because of Eq. 11e, the practical calculations prove somewhat difficult for a navigator whose calculator has died and has only tables at his disposal. Equations 11d. and 11e. can be rendered superfluous by means of a Mid-Latitude (Middle Latitude) approximation in lieu of the mean-latitude $\underline{\varphi}$. Hence, by defining a latitude φ by:

13. $\psi = \dfrac{\varphi_A + \varphi_B}{2}$, where it is assumed that $\psi \cong \underline{\varphi}$.

The algorithm defined by 11 then reduces to:

14. a. $\text{Dep} = \mathbb{D} \cdot \sin C$
   b. $\tilde{\varphi}_B - \varphi_A = \mathbb{D} \cdot \cos C$
   c. $\text{Dep} = \cos \psi (\tilde{\lambda}_B - \lambda_A)$
   d. $\tan C = \dfrac{\text{Dep}}{\tilde{\varphi}_B - \varphi_A}$

The geometric interpretation is so simple that it almost defies further explanations. In particular, it gives rise to one of the most simple and most accurate plotting sheets a navigator can use for plotting relatively short distances. It also allows him to solve the Mid-Latitude problem by simply using a compass, protractor and a straight-edge ruler, one that preferably has a millimeter scale on it (see Fig. 1.2.1).

The resulting errors at the end of point B can readily be evaluated by employing the formulae of (11) and (14) above. In particular we find that:

$$\tilde{\varphi}_B = \varphi_B \text{ and } \Delta \varphi_B = 0, \text{ and } \Delta \lambda_B = \mathbb{D} \cdot \sin C \left[\sec \underline{\varphi} - \sec \dfrac{(\varphi_A + \varphi_B)}{2} \right].$$

In the next section, we will see how the actual mapping from the plotting sheet on to the globe and from there on to the Mercator-Chart can be depicted. I will also furnish the reader with some error estimates for whenever linear approximations or plotting sheets are being used.

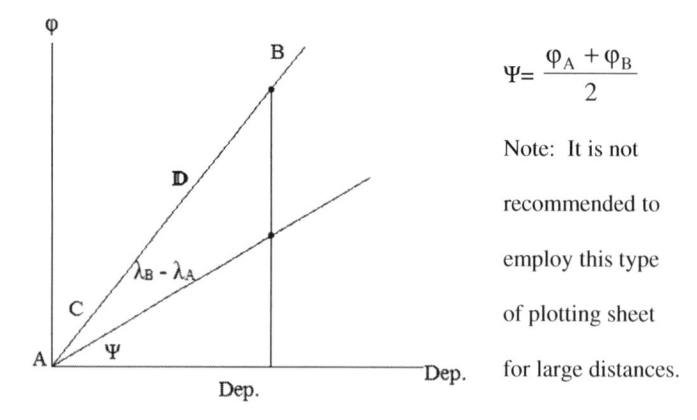

$\Psi = \dfrac{\varphi_A + \varphi_B}{2}$

Note: It is not recommended to employ this type of plotting sheet for large distances.

Fig. 1.2.1

1.3 Approximations of Loxodromes by Straight Lines on the Plotting Sheet

The objective in devising plotting sheets is always to map the meridians and circles of equal latitude in the vicinity of a point A with coordinates λ and φ (representing longitude and latitude respectively) on to a flat sheet of paper with a rectangular system of coordinates. This graticule consists of straight and perpendicular lines (see Fig. 1.3.1a).

Furthermore, it is always assumed that any straight line on such a plotting sheet represents a Loxodrome so closely that said line may actually represent a Line of Position (LOP). If that is the case, then, two such non-parallel lines will intersect and determine or "fix" the position of the vessel.

In practice, the navigator chooses one set of coordinates with a scale representing one minute of a degree of latitude (one Nautical Mile) on the globe. Having done so, the remaining coordinate axis representing the longitude will have to use a different scale.

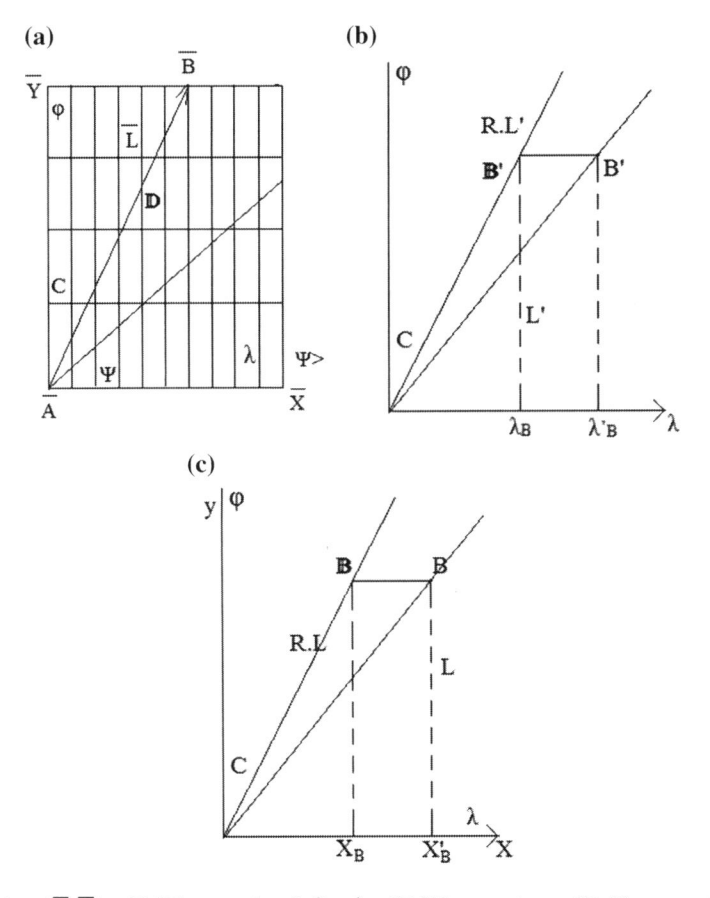

Fig. 1.3.1 **a** $(\overline{X}, \overline{Y})$ – PLOT—mapping. **b** (λ, φ) – PLOT—mapping. **c** (X, Y)—mercator chart

In this section, we will investigate the error resulting from the use of such plotting sheets. We will also show the actual position of the vessel on the globe when its position is fixed on a plotting sheet. In other words, if the navigator moves his or her pencil along a straight line on the plotting sheet and then reads off the latitude and longitude, we will be able to show the corresponding position on the globe or the Mercator-Chart.

In order to show this in detail, we will have to use three plotting sheets. Firstly, we will map our original plotting sheet (see Fig. 1.3.1a) with the straight line $\bar{L}$ on to an intermediate sheet that relates the longitude and latitude on the globe, but depicts it by using a rectangular coordinate system (see Fig. 1.3.1b). Secondly, we will then map the image L' of $\bar{L}$ together with the curve representing the Loxodrome on the coordinate system (λ, φ) on to the Mercator-Chart (see Fig. 1.3.1c).

Mathematically, all of this amounts to using the two mappings defined by:

1. $\bar{X} - \bar{X}_A = \cos\psi \cdot (\lambda - \lambda_A); \bar{Y} - \bar{Y}_A = \varphi - \varphi_A;$ and $\varphi_A \leq \psi \leq \varphi_B$
2. $X - X_A = n^{-1} \cdot (\lambda - \lambda_A); Y - Y_A = n^{-1} \cdot [\mathbb{R}(\varphi) - \mathbb{R}(\varphi_A)]$

The mapping defined by (1) of our plotting sheet (3a.) maps $\bar{L}$ on to L', and (2) maps L' and the Rhumb-Line RL' on to L and RL. respectively. Since the equation of $\bar{L}$ on $(\bar{X}, \bar{Y})$ is :

3. $\bar{L} : \bar{Y} - \bar{Y}_A = \cot C \cdot (\bar{X} - \bar{X}_A)$. Its image L' on (λ, φ) has the equation:
4. $L' : \varphi - \varphi_A = \cot C \cdot \cos\psi \cdot (\lambda - \lambda_A)$, and the equation for the Rhumb-Line RL' is according to (2) of Sect. 1.2:
5. $\mathbb{R}(\varphi) - \mathbb{R}(\varphi_A) = \cot C \cdot (\lambda - \lambda_A)$.

Next, the mapping defined by (2) maps L', given by (4), on the curve L on the Mercator-Chart and is given by:

6. $L : X - X_A = n^{-1} \cdot (\lambda - \lambda_A)$
   $Y - Y_A = n^{-1} \cdot \{\mathbb{R}[\varphi_A + \cot C \cdot \cos\psi \cdot (\lambda - \lambda_A)] - \mathbb{R}(\varphi_A)\}$

Similarly, the Rhumb-Line RL' is mapped on to the straight line RL, given by:

7. $RL : X - X_A = n^{-1} \cdot (\lambda - \lambda_A)$
   $Y - Y_A = n^{-1} \cdot \cot C \cdot (\lambda - \lambda_A)$

Hence, tracing a point $\bar{B}$ on the plotting sheet (see Fig. 1.3.1) its image B on the Mercator-Chart moves along the curve L relative to the Rhumb-Line. Although the relations (6) and (7) enable us to trace the error in approximating the Rhumb-Line

1.3 Approximations of Loxodromes by Straight Lines on the Plotting Sheet

by a straight line $\bar{L}$ on the plotting sheet as a function of the longitude λ, we still have not solved the problem stated at the beginning of this section. To solve that problem explicitly, we proceed as follows.

Firstly, let us fix a point $\bar{B}$ on our plotting sheet that is $\mathbb{D}$ Nautical Miles or minutes of a degree away from point A and lines on the line $\bar{L}$ (see Fig. 1.3.1a). Then we conclude that:

$\bar{X}_B - \bar{X}_A = \mathbb{D} \cdot \sin C$ and $\bar{Y}_B - \bar{Y}_A = \mathbb{D} \cdot \cos C$. Substituting these expressions into (7) yields:

8. $\varphi'_B - \varphi_A = \mathbb{D} \cdot \cos C$

9. $\lambda'_B - \lambda_A = \mathbb{D} \cdot \sin C$

Next we specify a point B' on the RL' that is also $\mathbb{D}$ minutes of a degree away from A. According to formula (3) of Sect. 1.2, the latitude of this point also satisfies the relation $\varphi_B - \varphi_A = \mathbb{D} \cdot \cos C$. Hence we may conclude that:

10. $\varphi_B = \varphi'_B$, i.e., $\Delta \varphi_B = 0$.

By using this result in Eq. 5. we find that:

11. $\lambda_B - \lambda_A = \tan C \cdot [\mathbb{R}(\varphi'_B) - \mathbb{R}(\varphi_A)] = \mathbb{D} \cdot \sin C \cdot \cos \underline{\varphi}^{-1}$.

Therefore the coordinates λ_B, φ_B of point B' on the Rhumb-Line that correspond to the point $\mathbb{B}$ on the Mercator-Chart have also been found. It follows that the error in longitude $\Delta \lambda_B$ at B' is given by:

12. $\Delta \lambda_B = \lambda'_B - \lambda_B = \mathbb{D} \cdot \sin C \cdot (\sec \psi - \sec \underline{\varphi})$.

In cases where we specify ψ as follows, we obtain the following types of plotting sheets:

13. a. $\psi = \underline{\varphi}$, the exact, i.e. the Mercator plotting sheet results.
    b. $\psi = \dfrac{\varphi_A + \varphi_B}{2}$, the Mid–Latitude plotting sheet results.
    c. $\psi = \varphi_A$, the most widely used commercial plotting sheet results

Similarly, to the derivation of (10) and (12), we also derive from (6) and (7) the expressions for the errors ΔX_B and ΔY_B. For we deduce readily for the values of X_B and Y_B on the curve L that:

14. $X_B - X_A = n^{-1} \cdot \mathbb{D} \cdot \sin C \cdot \sec \psi$
    $Y_B - Y_A = n^{-1} \cdot \mathbb{D} \cdot \cos C \cdot \sec \underline{\varphi}$.

For the values $X_B^{\bar{R}}$ and $Y_B^{\bar{R}}$ on the Rhumb-Line, we find that:

15. $X_B^R - X_A^R = n^{-1} \cdot \mathbb{D} \cdot \sin C \cdot \sec \underline{\varphi}$
    $Y_B^R - Y_A^R = n^{-1} \cdot \mathbb{D} \cdot \cos C \cdot \sec \underline{\varphi}$

It follows then from (14) and (15) that:

16. $\Delta X_B = X_B - X_B^R = n^{-1} \cdot \mathbb{D} \cdot \sin C \cdot (\sec \psi - \sec \underline{\varphi})$
    $\Delta Y_B = Y_B - Y_B^R = 0$

Admittedly, the errors due to the use of commercial or homemade plotting sheets are relatively small in comparison to all the other errors inherent in Celestial Navigation. But they are definitely a limiting factor in the accuracy of that method.

Furthermore, if navigators require a higher degree of accuracy in terrestrial navigation, they would have to take the ellipticity of the Earth into account and distinguish between both geographic and geocentric coordinates. The latter implies that we will have to adjust the Rhumb-Line function (4) of Sect. 1.2. In that case, if "e" denotes the ellipticity defined by:

$e = \dfrac{a-b}{a} \cong \dfrac{1}{291}$ where "a" denotes the semi-major axis and "b" the semi-minor axis of the Earth, then the modified Rhumb-Line function is given by:

$$\tilde{R}(\varphi) = \tfrac{1}{\sin 1'} \cdot \left(\ln[\tan(45 + \tfrac{\varphi}{2})] - e \cdot \sin \varphi\right) \qquad [8]$$

In the next section, we will consider some numerical examples and I will also elaborate on what is referred to as "Flattening the Earth" mathematically. I will also present the missing derivation of the Transverse Mercator Projection.

1.4 Applications and Numerical Examples

In terrestrial navigation, the two most frequently encountered problems in Rhumb-Line navigation are:

A. Given the coordinates λ_A, φ_A of departure A of a voyage, the course C to be followed, and the distance $\mathbb{D}$ to be traveled along the Loxodrome C—find the coordinates λ_B, φ_B of the destination. And…

B. Given the coordinates λ_A, φ_A of a departure A, and the coordinates λ_B, φ_B of the point of destination B—find the course C, and the distance $\mathbb{D}$ along the Loxodrome to be followed.

Navigators have two options[1] for solving these problems:

(i) They can use formula (11) of Sect. 1.2. to find the exact solution. Or…
(ii) They can choose an approximate method as defined by the relations (14) of Sect. 1.2. with a specific value for ψ (see also Sect. 1.3 #13.).

---

[1] Not counting the option of using GPS.

1.4 Applications and Numerical Examples

In the case that they choose (ii), they may either use an approximate plotting sheet, compass, protractor, and straight edge, or they can use a trigonometric table or calculator to solve the relations Sect. 1.2 #14. In either case, they can then employ the formulae provided for the ensuing error of Sect. 1.3.

Next, let's consider some numerical examples that will illustrate the use of our formulae.

Example #1 This first example corresponds with problem A above and pertains to a flight I made in a single engine float plane over Lake Huron to Georgian Bay. With the following data provided: Way-Point A with $\lambda_A = -81°\ 25'$ W, and $\varphi_A = 45°\ 25'$ N, and C = 40 T, and $\mathbb{D}$ = 38 N.M. compute the λ_B and φ_B of Way-Point B.

Solution:

Using formula 11b. Section 1.2 yields the following—

$$\varphi_B = 45°.416666 + 38' \cdot \cos 40 = 45°.901828147$$

Next calculate $\mathbb{R}(\varphi_B)$, and $\mathbb{R}(\varphi_A)$ by employing 11e. Section 1.2 and $\dfrac{1}{\sin 1'} = 3437.749237$.

This results in: $\mathbb{R}(\varphi_A) = \dfrac{1}{\sin 1'} \cdot \ln[\tan(45 + 22.70833)] = 3065.425668$, and

$\mathbb{R}(\varphi_B) = \dfrac{1}{\sin 1'} \cdot \ln[\tan(45 + 22.95091407)] = 3107.075933$. Hence—

$\mathbb{R}(\varphi_B) - \mathbb{R}(\varphi_A) = 41.65026625 = 0°.694171104$, and

$\varphi_B - \varphi_A = 0°.48576214$.

Formula 11a. Section 1.2 yields: Dep. = $38' \cdot \sin 40 = 24'.422592917$, and expression 11d. Section 1.2 yields: $\cos \underline{\varphi} = 0.699772919$.

Finally, the application of formula 11c. Section 1.2 produces:

$$\lambda_B = -81°.41666 + 24'.422592977 \cdot 1.429035009 = -80°.83498699 =$$
$$= 80°\ 50'W.$$

Suppose I used a commercial plotting sheet with $\psi = \varphi_A$. In this case Formula (14), Sect. 1.2 yields the same value for φ and Dep. as above. However, in this case, according to Formula 11c. Section 1.2, the approximation of λ_B now turns out to be different, namely:

$$\tilde{\lambda}_B = -81°.41666 + 24.422592917 \cdot \sec 45.416666 =$$
$$= -80°.83678168.$$

Hence the errors are: $\Delta\lambda_B = -0'.108$ and $\Delta\varphi_B = 0$
(Note: that because of $\Psi = \varphi_A$, the displacement of B is to the left and not to the right as shown in Fig. 1.3.1 where it was assumed that $\Psi > \varphi_A$.)

Example #2 On one trip to Newfoundland, I flew over the Gulf of Saint Lawrence and used the following two way-points: A $\sim$ $-62°$ 12' W, 49°39' N and B $\sim$ $-55°49'$W, 50°20'N. I chose to calculate the distance $\mathbb{D}$ and course C by employing Formulae 11, Sect. 1.2.

Solution:

Calculate $\lambda_B - \lambda_A = 6°23' = 6°.38333$ and $\varphi_B - \varphi_A = 0°41' = 0°.68333$
Calculate $\mathbb{R}(\varphi)$ by using Formula 11e. Section 1.2, i.e.,

$$\mathbb{R}(\varphi_A) = \frac{1}{\sin 1'} \cdot \ln[\tan(45° + 24°.825)] = 3441.923282 \text{ and}$$

$$\mathbb{R}(\varphi_B) = \frac{1}{\sin 1'} \cdot \ln[\tan(45° + 25°.16666)] = 3505.698252, \text{ hence}$$

$$\mathbb{R}(\varphi_B) - \mathbb{R}(\varphi_A) = 63'.77496974 = 1°.06291.$$

By using the formula 11d. Section 1.2, we find that: $\cos \underline{\varphi} = 0.64288545$.

Next, we calculate the departure dep. by employing 11c. Section 1.2 and obtain:

Dep. $= 0.642885445 \cdot 6.3833 \cdot 60 = 246.224997$ NM

We can now calculate C by using formulae 11a. and 11b. Section 1.2 deducing that:

$$\tan C = \frac{246.224997}{41} = 6.005487733, \text{ or explicitly,}$$
$$C = 80°.54616818 = 80°32'.77$$

Finally by substituting this value in 11b. Section 1.2, we find that:

$$\mathbb{D} = \frac{41'}{\cos 80.54616818} = 249.6152022 \text{ NM}$$

In the case that the navigator used a plotting sheet with $\Psi = \varphi_A$, i.e., formula (14). Section 1.2, instead, the answer would have been:

$\widetilde{\text{Dep}}. = \cos 49.65 \cdot 6.38333 \cdot 60 = 247.9752853$, and...

$$\widetilde{C} = 80.6116999, \text{ and } \mathbb{D} = \frac{41'}{\cos 80.6116999} = 251.341768 \text{ NM}$$

1.4 Applications and Numerical Examples

(Note that in the above example the difference in latitudes φ_B and φ_A is less than one degree which accounts for the relative small error in $\tilde{C}$ and $\mathbb{D}$. However, in general, the errors are much more pronounced and suggest that such plotting sheets should be used for relatively small distances—less than 60 NM. Furthermore, in the above situation, I chose to follow the Loxodrome since it only deviated insignificantly from the Great Circle route.)

In the next section we will learn how to follow a Great-Circle route by computation. In particular, we will see how a Great-Circle route can be approximated by sections that, in turn, can be navigated by using Rhumb-Lines without significantly increasing the total distances. We will also encounter a different type for conformal projection, i.e., charts on which great circles appear as slightly deformed straight lines. Since the scientific part of cartography consists in devising suitable projections, and hence charts, this subject should be of interest to any serious navigator. Remember, the first accurate charts for seafaring were devised by the Captains of ocean-going vessels, Captains like F. Magellan, Cook, and Vancouver, to name a few. [14]

In this section, I have also referred to the publication entitled "Flattening the Earth" in which its author has quoted the formulae for the Transverse Mercator Projection, but has not provided its derivation. In passing from Sects. 1.4 to 1.5, I will give this mathematical derivation and also explain what the actual flattening of the Earth entails geometrically. [39]

In most references to this subject, the expression "flattening of the Earth" merely means projecting the surface of the Earth on a flat sheet of paper. This, of course, can be done in many different ways. However, for the mathematician, this can only mean approximating a small area of the surface of the Earth by a similar area on a tangent plane since it is geometrically impossible to flatten a piece of the surface of a sphere. Therefore, from the geometrical point of view, the question which arises is: what should the size of the area be so that the laws of Spherical Geometry can be approximated by the laws of Euclidean Geometry?

This question is more relevant than you, at first, might think. We don't have to go back very far in history to show that men used to believe that the Earth was flat and not a geoid despite the observations and teachings of men like Archimedes. (In fact a Flat Earth Society still exists, today.) Similarly, people still refuse to accept that we really do not know the 'real' geometry of our universe. The discoveries of scientists like Riemann, Einstein and Lobachevsky all point to a concept of 'curvature'. Admittedly, for the distances generally considered by astronomers, the concept of Euclidean Space suffices as a good approximation. (However, for anyone trying to explain the origin of the universe, it might first be necessary to explain the configuration of the space.)

In order to understand the concept of approximating the Spherical Geometry by the Euclidean Geometry, let us consider a small spherical triangle on the surface of the Earth with sides of about five to ten Nautical Miles long. The application of the COS-TH of Spherical Trigonometry yields:

$$\cos x = \cos y \cdot \cos z + \sin y \cdot \sin z \cdot \cos X,$$

where "x", "y", and "z" denote the sides of said triangle in minutes of a degree and "X" denotes the enclosed angle in degrees.

By converting the minutes in the radians and expressing the trigonometric functions in terms of their Taylor-Series expansion we obtain the relationship:

$$1 - \frac{x^2}{2}\sin^2 1' \cong (1 - \frac{z^2}{2}\sin^2 1') \cdot (1 - \frac{y^2}{2}\sin^2 1') + y \cdot z \cdot \cos X \cdot \sin^2 1',$$

where we have neglected all higher than second order terms.

By truncating again, higher than second order terms, we obtain the simple relation:
$x^2 = y^2 + z^2 - 2 \cdot y \cdot z \cdot \cos X$, where x, y, z are expressed in Nautical Miles and X in degrees.

Similarly, by invoking the SIN-TH of Spherical Trigonometry we find that:

$$\frac{\sin X}{\sin x} = \frac{\sin Y}{\sin y} = \frac{\sin Z}{\sin z}$$

We deduce:

$$\frac{\sin X}{x} = \frac{\sin Y}{y} = \frac{\sin Z}{z}$$

The first equation above is the Cosine Theorem (COS-TH.) and the last the set of equations which constitute the Sine Theorem (SIN-TH.) of Euclidean Geometry.

By noting that the maximum error in the above approximations is of the order of:

$$\frac{x^2}{2} \cdot \sin^2 1' = \frac{10^2}{2 \cdot 10^{-8}} \cong 10^{-6}$$

Therefore, we can conclude that for distances of less than 10 NM the error would be in the order of linear millimeters only. No measurable differences in those geometries can be detected by optical or mechanical means. What about using Laser techniques? On the whole, it would not be possible to verify that the sum of the three angles must satisfy the inequality: $X + Y + Z > 180°$. (This is one of the reasons that the great mathematician G.F. Gauss could not prove mathematically that the Earth is a geoid. Nor could Lobachevsky prove that his Non-Euclidean Geometry represents our true space that must have a curvature.)

Before concluding Sect. 1.4, let's take a look at the mathematical derivation for the Transverse Mercator Projection.

For certain applications, such as world maps and other special maps depicting large north-south sections of the globe, it is desirable to have charts that are based on a conformal cylindrical projection that do not have the same distortions along the latitude scale that the standard Mercator-Chart has. One of these projections is based on the cylindrical projection that utilized a meridian as a tangent circle instead of the equator (see Fig. 1.4.1a, b). This can readily be accomplished by a rotation of the cylinder or, mathematically speaking, by the use of another suitable coordinate system. However, everything said and expressed in mathematical terms

1.4 Applications and Numerical Examples

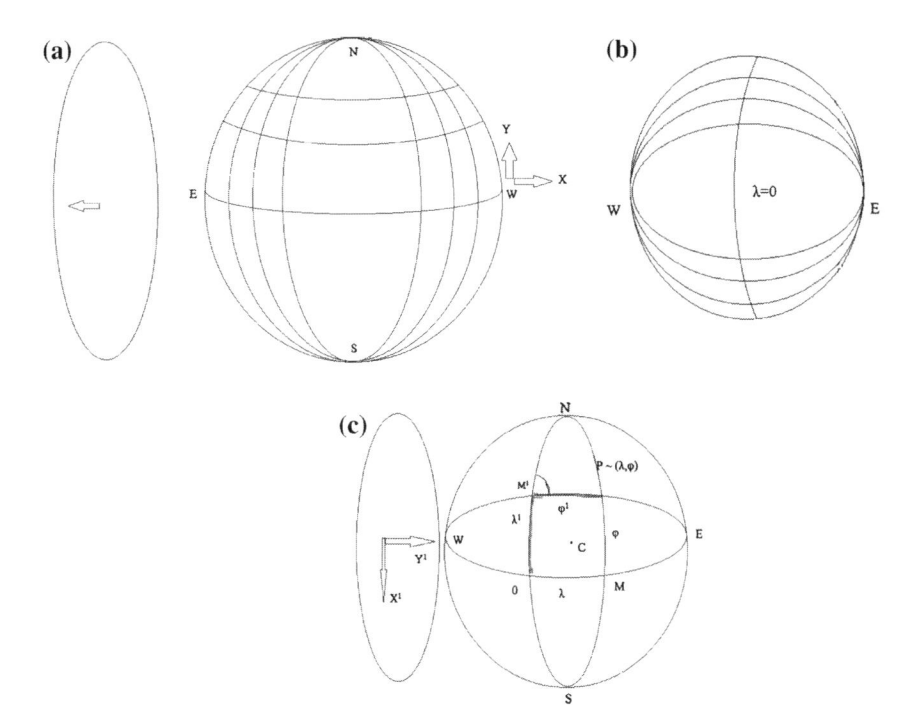

Fig. 1.4.1

about the standard Mercator Projection remains valid for the Transversal Mercator Projection if expressed in those new coordinates λ^1 and φ^1, i.e., the Eqs. 1. of Sect. 1.1. remain invariant. Therefore, we have again:

$$\Delta X = -\alpha' \cdot \Delta x' = \cos \varphi' \cdot \Delta \lambda'$$
$$\Delta Y = \alpha \cdot \Delta y' = \Delta \varphi', \text{ and also}$$
$$\alpha^1 = n \cdot \cos \varphi'.$$

Hence, again we have:

$$x' - x'_A = -n^{-1}(\lambda' - \lambda'_A)$$
$$y' - y'_A = n^{-1}[\mathbb{R}(\varphi') - \mathbb{R}(\varphi'_A)] = n^{-1} \cdot \frac{1}{\sin 1'} \cdot \ln[\tan(45° + \frac{\varphi'}{2})/\tan(45° + \frac{\varphi'_A}{2})]$$

By employing the trigonometric identity:

$$\tan^2(45 + \frac{\beta}{2}) = \frac{1 + \sin \beta}{1 - \sin \beta}, \text{ and the fact that } \ln x^n = n \cdot \ln x, \text{ we deduce that:}$$
$$x' - x'_A = -n^{-1}(\lambda' - \lambda'_A),$$

$$y' - y'_A = n^{-1} \cdot \frac{1}{\sin 1'} \cdot \frac{1}{2} \cdot \ln\left[\frac{1 + \sin \varphi'}{1 - \sin \varphi'} \bigg/ \frac{1 + \sin \varphi'_A}{1 - \sin \varphi'_A}\right]$$

By identifying the new coordinates λ^1, φ^1 in relation to our standard coordinates λ, φ we are able to deduce their mathematical relationship-Transformation by merely employing the SIN-TH. to the spherical triangles M'NP and MEP and by applying COS-TH. also to triangle MEP. Accordingly we have:

$$\frac{\sin \lambda}{\sin \varphi^1} = \frac{1}{\cos \varphi} \text{ or } \sin \varphi^1 = \sin \lambda \cdot \cos \varphi \text{ and}$$

$$\frac{\sin \lambda^1}{\sin \varphi} = \frac{1}{\cos \varphi^1} \text{ and } \cos \varphi = \sin \lambda \cdot \sin \varphi^1 + \cos \lambda \cdot \cos \varphi^1 \cdot \cos \lambda^1.$$

Substituting in the corresponding expressions yields:

$$\tan \lambda^1 = \sec \lambda \cdot \tan \varphi$$

Explicitly, we have found the Transverse Mercator Projection as:

$$x - x_A = y' - y'_A = n^{-1} \cdot \frac{1}{\sin 1'} \cdot \frac{1}{2} \cdot \ln\left[\left(\frac{1 + \sin \lambda \cdot \cos \varphi}{1 - \sin \lambda \cdot \cos \varphi}\right) / \left(\frac{1 + \sin \lambda_A \cdot \cos \varphi_A}{1 - \sin \lambda_A \cdot \cos \varphi_A}\right)\right],$$

$$y - y_A = -(x' - x'_A) = n^{-1} \cdot \tan^{-1}(\sec \lambda \cdot \tan \varphi) - n^{-1} \cdot \tan^{-1}(\sec \lambda_A \cdot \tan \varphi_A).$$

Now, let's take a look at the Great Circle Route and how to compute the aspects of following the shortest distance route between two distinct points on the globe.

1.5 Gnomonic or Great-Circle Navigation

For large distance navigation, it is important to choose the route from A to B so as to minimize the actual distance traveled. As we all know, the shortest distance between two points on a sphere is provided by a great circle that passes through those points. In reality, such a great circle is the intersection of the sphere by a plane passing through those two points and the center of the sphere.

Since any curve on our sphere can be described by a function representing one of the coordinates, say the latitude φ, in terms of the other coordinate, here the longitude λ, we can obtain a representation like: $\varphi = f(\lambda); \lambda_A \leq \lambda \leq \lambda_B; \lambda = f^{-1}(\varphi)$.

For example, in the case of the Rhumb-Line, we derived the equation of the Loxodrome (see formula 2 in Sect. 1.2). Again, the departure point A has the two coordinates, longitude λ_A and latitude φ_A. The destination point B has the two coordinates λ_B and φ_B. Then, the problems associated with great-circle navigation are:

1. Find the shortest distance D between A and B.
2. Calculate the initial course C at A.
3. Find the λ_M at which the maximum latitude occurs.
4. Calculate the maximum latitude for all $\lambda_A \leq \lambda \leq \lambda_B$.
5. Find the latitude φ for any given value λ, satisfying $\lambda_A \leq \lambda \leq \lambda_B$.

1.5 Gnomonic or Great-Circle Navigation

Please note that tasks #3 and #4 are very important because the navigator has to be sure that the great-circle route does not endanger the vessel in the vicinity of regions of ice.

Once a formula for $\varphi = f(\lambda)$, the equation of the great-circle route, has been found, the navigator can pick any point along this route by specifying λ and then calculating for φ, thereby establishing way-points along this route, which, in turn, are navigated by means of Rhumb-Line navigation. To physically chart this, we could use a chart that depicts any great-circle as a straight line. But remember, such charts are no longer based on conformal mapping and entail meridians that are no longer parallel. Such charts are referred to as GNOMONIC charts and are based on azimuthal projections of the points on the sphere from the center of the sphere on a tangent plane to the sphere at a fixed points, λ_0 and φ_0. (Gnomonic charts are commercially available, but for navigators who wishes to make their own plotting sheets or maps, the formulae for the necessary projections are provided without further proofs at the end of Sect. 1.6).

Besides the Gnomonic charts, a navigator also has access to other types of conformal charts which depict a great-circle by curves which are approximately straight lines. Such charts are known as Lambert Conformal Conic Projection Charts and are widely used by Aviators. These projections are made by projecting part of the sphere on a cone that intersects the sphere at two distinct circles of equal latitude. Since this projection is also conformal, distortions are minimal. The Rhumb-Lines appear as curves and great-circles are almost straight lines. The aviation navigator also uses flight computers and plotters suitable for position plotting based on linear approximations. Because of the versatility of Lambert Charts, the underlying mathematics of this type of projection is even more difficult than the derivation of the mathematical formulae of the Mercator Projection. (All the relevant mathematical formulae are provided in Sect. 1.6 without further proof.) [55], [2], [8], [16]

The current objective is to provide straight forward formulae which can be evaluated on any scientific calculator. Like the pocket calculator which is inexpensive and fits neatly into a shirt pocket, these formulae also do not occupy much space.

Let's go back to our task of deriving formulae for the solutions of the five problems listed above. First of all, let us depict a Great-Circle course on a sphere-globe.

In order to derive the required formulae it is necessary to employ some of the standard formulae of Spherical Trigonometry. (The formulae are listed in an Appendix to the first part of this book. However, for those not interested in the underlying mathematics, evaluating them on the calculator will suffice.)

First we apply the COS-TH. to the spherical triangle defined by N, (λ_1, φ_1), and (λ_2, φ_2) to find the distance D as:

1. $\cos D = \sin \varphi_1 \cdot \sin \varphi_2 + \cos \varphi_1 \cdot \cos \varphi_2 \cdot \cos(\lambda_2 - \lambda_1)$.

Similarly, we find that:

$\sin \varphi_2 = \sin \varphi_1 \cdot \cos D + \cos \varphi_1 \cdot \sin D \cdot \cos C$.

By solving cos C we discover:

2. $\cos C = \dfrac{\sin \varphi_2 - \sin \varphi_1 \cdot \cos D}{\cos \varphi_1 \cdot \sin D}$. Hence we have found the shortest distance D between A and B, and have calculated the initial course C at A.

Applying the COS-TH. to the triangle defined by $(\lambda_0, 0)$, $(\lambda_1, 0)$ and (λ_1, φ_1), we find:

(i) $\cos(\lambda_1 - \lambda_0) = \cos \varphi_1 \cdot \cos d + \sin \varphi_1 \cdot \sin d \cdot \cos C$, and also:
(ii) $\cos d = \cos \varphi_1 \cdot \cos(\lambda_1 - \lambda_0)$, from which it follows that:

$\dfrac{\cos(\lambda_1 - \lambda_0)}{\cos d} = \dfrac{1}{\cos \varphi_1}$, and also $\dfrac{\cos(\lambda_1 - \lambda_0)}{\cos d} = \cos \varphi_1 + \sin \varphi_1 \cdot \tan d \cdot \cos C$.

Hence: $\cos \varphi_1 + \sin \varphi_1 \cdot \tan d \cdot \cos C = \dfrac{1}{\cos \varphi_1}$, therefore we find that:

$\sin \varphi_1 = \cos \varphi_1 \cdot \tan d \cdot \cos C$, or
3. $\tan d = \tan \varphi_1 \cdot \sec C$.

Note that the distance d serves only as an auxiliary quantity. We also conclude from (ii) that:

4. $\cos(\lambda_1 - \lambda_0) = \cos d \cdot \sec \varphi_1$, i.e. the equation for finding λ_0 and, therefore, λ_M, the longitude at which the maximum of the latitude occurs.

The application of the SIN-TH. then yields:

$\dfrac{\sin \alpha}{\sin |\varphi_1|} = \dfrac{\sin 90°}{\sin d}$, and hence the formula for α:

5. $\sin \alpha = \dfrac{\sin |\varphi_1|}{\sin d}$.

All that remains is to derive the actual "Equation of the Great-Circle Route", $\varphi = f(\lambda)$. Now, let's introduce another auxiliary variable ℓ, representing the length of the great-circle arc extending from λ_0 to λ. The application of the COS-TH. to the spherical triangle N, $(\lambda_0, 0)$, (λ, φ) yields:

$\cos \ell = \cos 90° \cdot \cos(90° - \varphi) + \sin 90° \cdot \sin(90° - \varphi) \cdot \cos(\lambda - \lambda_0)$, or in short:

(iii) $\cos \ell = \cos \varphi \cdot \cos(\lambda - \lambda_0)$, and
(iv) $\cos \varphi = \cos \ell \cdot \cos(\lambda - \lambda_0) + \sin \ell \cdot \sin(|\lambda - \lambda_0|) \cdot \cos \alpha$.

Combining (ii) and (iii) yields: $\cos \varphi \cdot \sin^2(|\lambda - \lambda_0|) = \sin \ell \cdot \sin(|\lambda - \lambda_0|) \cdot \cos \alpha$, and hence:

1.5 Gnomonic or Great-Circle Navigation

Fig. 1.5.1

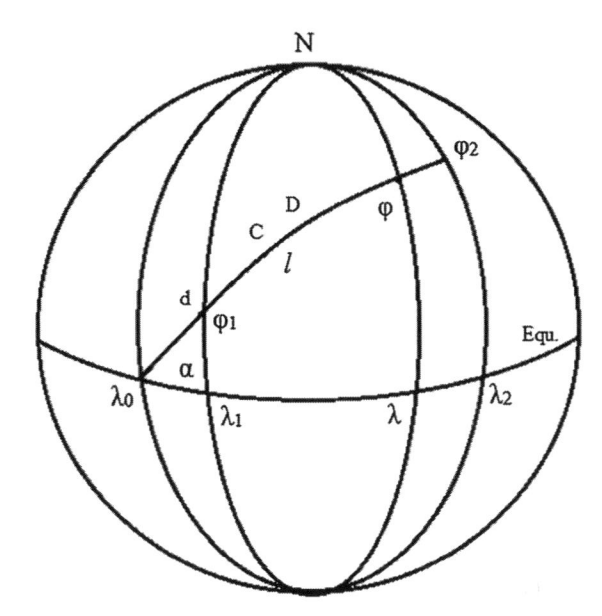

(v) $\cos|\varphi| = \dfrac{\sin \ell \cdot \cos \alpha}{\sin(|\lambda - \lambda_0|)}$.

Next we apply the SIN-TH. to the same triangle to obtain: $\dfrac{\sin 90°}{\sin \ell} = \dfrac{\sin \alpha}{\sin|\varphi|}$.

This, then, yields:

(vi) $\sin \ell = \dfrac{\sin|\varphi|}{\sin \alpha}$. By combining (v) and (vi) we obtain:

$\cot|\varphi| = \cot \alpha \cdot \dfrac{1}{\sin(|\lambda - \lambda_0|)}$, the desired equation $\varphi = f(\lambda)$. The later can be written as:

6. $\tan|\varphi| = \tan \alpha \cdot \sin(|\lambda - \lambda_0|)$, the Equation of Great-Circle navigation.

(Note that $\varphi = \pm |\varphi|$, depending on the signs of φ_1 and φ_2 and the location of λ.[2])

With the help of relation (6), we can establish WAY-POINTS along our great-circle route in order to navigate from way-point to way-point employing Rhumb-Line navigation. Equation 6 also confirms what we can readily deduce

---

$${}^2\varphi = \begin{cases} |\varphi| & \begin{cases} \text{if } \varphi_1, \varphi_2 > 0 & \forall \lambda \in [\lambda_1, \lambda_2] \\ \text{if } \varphi_1 < 0, \varphi_2 > 0 & \forall \lambda \in [\lambda_0, \lambda_2] \\ \text{if } \varphi_1 > 0, \varphi_2 < 0 & \forall \lambda \in [\lambda_1, \lambda_0] \end{cases} \\ -|\varphi| & \begin{cases} \text{if } \varphi_1, \varphi_2 < 0 & \forall \lambda \in [\lambda_1, \lambda_2] \\ \text{if } \varphi_1 < 0, \varphi_2 > 0 & \forall \lambda \in [\lambda_1, \lambda_0] \\ \text{if } \varphi_1 > 0, \varphi_2 < 0 & \forall \lambda \in [\lambda_0, \lambda_2] \end{cases} \end{cases}$$

Also note: $\varphi < 0$ for all latitudes south, and positive for all latitudes north and $\lambda < 0$ for all longitudes west, and positive for all longitudes east.

from Fig. 1.5.1, namely—the maximum of the latitude φ_M occurs either at λ_1, λ_2 or at $\lambda_M = \lambda_0 \pm 90°$.

The maximum value of φ attained on the interval $\lambda_0 \leq \lambda \leq \lambda_2$ is either φ_1, φ_2 or $\varphi_M = \alpha$. Analytically, these results are verified by differentiating (6) that yields: $\frac{1}{\cos^2 \varphi} \cdot \frac{d\varphi}{d\lambda} = \pm \cos(\lambda - \lambda_0) \cdot \tan \alpha$. Hence the relative *extremum* occurs at $\lambda_M - \lambda_0 = \pm 90°$, i.e., $\lambda_M = \lambda_0 \pm 90°$. Substituting in (6) yields: $\tan \varphi_M = \tan \alpha$, i.e. $\varphi_M = \alpha$. Therefore, we have solved all relevant problems (1–5) of great-circle navigation. Next, let's take a look at a practical example—a trip from Baja California (Mexico) to Honolulu (Hawaii).

1.6 Numerical Examples and More Chart Projections

In this section, let's look at an actual journey from the west coast of Baja California, Mexico to Honolulu, Hawaii.

The coordinates of the point of departure A $\sim (\lambda_1, \varphi_1)$ are:

$$\lambda_1 = -110°40'.659 = -110°.67765 \text{ and } \varphi_1 = 23°.45'.970 = 23°.76616.$$

The coordinates of the destination B $\sim (\lambda_2, \varphi_2)$ are given as:

$$\lambda_2 = -155°04'.980 = -155°.083 \text{ and } \varphi_2 = 19°50'.808 = 19°.8468$$

First, let's look at the problems of determining the actual distance D; the initial course C to be steered; and the maximum latitude to be reached, together with its longitude $\lambda_0 - 90°$.

The application of formula (1) Section 1.5, yields:

$\cos D = \sin 23.76616 \cdot \sin 19.8468 + \cos 23.76616 \cdot \cos 19.8468 \cdot \cos -44.40535 =$
    0.751812073, hence,
$D = \cos^{-1} 0.751812073 = 41°.252410 = 2475'.1446$, i.e.
$D = 2475.1446$ Nautical Miles.

Next, evaluate formula (2), Sect. 1.5, to find the course C as:

$$\cos C = \frac{\sin 19.8468 - \sin 23.76616 \cdot \cos 41.252410}{\cos 23.76616 \cdot \sin 41.252410} = 0.060521631, \text{ hence,}$$

$C = -[(\cos^{-1} 0.060521631 - 360°)] = -[(86°.53024562 - 360°)] = 273°.4697544$, i.e.
$C = 273°28'.2$

1.6 Numerical Examples and More Chart Projections

Our auxiliary variable d is to be found by applying formula (3) Sect. 1.5, to obtain:

$$\tan d = \frac{\tan 23.76616}{\cos 86.53024562} = 7.275865248, \text{hence,}$$

$$d = \tan^{-1} 7.275865248 = 82°.17425821 = 4930.4555 \text{ NM}$$

The longitude λ_0 is readily found by applying formula (4) Sect. 1.5, resulting in:

$$\cos(\lambda_1 - \lambda_0) = \frac{\cos 82.17425821}{\cos 23.76616} = 0.148777317, \text{ and hence:}$$

$$\lambda_1 - \lambda_0 = \cos^{-1} 0.148777317 = -81°.44392306, \text{ i.e.}$$

$$\lambda_0 = -29°.233726694 \text{ and } \lambda_M = -119°23372694.$$

Next, the application of formula (5) Section 1.5, yields:

$$\sin \alpha = \frac{\sin 23.76616}{\sin 82.1742582} = 0.406793395, \text{ and therefore:}$$

$$\alpha = \sin^{-1} 0.406793395 = 24°.00355942, \text{ i.e. } \varphi_M = 24°$$

It follows then that: $\tan \alpha = 0.445303125$.

Finally, formula (6) Section 1.5, yields the equation for the intermediate points as:

$$\tan \varphi = 0.445363125 \cdot \sin(\lambda_0 - \lambda).$$

Secondly we want to establish way-points along the great-circle route. Here, the objective is to obtain the coordinates λ, φ along the route subject to the condition that the distance between any two successive way-points does not exceed 300 NM. Therefore, let's define a grid along the equator defined by: $\lambda^i = \lambda_1 - (i-1)5°$, $1 \leq i \leq 9$. Hence, $\lambda_0 - \lambda^i = 81°.44392306 + (i-1)5°$, $1 \leq i \leq 10$ and then compute the corresponding values $\varphi^i = \varphi(\lambda^i)$ by employing Eq. 6. Sect. 1.5.

All the relevant values of these calculations are listed as follows:

i	λ^i	$\tan \varphi^i$	φ^i
1	−110.67765	0.440347223	23.76616
2	−115.67765	0.444443725	23.96255024
3	−120.67765	0.445161726	23.9967980
4	−125.67765	0.442489777	23.86890024
5	−130.67765	0.436450214	23.57887314
6	−135.67765	0.427089002	23.12679508

(continued)

(continued)

i	λ^i	$\tan \varphi^i$	φ^i
7	−140.67765	0.41.4477384	22.512901
8	−145.67765	0.398711343	21.73773069
9	−150.67765	0.379910868	20.80232838
10	−155.083	0.360945109	19.8467999

As has been mentioned in Sect. 1.5, the avid navigators who like to construct their own "exact" plotting sheets or charts can employ the following mathematical projection formulae for the GNOMONIC and LAMBERT CONFORMAL CONIC charts. These formulae are stated without proofs which have been left to the reader to verify. [1], [2], [8], [14], [16], [32], [39], [40]

1. Gnomonic Projection—Non-Conformal Projection.

$$x = c \cdot \frac{\sin(\lambda - \lambda_{sp}) \cos \varphi}{\cos a}$$

$$y = c \cdot [\frac{\cos \varphi_{sp} \sin \varphi - \sin \varphi_{sp} \cos \varphi \cos(\lambda - \lambda_{sp})}{\cos a}]$$

$\cos a = \sin \varphi_{sp} \cdot \sin \varphi + \cos \varphi_{sp} \cdot \cos \varphi \cdot \cos(\lambda - \lambda_{sp})$, where "c" denotes the scale factor of the chart and λ_{sp}, φ_{sp} are the spherical coordinates of the center of projection and "a" denotes the angular distance of the point (λ, φ) on the sphere from the center of projection.

Example: Choose $(\lambda_{sp}, \varphi_{sp}) = (0, 0)$, then: $x = c \cdot \tan \lambda$, and $y = c \cdot \frac{\tan \varphi}{\cos \lambda}$

The corresponding formulae for the Lambert Conformal Conic Projections are a little more complicated and more difficult to derive. Therefore, no attempt has been made to derive them here. It will suffice to merely state them as follows:

2. Lambert Conformal Conic Projections— [2], [14], [39], [40]

$x = p \cdot \sin[n(\lambda - \lambda_0)]$,
$y = p_0 - p \cdot \cos[n(\lambda - \lambda_0)]$

$$\text{with } n = \frac{\ln(\cos \varphi_1 \cdot \cos \varphi_2)}{\ln\left[\tan(45 + \frac{\varphi_2}{2}) \cdot \cot(45 + \frac{\varphi_1}{2})\right]}$$

$$p = F \cdot \cot^n\left(45 + \frac{\varphi}{2}\right)$$

$$p_0 = F \cdot \cot^n\left(45 + \frac{\varphi_0}{2}\right), \text{ and}$$

$$F = \frac{\cos \varphi_1 \cdot \tan^n\left(45 + \frac{\varphi_1}{2}\right)}{n}$$

1.6 Numerical Examples and More Chart Projections

... where λ_0 and φ_0 are the coordinates of the point of reference; where φ_1 and φ_2 are the coordinates of the two standard parallels; and where λ and φ are the spherical coordinates of the point $P \sim (\lambda, \varphi)$ to be projected on the chart $Q \sim (X, Y)$.

For the purpose of numerical evaluations, the following useful data is hereby included:

$\sin 1' = \dfrac{1}{3438}$,

1 NM = 1.852 km,

1 km = 3280.83999 ft = 0.5399568 NM

1 knot = 1.68780986 ft/s. = 0.514444 m/s.

1.9438449 knots = 3.6 km/h

International Spheroid: $R_M = \dfrac{2a+b}{3} = 3440.193$ NM = Mean Radius

$f = \dfrac{a-b}{a} = \dfrac{1}{297} =$ Ellipticity, and

$e = \sqrt{2f - f^2} = 0.081991889 =$ Exc.

Any treatise on terrestrial navigation would not be complete without solving the very important problem of determining a vessel's position at sea (or air) by the use of land or sea marks whose distances and/or angles relative to magnetic north (or a celestial object) can be measured. This measurement gives rise to the concept of Lines of Position (LOP). In the next chapter we will investigate Lines of Position and the definition of land and sea marks that will include time dependent (moving) objects such as substellar points of celestial bodies.

Chapter 2
Astro-navigation

2.1 Lines of Position, Position Fix, Navigational Triangle and Fix by Computation

If it were possible to determine the velocity of a vessel relative to the surface of the Earth, we would know its position at any instant by using the formulae of the previous section in all cases with a v constant. However, we seldom have accurate values for v available mainly because we can only estimate wind and current velocities to establish a valid drift angle.[1] This is particularly true of sailing vessels and float planes that travel at low speeds and present great lateral resistance. It should not surprise the reader that the actual error incurred in DEAD-RECKONING can be of the order of 25% (or even higher) of the distance traveled in 24 h of sailing. Because of this error, it is imperative that a navigator checks the actual position of the vessel at least once every 24 h and correct the course accordingly. [24]

In what follows, it is assumed that the navigator has access to a straight edge, a compass, a protractor, and a local chart since the plotting sheets will not reveal land masses and other obstacles. Of course, a pocket calculator and a chronometer are always indispensable. Moreover, it is assumed that the navigational aids are within line of sight. This would most likely occur if the vessel were moving along a coast line or an interior lake close to shore.

The actual procedure for obtaining a terrestrial fix consists in establishing two or more lines of position (LOPs) which are simply obtained by noting the angle between the north direction of the magnetic compass and the line of sight towards the navigational aid, as for instance, a tower, lighthouse or buoy that can be identified on a chart. This angle will be the AZIMUTH of the navigational aid at the actual position of the vessel. This azimuth is then transferred to the chart and the

---

[1] Except in the case of the INERTIA NAVIGATION SYSTEMS. See appendix at the end of this section.

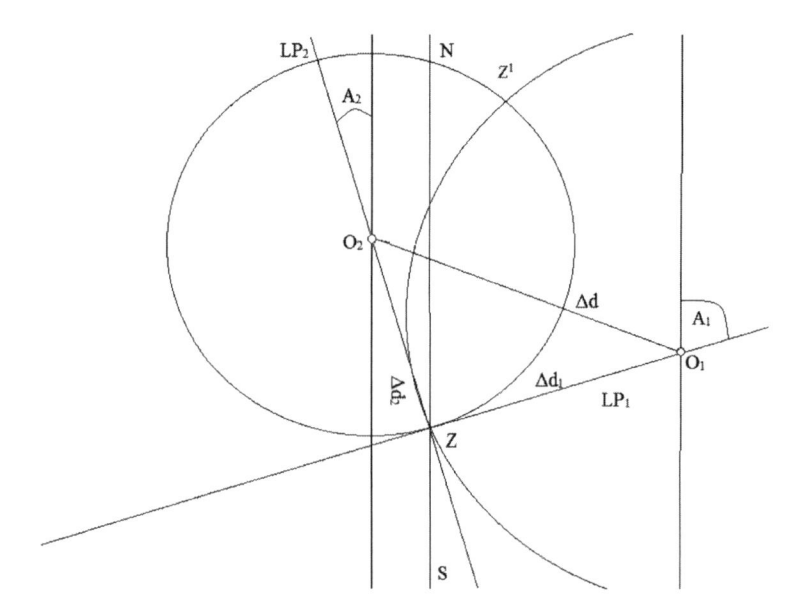

Fig. 2.1.1

resulting LOP is then drawn at this angle relative to the magnetic meridian through the position of the aid on the chart. Similarly, a second LOP is established simultaneously. The case of establishing a fix based on LOPs taken at different times is easily accomplished by advancing the first LOP parallel to itself in accordance with the velocity v of the vessel and the elapsed time. It is obvious that the two LOPs will always intersect at a distinct point Z providing that their azimuths are distinct.

Figure 2.1.1 depicts the procedure outlined above:

ON A TANGENT PLANE AT Z constituting an approximation only.

Here we can apply the plane Euclidean Geometry with the following properties:

(i) Parallels cross at infinity,
(ii) The sum of the angles in a triangle is always 180°,
(iii) The shortest distance between two points is the line segment of the straight line passing through the two points.

(Note that the north pole is actually located at infinity. Also note that all distances in Fig. 2.1.1 are linear distances.) [2, 8]

Next let's look at the true picture on the globe. Here the corresponding points N, O_2, and O_1 form a spherical triangle with two sides, co-latitude of O_1 and O_2, and hour angle t are known. Hence we can calculate the angular distance d between the two points, $O_1 \sim (\lambda_1, \varphi_1)$ and $O_2 \sim (\lambda_2, \varphi_2)$. Besides the triangle $N\hat{O}_2O_1$, we also have another triangle, namely, $O_2\hat{Z}O_1$ (see Fig. 2.1.3iv). Furthermore, we also have two navigational triangles $N\hat{O}_2Z$ and $N\hat{Z}O$ (see Fig. 2.1.3ii, iii).

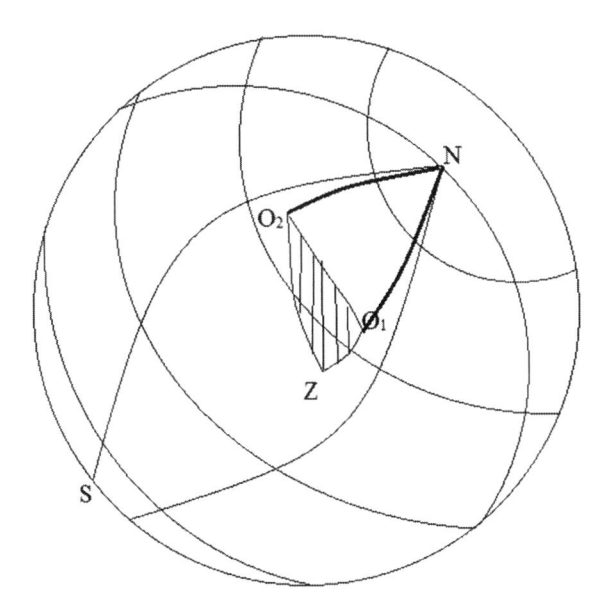

2.1 Lines of Position, Position Fix, Navigational Triangle ...

Fig. 2.1.2

Since the task of fixing the position of our vessel actually amounts to finding the coordinates (λ, φ) of our zenith Z, we must establish the elements in spherical triangles that determine these coordinates in the actual calculations.

ON THE ACTUAL SPHERE where we use Spherical-Geometry (Non-EUCLIDEAN Geometry) with the following properties:

(i) Great Circles always cross at two distinct points,
(ii) The sum of the angles in a spherical triangle is always greater than 180°,
(iii) The shortest distance between two points is the length of the arc of the great-circle that passes through those points.

Because of the differences in the two geometries it is no surprise that we have to consider another approach to the solution of the approximation defined by Fig. 2.1.1, namely, instead of merely measuring the azimuths A_1 and A_2 in Fig. 2.1.1, we must also introduce the linear distances and then use the angular distances d, d_1, and d_2 in order to solve the exact problem depicted by Figs. 2.1.2 and 2.1.3.[2]

By looking at Fig. 2.1.3i, we readily see that with φ_1, φ_2, λ_1, and λ_2 given we can calculate Δd and α_1. Next we deduce from Fig. 2.1.3iv that with Δd_1 and Δd_2 given we can then calculate α_2 and therefore, compute φ and λ. The actual formulae used are the COS-TH. and the SIN-TH. of spherical trigonometry.

---

[2]In this chapter I merely want to state qualitatively the exact solution to a terrestrial fix by employing the angular distances outlined above.

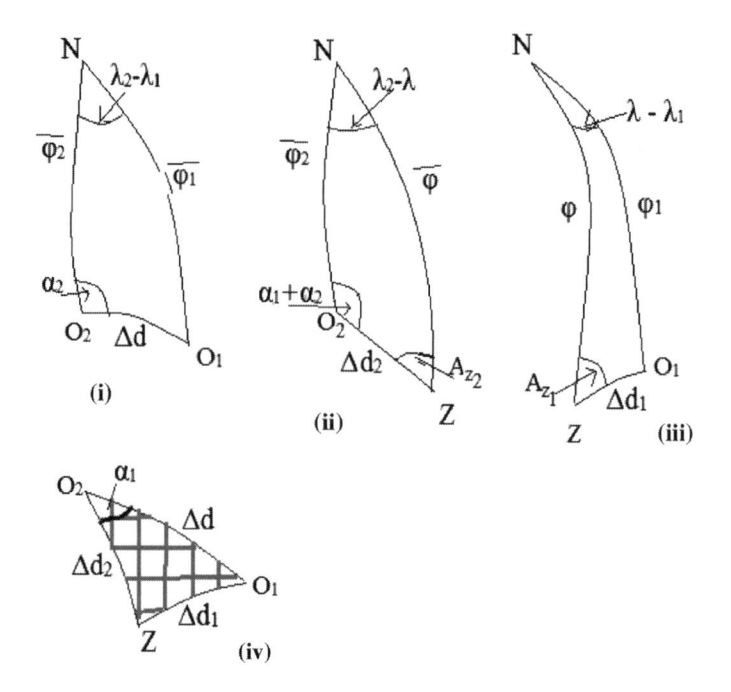

Fig. 2.1.3

The explicit formulae for finding φ and λ of Z and all other relevant formulae will be deduced in the next chapter. The only differences will be that the navigational aids will be replaced by the substellar points of the heavenly aids (stars, planets, sun and moon) and that the coordinates of $\varphi_{1,2}$, $\lambda_{1,2}$ of the substellar points depend on the precise time at which the angular distances of these points are measured. The next chapter will also take the movement of the vessel into account. Basically, we have made the transition from the Euclidean Geometry to the Elliptic Geometry and thereby laid the foundation of exact navigation.

In order to understand the significance of the most recent developments in navigation, let's take a moment and reflect on the historical mile-stones of the achievements of navigators and astronomers. This may give us a better understanding of what future developments may have in store.

It was certainly a Quantum-Leap in the making of men to accept the shape of the Earth as what it actually is—a GEOID—and for most practical applications a SPHEROID. Remember, it took men more than sixteen centuries, from Archimedes to Columbus' times[3] to fully integrate the true shape of the Earth into the concept of navigation. Perhaps now is the time to ask the men of vision how much longer it will take them to readily accept that the physical Universe = Space + Energy itself

---

[3] According to Joshua Slocum, there was at least one world leader in the 19th Century who still believed that the Earth was flat.

requires the adoption of the concept of another Non-Euclidean Space as conceived a long time ago by some mathematicians and subsequently employed by Albert Einstein in his treatise of The General Theory of Relativity.

As for mathematicians, it should be pointed out that Riemann had developed the necessary calculus for a Non-Euclidean Geometry (also used subsequently by Albert Einstein). The Russian mathematician Lobachevski had also shown earlier that the formulae of Euclidean Geometry constituted an approximation to his Hyperbolic Geometry in an infinitesimally small neighborhood of a point on the hyperbolic surface. Therefore, we can only conjecture that approximating the true space by Euclidean Space is very similar to approximating the spheroid-Earth by its tangent planes.

Unfortunately, the actual distances involved are of the order of 10^6 AU, and therefore a physical proof of the Non-Euclidean Geometry of our space can still not be rendered by experimental means currently available. The argument most frequently used by astronomers and astrophysicists is why bother looking for the actual configuration of space if the Euclidean Space constitutes a sufficiently close approximation to reality. The argument against this is simply why are you trying to prove the BIG-BANG theory if you still don't know which type of space was created at the beginning of our universe. The other logically correct argument or opting for the true space is: "If the proponents of the BIG-BANG theory assume that at the beginning there was only one singularity containing all the energy of the universe, the argument against it is that a singularity can only exist if a space is in existence, or if the space, itself, was a singularity at the beginning of the universe. The arguments rest on the abstract philosophical concept of a point that has no physical dimensions whatever. Hence, no physical proof of this theory may exist."

Conclusions

In the foregoing sections, I have chosen to use the analogy of approximating the Non-Euclidean Geometry, i.e., the Elliptic and Spherical Geometry by the Euclidean Geometry thereby employing the tangent plane on the sphere in order to arrive at the basic equations for exact navigation. This approach also shows that there is, from a mathematical point of view, a difference in the geometry between approximate terrestrial and celestial navigation.

In the sections to come, we will explore some of the practical applications and approximations which were made in order to enable the mariners and navigators for the past to arrive at numerical results for the position of their vessels. It should be noted that a majority of these men were ill prepared for the mathematical task they encountered in navigation.

Things have drastically changed. Today, children learn to use computers. Across the world, people use programs they don't have to understand on computers for all kinds of things. For instance, some elementary school students have been taught to execute a simple program on their calculators involving trigonometric and logarithmic expressions without these students actually having the knowledge of trigonometry or logarithms. Most of these same students even know how to use a GPS. Therefore, why write a book on the Science of Navigation when children, today, already have access to such a navigational tool? The answer is simple—what are those children or any adult going to do if that GPS goes dark?

Appendix

A more sophisticated system of terrestrial navigation had been put to use about sixty years ago by the U.S. Navy and by its Russian counterpart. Used chiefly on board submarines, it basically consisted of two components: electric gyroscopes and an analogue computer. The gyroscopes measure the forces acting on the vessel. The analogue computer integrated the system of differential equations of motion whenever the vessel was in motion. Therefore, based on the input (the initial position and the initial velocity), the analogue computer provided the position and velocity at any given later time.

Such navigational systems were also referred to as Inertia Platforms and were characteristically heavy and bulky given the state of technology a half century ago. With the passing of time, the advent of microprocessors and sensor chips, such systems are evolving into something small enough to be carried aboard private vessels. It is only a matter of time before these platforms are marketed to owners of pleasure crafts.

Although the inertia navigational systems were developed long before GPS became available, they have not become obsolete. On the contrary, they actually constitute another independent terrestrial navigation system with the tremendous advantage of being independent of a land or air based system of navigation.

With all the terrorist activity reported in today's headlines, it should be obvious to all the belligerent nations of the world how utterly vulnerable any kind of GPS system really is. Methods for the shooting down of artificial satellites have been perfected to the point that virtually anyone with a modicum of training can push a red button and bring them down. Anyone interested in computing the orbits of Celestial Objects or intercepting Artificial Celestial Objects need only apply the basics of Celestial Mechanics to determine the critical parameters of their orbits. [54, 56]

2.2 Celestial Sphere, Equatorial and Horizon System of Coordinates, Navigational Triangle and the Ecliptic Coordinate System

In the previous chapter we used LONGITUDE and LATITUDE for the coordinates of a point on our sphere without explicitly defining the underlying coordinate system. Aside from the actual shape of the Earth, we also employed the physical reference system of its axis of rotation which passes through the center of the Earth and exits the sphere at the two points referred to as the North and South poles, respectively. (Those two points are not to be confused with the magnetic poles.) Then, the underlying coordinate system is based on the great-circle defined by the intersection of the plane which passes through the center of the Earth and which is perpendicular to the axis of rotation. This great-circle is called the EQUATOR and is assigned the latitude zero.

2.2 Celestial Sphere, Equatorial and Horizon System ...

If we then pass a plane parallel to the equatorial plane through a given point, we obtain a small circle of equal latitude. By assigning the angular distance of this circle from the equator as latitude, we have defined φ of said point. The angular distances between two circles of constant latitude are measured along any great-circle which passes through the north and south poles and is called a MERIDIAN.

Next, let us define longitude zero as the longitude corresponding to the point of intersection of the meridian that passes through GREENWICH, England with the equator. The longitude of any point on the sphere (Earth) is then defined by measuring the angular distance of the point of intersection of the meridian that passes through said point with the equator from the Greenwich meridian and is assigned negative values for all points ($|\lambda| \leq 180°$) lying West (left) of the Greenwich meridian and positive values for all other points lying East of said meridian. In addition, we also assign negative values for φ for points lying South of the equator (see Fig. 2.2.1).

Now, let's define SUBSTELLAR points (SP) and their coordinates on the sphere. For the sake of simplicity, let's assume that the observer is located at one of the poles. He or she, then, will see all heavenly bodies of the semi-sphere above them. If they connects those objects, like the stars, planets, Moon and Sun by means of imaginary lines to the center of the Earth, those straight lines will intersect the surface of the Earth at well-defined points at any instant. We shall refer to those well defined points on the surface of the Earth as Substellar Points (SP). Since these points lie on the surface of the globe, it is only logical to use the same system of coordinates that we use to locate any point on the Earth, except that we will use different names for these coordinates. Longitude will be called Greenwich Hour Angle, and Latitude will be

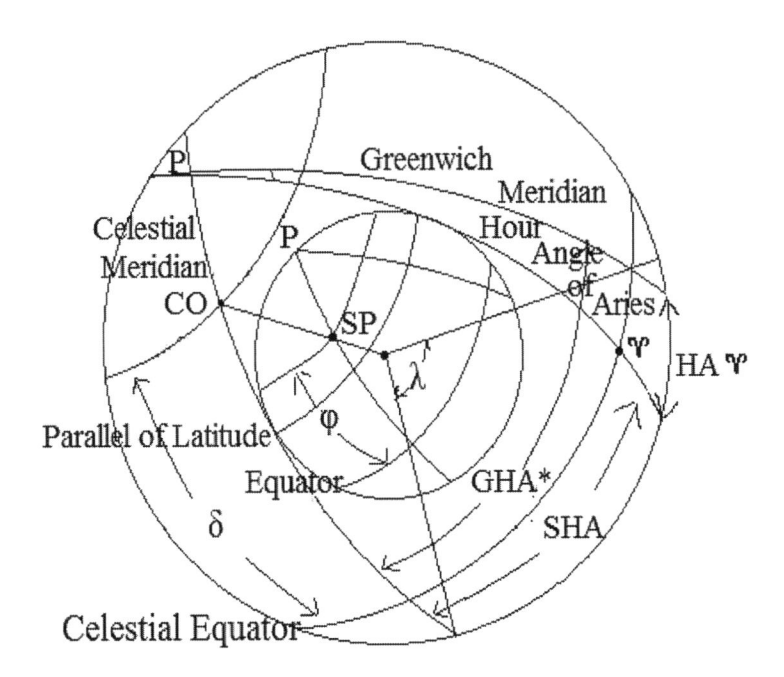

Fig. 2.2.1

called Declination. Accordingly, we will also use different symbols and denote the declination by "δ" and the Greenwich Hour Angle by "GHA".

It should be clearly understood that these coordinates are defined at the instant of observation and, therefore, depend on the rate of rotation of the Earth.

Because the distances of the stars are so great (light years) in comparison to the diameter of the Earth (the same is also partly true for the planets, the Sun, and less true for the Moon), we may actually assume in the above description that the observer is located at the center of the Earth, carrying with him his tangent plane. In the next chapter I will introduce corrections as in the cases of the Moon, Sun, and all the navigational planets because their distances from the center of the Earth are finite and not infinite as with the stars. However, for the moment, we will assume that all celestial objects are at infinite distances away from the center of the Earth. Therefore, linear distances will not enter the calculations explicitly in astronavigation and only angular will be considered here. Because of this, we may also picture the substellar points as interstellar points on a fictitious sphere of arbitrary radius, henceforth referred to as the CELESTIAL SPHERE (see Fig. 2.2.1).

By extending the plane of the terrestrial equator to intersect with this sphere, we obtain that great-circle on the celestial sphere referred to as the CELESTIAL EQUATOR. If we also extend the axis of rotation of the Earth to intersect with the celestial sphere, we obtain the CELESTIAL POLES—north and south, respectively.

Moreover, the observer will also notice that all COs observed will rotate at a fixed rate relative to the observer's position and except for the Sun, Moon, and planets, those COs will not sufficiently alter their relative position to each other. Therefore, the observer may conceive that all the stars are located fixed on the Celestial Sphere which, in turn, rotates about the extended axis of the Earth.

This model implies that the coordinates of the stars remain quasi-fixed[4] if we merely refer their longitude to a "fixed" point[5] on the Celestial Equator. This particular point is the position of a fictitious star called ARIES (Υ) corresponding to the vernal equinox.

Depending on whether the hour angle of the star is measured westward or eastward relative to Aries, we refer to it as SIDEREAL HOUR ANGLE (SHA) or RIGHT ASCENSION (RA). For now, we will refer to the celestial coordinates of a CO as DECLINATION and SIDEREAL HOUR ANGLE. It follows then, that the coordinates of any substellar point can be expressed in terms of its declination, the GHA of Aries, and its SHA. The latter also makes it possible to determine longitude in terms of exact time measurements at the instant of observation. [2, 16, 19]

In the previous chapter, we employed the concept of the Navigational Triangle without reference to the celestial sphere. Now, it's time to elaborate on this fundamental tool of astronavigation and apply it to the transformation of coordinate systems.

---

[4]The fact that the stars change their position on the Celestial Sphere will be explained and incorporated in the actual calculations in a later chapter.

[5]This fixed point, called Aries, is also a variable point as will be explained later.

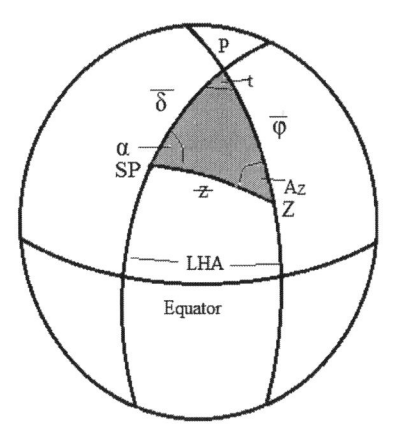

2.2 Celestial Sphere, Equatorial and Horizon System ...

Fig. 2.2.2

Let's recall that the zenith of the observer on the Earth is the intersection of the straight line passing through the position of the observer and the center of the Earth with the celestial sphere. This allows us to define the Navigational Triangle corresponding to the observer's position Z, the North Pole of the celestial sphere, and the position SP of the celestial object on the celestial sphere as the spherical triangle ZP̂SP spanned by these points—see Fig. 2.2.2.

Since an observer at Z will orientate him or herself with reference to their natural horizon, i.e., relative to the tangent plane at Z to the sphere, and since the distance R (the radius of the Earth) is immaterial for the above reasons, we shall adopt a new coordinate system based on the concept of the horizon plane. The intersection of this horizon plane with the local meridian circle is depicted in Fig. 2.2.3 and the mathematical definition therefore is:

> The Horizon Plane is the plane that passes through the center of the Earth and is orthogonal to the line drawn through the center of this sphere and the zenith.

We now proceed to redraw the navigational triangle with reference to the horizon plane and call the resulting system of coordinates the Horizon System with the two important coordinates of azimuth angle A_Z and the altitude $a = 90° - z$ (see Fig. 2.2.4).

$$Z_n = \begin{cases} A_Z \text{ if SP is East} \\ 360° - A_Z \text{ if West} \end{cases} \quad t = \begin{cases} LHA \text{ if SP is West} \\ 360° - LHA \text{ if East} \end{cases}$$

Since all observations are carried out on the platform referred to as horizon and all ephemeral data is provided with reference to the equatorial system of coordinates, it is imperative to know how the coordinates of one system can be calculated once the coordinate of the other system are known.

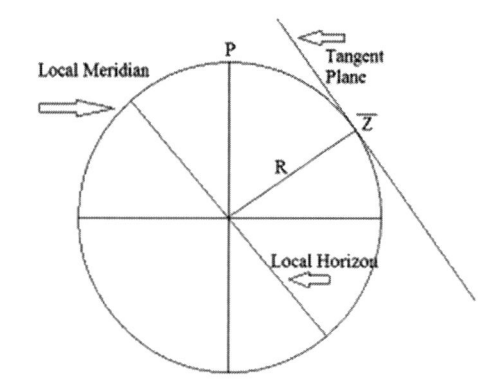

Fig. 2.2.3

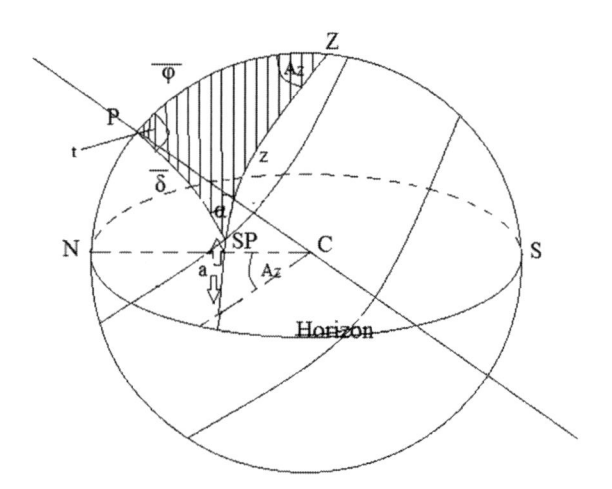

Fig. 2.2.4

a: altitude	α: parallatic angle	$\bar{z} = 90° - a$ Zenith distance
A_z: azimuth angle	Z_n: true azimuth	$\bar{\delta} = 90° - \delta$ polar distance
t: LHA (SP)	$\bar{\varphi} = 90° - \varphi$ colatitude	LHA = Local Hour Angle
	δ = declination of SP	[26]

Here we shall derive the corresponding equations by employing the fundamental equations of spherical trigonometry (FE) to our navigational triangle:

FE: (i) COS-TH. $\cos b = \cos a \cdot \cos c + \sin a \cdot \sin c \cdot \cos B$.
 (ii) SIN-TH. $\sin a \cdot \sin B = \sin b \cdot \sin A$.
 (iii) ANALOGUE FORMULAE:

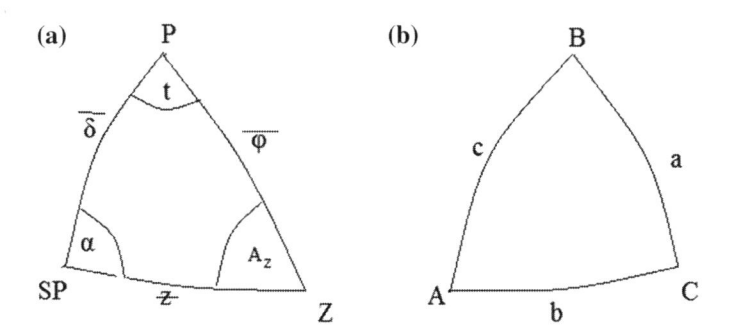

2.2 Celestial Sphere, Equatorial and Horizon System ...

Fig. 2.2.5

$$\sin a \cdot \cos C = \cos c \cdot \sin b - \sin c \cdot \cos b \cdot \cos A$$
$$\sin a \cdot \cos B = \cos b \cdot \sin c - \sin b \cdot \cos c \cdot \cos A$$

(see Sect. 4.2) [7, 41, 42]

First let us consider the problem of calculating the azimuth Z_n and altitude a when the declination δ, local hour angle t, and the latitude of φ of the observer are known.
Solution:
Application of the COS-TH. to Fig. 2.2.5b yields:

(1)   (i) $\sin a = \sin \varphi \cdot \sin \delta + \cos \varphi \cdot \cos \delta \cdot \cos t$
        Application of the SIN-TH. yields:
  (ii) $\cos a \cdot \sin A_Z = \cos \delta \cdot \sin t$
        Application of ANALOGUE FORMULA yields:
  (iii) $\cos a \cdot \cos A_Z = \sin \delta \cdot \cos \varphi - \cos \delta \cdot \sin \varphi \cdot \cos t$

Here we have also used $z = 90° - a$, $\bar{\delta} = 90° - \delta$, and $\bar{\varphi} = 90° - \varphi$. Furthermore, we have used $A_z = Z_n$ if $Z_n \leq 180°$, and $A_z = 360° - Z_n$ if $Z_n \geq 180°$.

Now, let's consider the problem of finding the declination δ and the local hour angle t when Z_n, a, and φ are given.
Solution:
Application of the COS-TH. yields:

(2)   (i) $\sin \delta = \sin \varphi \cdot \sin a + \cos \varphi \cdot \cos a \cdot \cos A_z$
        Application of the SIN-TH. yields:
  (ii) $\cos \delta \cdot \sin t = \cos a \cdot \sin A_Z$
        Application of the ANALOGUE FORMULA yields:
  (iii) $\cos \delta \cdot \cos t = \sin a \cdot \cos \varphi - \cos a \cdot \sin \varphi \cdot \cos A_z$.

In addition to those two coordinate systems, we still require a third one (the Ecliptic Coordinate System) in order to deal with the celestial mechanics of position astronomy. Therefore we need to introduce the pole K of the ecliptic as the intersection of the straight line that passes through the center of the sun and is perpendicular to the ecliptic with the celestial sphere (see Fig. 2.2.6a).

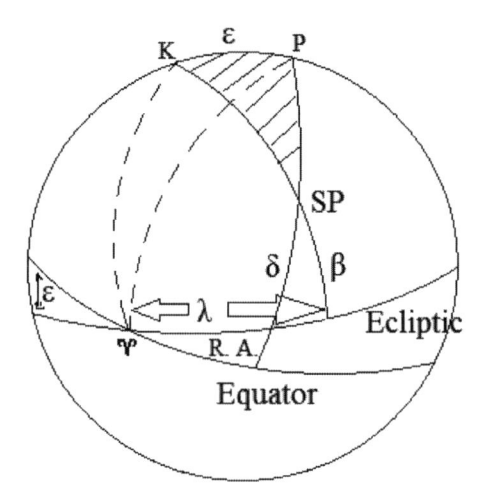

Fig. 2.2.6a

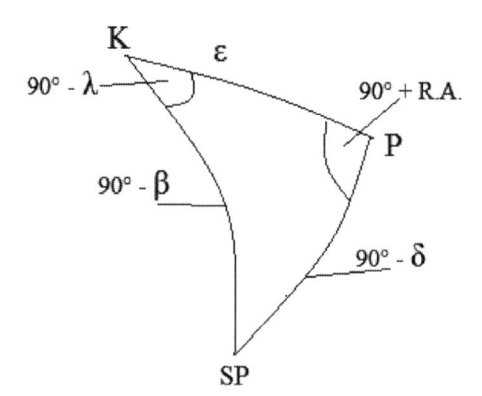

Fig. 2.2.6b

The ecliptic coordinates λ and β are then defined as follows:

(i) Ecliptic Longitude λ = length of arc on the ecliptic between the point of Aries ♈ and the intersection of the great-circle that passes through K and SP with the ecliptic.
(ii) Ecliptic Latitude β = length of arc on the great-circle that passes through K and SP between SP and the intersection of this great-circle with the ecliptic.

By identifying the elements of the triangle $\hat{KPSP}$ (see Fig. 2.2.6b), we may apply the FE of spherical trigonometry to this triangle to obtain the equations for λ and β for given values of RA and δ.

ε = Obliquity of Ecliptic
$\alpha_1 = 90° - \lambda$
$\alpha_2 = 90° + \text{RA}$

2.2 Celestial Sphere, Equatorial and Horizon System ... 41

(3)   (i) $\sin \beta = \sin \delta \cdot \cos \varepsilon - \cos \delta \cdot \sin \varepsilon \cdot \sin RA$
       (ii) $\cos \beta \cdot \cos \lambda = \cos \delta \cdot \cos RA$
       (iii) $\cos \beta \cdot \sin \lambda = \sin \delta \cdot \sin \varepsilon + \cos \delta \cdot \cos \varepsilon \cdot \sin RA$

By applying the FE of spherical trigonometry once more to the triangle above, but by starting with a different side of it, we find the formulae for calculating the RA, and δ for given values of λ and β as follows:

(4)   (i) $\sin \delta = \cos \varepsilon \cdot \sin \beta + \sin \varepsilon \cdot \cos \beta \cdot \sin \lambda$
       (ii) $\cos \delta \cdot \cos RA = \cos \beta \cdot \cos \lambda$
       (iii) $\cos \delta \cdot \sin RA = - \sin \beta \cdot \sin \varepsilon + \cos \beta \cdot \cos \varepsilon \cdot \sin \lambda$

At this point, readers will probably ask themselves why we need three equations for only two unknowns? The equations under consideration are transcendental ones. In all these cases, we obtain more than one possible solution as the example of a simple quadratic equation which has two solutions, in general, demonstrates. Therefore, in order to eliminate the ambiguity in such transcendental equations, we will need more equations than unknowns and, perhaps, some other additional information.

In Chap. 4 we will take a closer look at the problem of the branches of the multi-valued functions $\sin^{-1}x$, $\cos^{-1}x$, and $\tan^{-1}x$. However, here, we need to state the results for the most frequently encountered multi-valued functions, $\sin^{-1}x$, $\cos^{-1}x$, without elaborating on their derivations.

In accordance with the results of Chap. 4, the branches of said functions as follows:

(5)   (i) First Branches

$$\sin^{-1} x = \begin{cases} \sin^{-1}|x|, & \text{if } 0 \leq x \leq 1 \\ -\sin^{-1}|x|, & \text{if } -1 \leq x \leq 0 \end{cases}$$

$$\cos^{-1} x = \begin{cases} \cos^{-1}|x|, & \text{if } 0 \leq x \leq 1 \\ 90° + \sin^{-1}|x|, & \text{if } -1 \leq x \leq 0 \end{cases}$$

(ii) Second Branches

$$\sin^{-1} x = \begin{cases} 90° + \cos^{-1}|x|, \text{if } 0 \leq x \leq 1 \\ -(90° + \cos^{-1}|x|), \text{if } -1 \leq x \leq 0 \end{cases}$$

$$\cos^{-1} x = \begin{cases} -\cos^{-1}|x|, \text{if } 0 \leq x \leq 1 \\ -(90° + \sin^{-1}|x|), \text{if } -1 \leq x \leq 0 \end{cases}$$

(Note: The functions $\sin^{-1}|x|$ and $\cos^{-1}|x|$ as provided on calculators are referred to as the Main Branches of said functions.)

2.3 Conclusions and Numerical Examples

In the previous section, I laid down the foundations for solving the actual task of determining the position of an observer by computation. In this sections I shall address the problems associated with the ambiguities inherent in solving the transcendental equations related to the navigational triangle.

Lets us consider two typical examples to demonstrate the problem of eliminating the ambiguities in the computations by means of employing all three equations in 1, 2 (Sect. 2.2), and by using the branches of the multi-valued trigonometric functions $\sin^{-1} x$ and $\cos^{-1} x$.

Example 1 Given the latitude $\varphi = 38°$, zenith distance $z = 70°$, and the true azimuth
$Z_n = 60°$. Find the declination δ and local hour angle (LHA).

Solution:

Application of formula 2(i), Sect. 2.2 yields:

$$\sin \delta = \sin 38 \cdot \sin 20 + \cos 38 \cdot \cos 20 \cdot \cos 60$$
$$= 0.210568626 + 0.370243945 = 0.580812571 \ldots \text{hence}$$

$\delta = \sin^{-1} 0.580812571 = 35°.50771487\ldots$
and since $0 < \delta < 90°$ no ambiguity occurs.
Next by applying formula 2(ii) we find that:
$$\sin t = \frac{\cos 20 \cdot \sin 60}{\cos 35.50771487} = 0.999705617 \text{ and therefore in accordance with formulae (5) we find that:}$$

$$t = \begin{cases} 88°.6097124 \\ 90° + 1.390287598 = 91°.3902876 \end{cases}$$

Here we encounter an ambiguity. Therefore, we apply formula 2(iii) to obtain:
$$\cos t = \frac{(\sin 20 \cdot \cos 38 - \cos 20 \cdot \sin 38 \cdot \cos 60)}{\cos 35.50771487} = -0.024262674 \ldots \text{hence}$$
$t = \cos^{-1}(-0.024262674)$.
By employing formulae (5), we find that:

$$t = \begin{cases} 90° + 1.390285248 = 91.39028525 \\ -91°.39028525 \end{cases}$$

But since $t > 0$, it follows that: $t = 91°.39028525$
Therefore <u>LHA = 268°.609714</u> because the CO lies to the East of the observer.
(*Note* this example relates directly to the identification of stars.)

2.3 Conclusions and Numerical Examples

Next, suppose the declination of a CO is known and the latitude of the observer is also known. We would now like to find the CO's position in the local sky at an instant when its local hour angle is prescribed. (This problem is directly related to the situation where the navigator is preparing to take sextant sights.)

The following example explains how the formulae of Sect. 2.2 are applied in order to solve the above problem.

Example 2 Here are the givens—declination $\delta = 35°.50771487$; latitude $\varphi = 38°$; and the local hour angle LHA $= 268°.6097148$. The problem is now to find the altitude a and the azimuth Z_n of this CO in the local sky.

Solution:

First recall that the $t = 360° - $ LHA $= 91°039028525$.
Then by applying formula 1(i), Sect. 2.2, we find that:
$\sin a = \sin 38 \cdot \sin 35.50771487 + \cos 38 \cdot \cos 35.50771487 \cdot \cos 91.39028525$
$\quad = 0.342020142$, and hence...
$a = \sin^{-1} 0.342020142 \cong 20°$

Again, we do not have an ambiguity since $0 < a < 90°$.

Next we apply formula 1(ii) to find A_z by...

$$\sin A_z = \frac{(\cos 35.50771487 \cdot \sin 91.39028525)}{\cos 20} = 0.866025413 \text{ and therefore}$$

$$A_z = \begin{cases} 60° \\ 120° \end{cases} \text{ in accordance with formulae (5).}$$

Here again, we encounter an ambiguity and therefore proceed to use formula 1(iii) thereby finding:

$$\cos A_z = \frac{(\sin 35.50771487 \cdot \cos 38 - \cos 35.50771487 \cdot \sin 38 \cdot \cos 91.39028525)}{\cos 20}$$

$\quad = 0.499999\ldots = 0.5$ and therefore...

$A_z = \cos^{-1} 0.5$ and in accordance with formulae (5) we find that ...

$$A_z = \begin{cases} 60° \\ -60° \end{cases}$$

Therefore, we conclude that $Z_n = A_z = 60°$.

In the next section we shall solve the exact navigational task of determining the position of the observer by observing the altitude of two or more distinct celestial objects simultaneously or at distinct instants, or by observing the altitude of one of the COs at distinct instants. I will provide the necessary formulae for determining the position of the observer when altitude and azimuth are known. Furthermore, we shall also solve the problem associated with the movement of the vessel between the taking of the sights at distinct instants.

Of key importance for the navigator is that the formulae provided in the subsequent section do not assume that the observer has prior knowledge of an assumed or an approximate position and therefore, the formulae are the only means of determining the position without dead reckoning. Later, I will also derive methods of determining approximate positions when an estimate of the true position is available.

It should be clearly understood that the latter is quite different from the problem of determining the exact position by two observations. Although it is highly improbable that the navigator has lost orientation completely, the need for another, independent navigational system still exists.

It should be stressed that ambiguity in calculations of the aforementioned type can be easily resolved by the method of quadrant orientation as explained in detail in the next section. Therefore, it is imperative that the users of the formulae provided herein are familiar with the underlying definition of azimuth and make the necessary adjustments in all relevant formulae in case they use a different definition for azimuth.

Finally, I will make an attempt to throw some more light on the limitations of the standard approximation method (Sumner's Method), since no error analysis seems to exist.

2.4 The Use of the Exact Equations[6] for Finding the Position at Sea or Air by Employing Two or More Altitude Measurements Together with the Corresponding Measurements of Time

In the previous sections, I have clearly made the distinction between finding an observer's position without prior knowledge of an approximate position and the problem of merely improving on a given or assumed position. However, the real importance of finding one's position without the knowledge of an approximate position lies in obtaining another independent system of navigation. So far, we have only addressed the independent navigational system based on Dead Reckoning, like Rhumb-Line and great-circle navigation.

From the mathematical point of view, the problem of finding the exact solution consists in finding the two points of intersection of the two circles of equal altitude on the celestial sphere and then eliminating one point as impossible by means of the orientation of the observer relative to the celestial object (see Fig. 2.4.1).

Although there is always more than one method for solving a given problem, I am going to use here the method based on splicing two navigational triangles together (see Fig. 2.4.2).

Another method consists in using the two equations for constant zenith distances and then solving those transcendental equations for their common points of intersection. (We will examine this approach when discussing some approximate methods in the next section.)

---

[6]Note that the Exact Equation does not necessarily mean exact physical solution—See appendix in this section.

2.4 The Use of the Exact Equations for Finding …

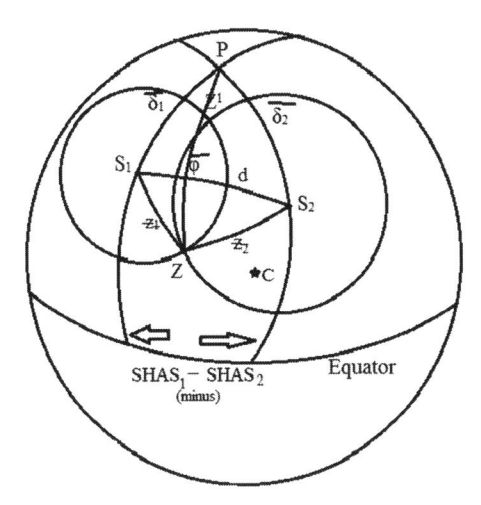

Fig. 2.4.1

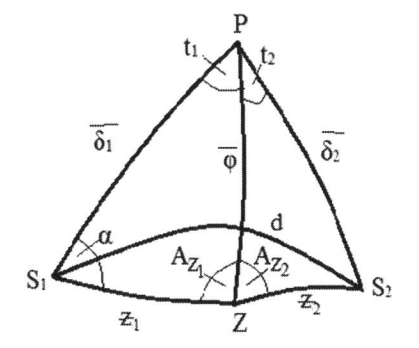

Fig. 2.4.2

By referring to Fig. 2.4.1, we encounter the two spherical triangles $P\hat{S}_1Z$ and $P\hat{S}_2Z$ already spliced together. By having spliced these two triangles together, we have created two more triangles, namely, $P\hat{S}_1S_2$ and $S_1\hat{Z}S_2$ (see Fig. 2.4.3).

As will be shown shortly, the solutions φ and t_1 will depend on the configuration and certain combinations of the two auxiliary angles, α_1 and α_2. Therefore, it will be imperative to employ the orientation of the observer Z relative to the two celestial objects S_1 and S_2. (Note—S_1 can be the same object but viewed at a different instant than S_2.)

To simplify matters, we will always assume (without loss of generality) that S_1 is always situated to the left of S_2, i.e., GHA (S_1) > GHA (S_2).

Then referring to the four quadrants I to IV (see Fig. 2.4.4), we can set up our orientation scheme (see Figs. 2.4.5a–2.4.7d).

Accordingly, there are $16 - 4 = 12$ relevant combinations possible. We have eliminated four combinations by the requirement GHA $(S_1) \geq$ GHA (S_2). However, in the actual calculations we will only have to distinguish between the two combinations:

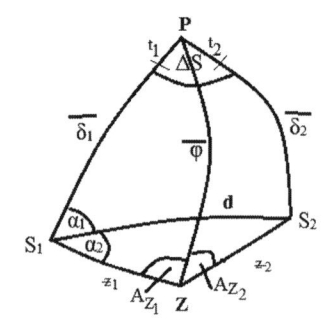

Fig. 2.4.3

$$\alpha = \alpha_1 + \alpha_2, \text{ and } \alpha = \alpha_1 - \alpha_2. \tag{26}$$

Later, I shall provide a simple box diagram for finding α by simply referring to the relative positions of S_1 and S_2 in the respective quadrant. Since the observer has to orientate him or herself with respect to True North, we shall again adopt the concept of the true azimuth Z_n, measured clockwise from N → E → S → W → N, i.e., we define

(1) $Z_n = \begin{cases} A_Z \text{ if } A_Z \text{ is East} \\ 360° - A_Z \text{ if } A_Z \text{ is West} \end{cases}$. Next let us carry out a detailed case study of those twelve cases under consideration:

(Note that Fig. 2.4.5a–d. depict the cases where the observer, when facing north, is viewing S_1 to the left (west) and S_2 to the right (east). Similarly, Fig. 2.4.6a–d. depict the cases where the observer is viewing S_1 and S_2 to his left (west); and Fig. 2.4.7a–d. the cases where S_1 and S_2 are seen to his right (east)—always recalling that GHA $(S_1) \geq$ GHA (S_2).

(It should also be noted that α, the crucial variable in our calculations, is actually the PARALLACTIC Angle α in our navigational triangle PŜZ (see Fig. 2.4.2 of this section). The angles α_1 and α_2 are only auxiliary parameters as is the variable (d).

With reference to our quadrant oriented diagram (see Fig. 2.4.4), we can now schematically summarize the results depicted by Figs. 2.4.5a–2.4.7d as follows:

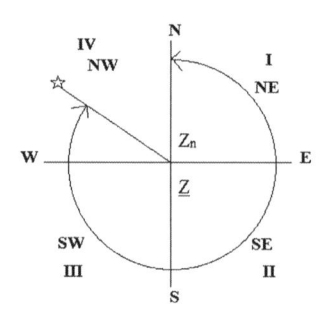

Fig. 2.4.4

2.4 The Use of the Exact Equations for Finding ...

(a)

$S_1 \in$ NW (IV), $S_2 \in$ NE (I)
$Z_{n_1} \geq 270°$, $Z_{n_2} \leq 90°$
$\alpha = \alpha_1 + \alpha_2$

(b)

$S_1 \in$ SW (III), $S_2 \in$ SE (II)
$180° \leq Z_{n_1} \leq 270°$, $90° \leq Z_{n_2} \leq 180°$
$\alpha = \alpha_1 - \alpha_2$

(c)

$S_1 \in$ NW (IV), $S_2 \in$ SE (II)
$Z_{n_1} \geq 270°$, $90° < Z_{n_2} \leq 180°$
$\alpha = \alpha_1 - \alpha_2$

(d)

$S_1 \in$ SW (III), $S_2 \in$ NE (I)
$180° < Z_{n_1} \leq 270°$, $Z_{n_2} \leq 90°$
$\alpha = \alpha_1 + \alpha_2$

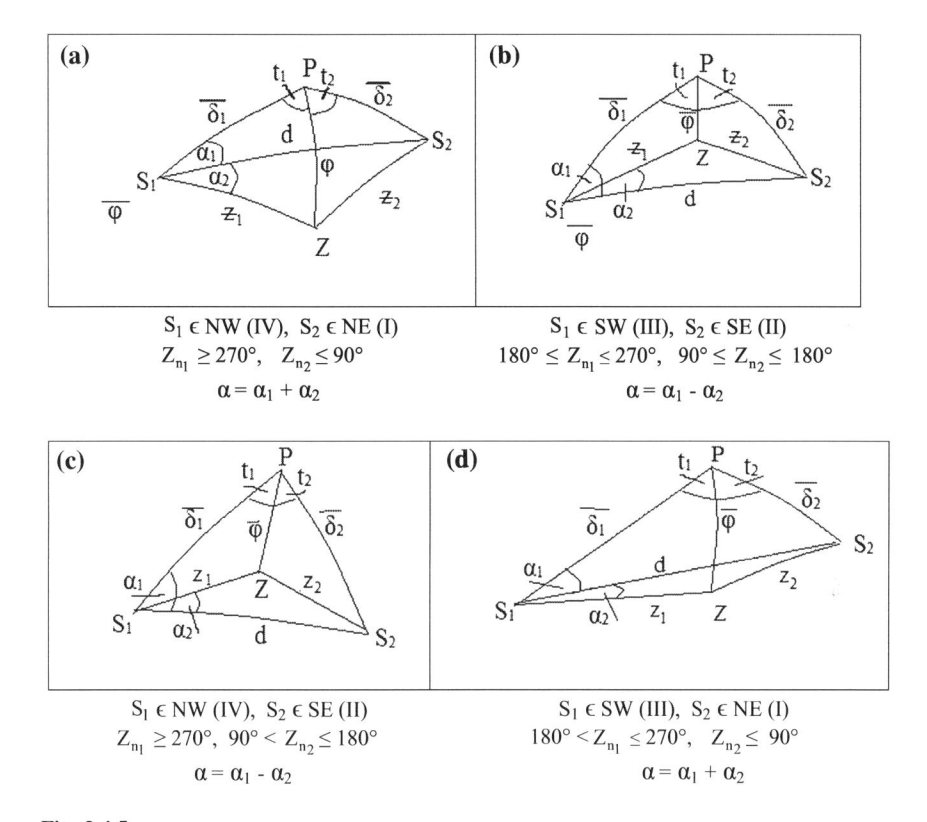

Fig. 2.4.5

S_1	IV	III	II	I	S_2
IV	$\alpha_1 + \alpha_2$	$\alpha_1 + \alpha_2$	XXXXXXX	XXXXXXX	IV
III	$\alpha_1 - \alpha_2$	$\alpha_1 - \alpha_2$	XXXXXXX	XXXXXXX	III
II	$\alpha_1 - \alpha_2$	$\alpha_1 - \alpha_2$	$\alpha_2 - \alpha_1$	$\alpha_2 - \alpha_1$	II
I	$\alpha_1 + \alpha_2$	$\alpha_1 + \alpha_2$	$\alpha_2 - \alpha_1$	$\alpha_2 - \alpha_1$	I

(Note that the combinations (IV,I), (IV,II), (III, I), and (III, II) do not appear because the conditions that GHA S_1 (T_1) > GHA S_2 (T_1), if not—see appendix of this section.[7])

In order to calculate the longitude λ of the observer, we require the hour angle t—see Fig. 2.4.2. It can easily be shown that, independent of the relative position of the observer with reference to the Greenwich Meridian, the longitude is given by:

---

[7]In all cases $T_1 \neq T_2$, Figs. 2.4.5a–2.4.7d depict the position of S_1 at T_1 and S_2 at T_2.

(a)

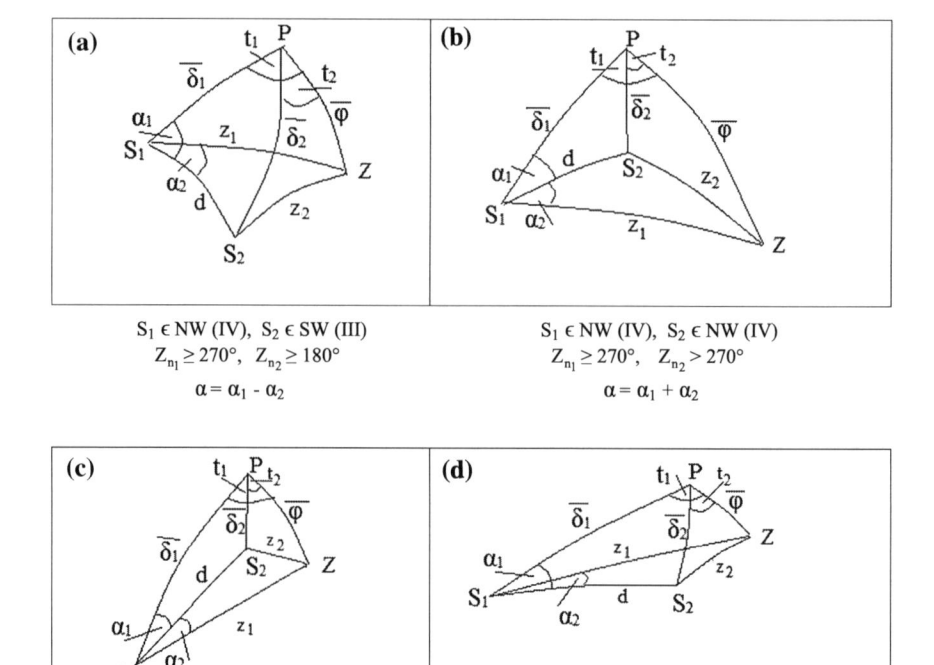

$S_1 \in$ NW (IV), $S_2 \in$ SW (III)
$Z_{n_1} \geq 270°$, $Z_{n_2} \geq 180°$
$\alpha = \alpha_1 - \alpha_2$

(b)

$S_1 \in$ NW (IV), $S_2 \in$ NW (IV)
$Z_{n_1} \geq 270°$, $Z_{n_2} > 270°$
$\alpha = \alpha_1 + \alpha_2$

(c)

$S_1 \in$ SW (III), $S_2 \in$ NW (IV)
$180° < Z_{n_1} < 270°$, $Z_{n_2} \geq 270°$
$\alpha = \alpha_1 + \alpha_2$

(d)

$S_1 \in$ SW (III), $S_2 \in$ SW (III)
$180° \leq Z_{n_1} < 270°$, $Z_{n_2} < 270°$
$\alpha = \alpha_1 - \alpha_2$

Fig. 2.4.6

(2) $\lambda = \begin{cases} t_1 - \text{GHA}(S_1) & \text{if } S_1 \text{ is West of the observer} \\ -t_1 + \text{GHA}(S_1) & \text{if } S_1 \text{ is East of the observer} \end{cases}$

We still need the angle subtended by S_1 and S_2 at the pole P, namely ΔS defined by:

(3) $\Delta S = \text{GHA } S_1 - \text{GHA } S_2$, valid in general.

In all cases where the altitudes a_1 and a_2 are measured simultaneously we simply have:

(4) $\Delta S = \text{SHA}(S_1) - \text{SHA}(S_2)$, valid if S_1 and S_2 are stars.

Since it is not always possible or practical to measure the required altitudes at the same instant, we require Eq. (3). In cases where GHA S_1 is measured at time T_1 and GHA S_2 at time $T_2 < T_1$, we like to reduce our calculations to the common time T_1 without employing approximations that are only valid for very short differences in time, like 4 min or less.

2.4 The Use of the Exact Equations for Finding ...

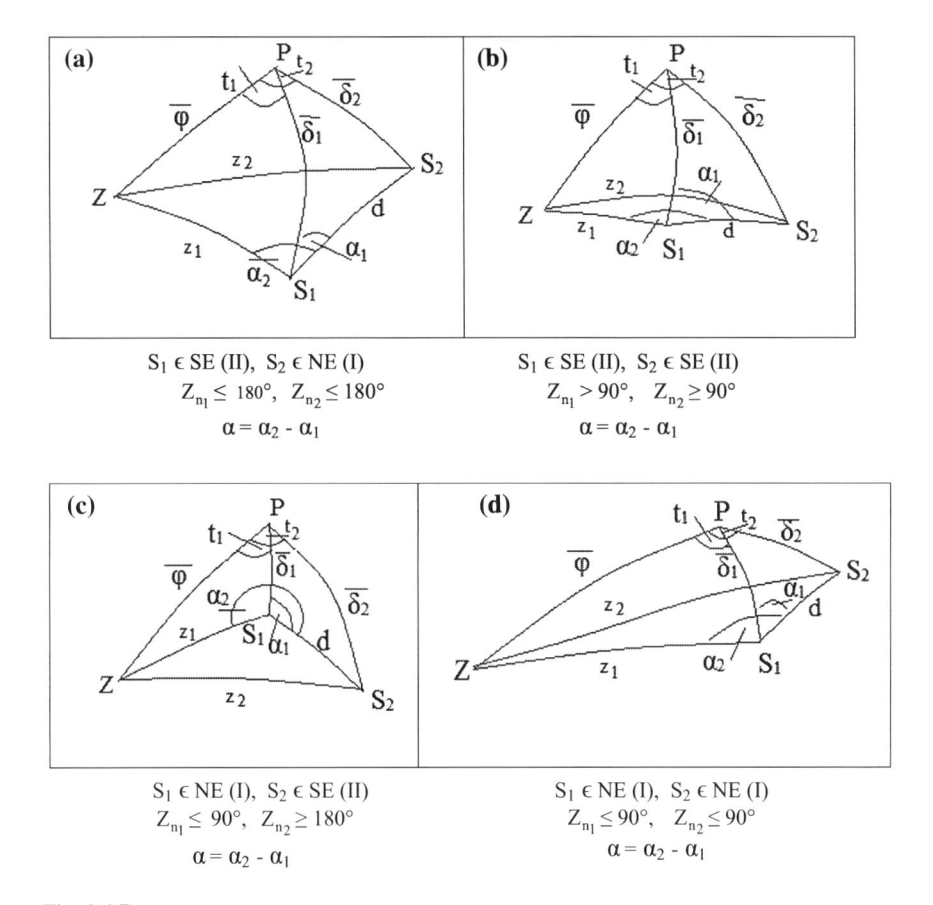

(a) $S_1 \in SE\ (II),\ S_2 \in NE\ (I)$
$Z_{n_1} \leq 180°,\ Z_{n_2} \leq 180°$
$\alpha = \alpha_2 - \alpha_1$

(b) $S_1 \in SE\ (II),\ S_2 \in SE\ (II)$
$Z_{n_1} > 90°,\ Z_{n_2} \geq 90°$
$\alpha = \alpha_2 - \alpha_1$

(c) $S_1 \in NE\ (I),\ S_2 \in SE\ (II)$
$Z_{n_1} \leq 90°,\ Z_{n_2} \geq 180°$
$\alpha = \alpha_2 - \alpha_1$

(d) $S_1 \in NE\ (I),\ S_2 \in NE\ (I)$
$Z_{n_1} \leq 90°,\ Z_{n_2} \leq 90°$
$\alpha = \alpha_2 - \alpha_1$

Fig. 2.4.7

It should be clear that we measure the time with our UTC chronometer. The corresponding time difference in Sidereal hours is therefore given by:

(5) $\Delta T_s = (T_1 - T_2) \cdot 1.002737962$. Therefore and in general, we have:
(6) $\Delta S = SHA\ S_1\ (T_1) - SHA\ S_2\ (T_2) + (T_1 - T_2) \cdot 1.002737962 \cdot 15°$.

However, in some applications, the use of a common time that is based on an averaging process of successive altitude measurements often eliminates the need for the use of two distinct time observations.

Next, we still have to consider the important case where the observer (vessel or airplane) is moving between the two sights. In this particular situation, we will have to calculate the new position Z' of the observer and therefore calculate the new zenith distance z'_2. However, all the other calculations remain the same once z_2 is replaced by z'_2.

The computation of z'_2 is based on the additional triangle Figs. 2.4.8 and 2.4.9, respectively, and requires the knowledge of direction (course) C and distance traveled, as well as the azimuth Z_n of S_2, as observed at the time T_2 when z_2 is being measured.

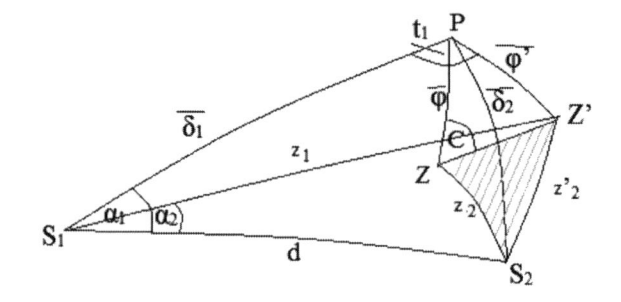

Fig. 2.4.8

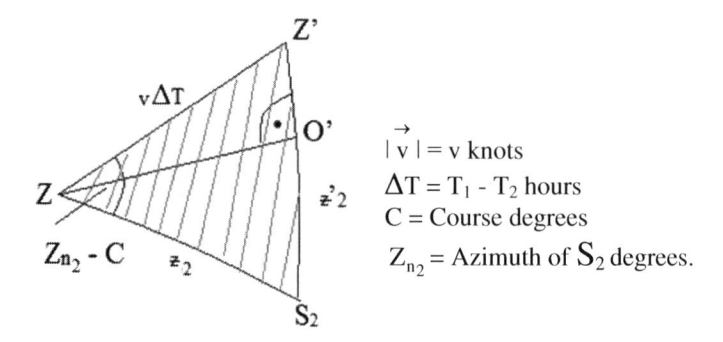

Fig. 2.4.9

$|\vec{v}| = v$ knots
$\Delta T = T_1 - T_2$ hours
C = Course degrees
Z_{n_2} = Azimuth of S_2 degrees.

Note that $\bar{\delta}_1$ is determined at T_1 and $\bar{\delta}_2$ at T_2, $T_1 > T_2$; also z_1 is measured at T_1 and z_2 at T_2. Furthermore, Z_{n_2} is also measured at T_2. Therefore, the necessary parameters used are z_1 and z'_2. Since $Z_{n_2} - C$ is known and $v\Delta T$ is also known, z'_2 can be computed easily by employing the COS-TH. of spherical trigonometry and obtaining:

(7) $\cos z'_2 = \cos v\Delta T/60 \cdot \cos z_2 + \sin v\Delta T/60 \cdot \sin z_2 \cdot \cos(Z_{n_2} - C)$, or
$a'_2 = \sin^{-1}(\cos v\Delta T/60 \cdot \sin a_2 + \sin v\Delta T/60 \cdot \cos a_2 \cdot \cos(Z_{n_2} - C))$,

since $z = 90° - a$.

It should be obvious that the proportions of the triangle $Z\tilde{S}_2Z'$ are highly exaggerated for reasons of clarity, for in real dimensions the arc extending from Z to Z' is relatively small in comparison to z_2 and, therefore, also very small in comparison to z'_2. Because of this, Fig. 2.4.9 suggests an approximation, namely the spherical triangle.

$Z\hat{O}'Z'$' by the plane triangle $Z\hat{O}'Z'$ with angles γ, $90°$ and β given by:
$\gamma = (Z_{n_2} - C) - 90°$ and $\beta = 180° - (Z_{n_2} - C)$—see Fig. 2.4.10 where $\overline{\Delta z_2} \cong \Delta z_2$, for sufficiently values of $v\Delta T$.
The application of the SIN-TH. of plane trigonometry yields:
$\sin \gamma / \overline{\Delta z_2} = 1/v\Delta T$ or

2.4 The Use of the Exact Equations for Finding ...

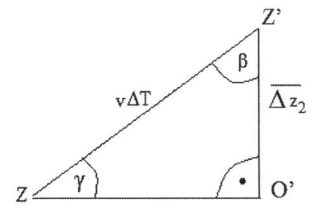

Fig. 2.4.10

$\Delta z_2 = v\Delta T \sin \gamma = -\cos(Z_{n_2} - C) \cdot v\Delta T$, since $\sin \gamma = -\cos(Z_{n_2} - C)$, hence:

(8) $a'_2 \cong a_2 + \cos(Z_{n_2} - C) \cdot v\Delta T/60$.

We now return to the task of finding the latitude φ and the longitude λ as defined by the spherical triangles $P\hat{S}_1Z$ and $P\hat{Z}S_2$ (see Fig. 2.4.2). As has already been pointed out, the solution of this problem is based on the calculation of the parallactic angle α, depicted in Fig. 2.4.2. In particular, we first calculate the interstellar distance d by employing:

$\cos d = \sin \delta_1 \cdot \sin \delta_2 + \cos \delta_1 \cdot \cos \delta_2 \cdot \cos \Delta S$, where ΔS is given by expression (6). Hence:

(9) $d = \cos^{-1}(\sin \delta_1 \cdot \sin \delta_2 + \cos \delta_1 \cdot \cos \delta_2 \cdot \cos \Delta S),^8 \; 0 \leq d \leq 180°$.

Next we calculate by employing again the COS-TH. to obtain:

(10) $\alpha_1 = \cos^{-1}\left(\dfrac{\sin \delta_2 - \sin \delta_1 \cdot \cos d}{\cos \delta_1 \cdot \sin d}\right)$ with $\cos \delta_1 \cdot \sin d \neq 0$, and $0 \leq \alpha_1 \leq 180°$.

Similarly we obtain α_2 by applying the COS-TH. to the triangle $S_1\hat{Z}S_2$ (see Fig. 2.4.3) as:

(11) $\alpha_2 = \cos^{-1}\left(\dfrac{\sin a'_2 - \cos d \cdot \sin a_1}{\cos a_1 \cdot \sin d}\right) = \cos^{-1} X_\alpha$, $\quad 0 \leq a_1 < 90°$,
$0 \leq \alpha_2 \leq 180°$, and a'_2 (given by formula (8)).

By employing the combinations provided by the box-diagram (see Fig. 5), we find:

(12) $\alpha = \begin{cases} \alpha_1 + \alpha_2 & \text{if} (IV, IV); (IV, I); (III, IV); (III, I); (S_1, S_2) \\ \alpha_1 - \alpha_2 & \text{if} (IV, III); (IV, II); (III, III); (III, II) \\ \alpha_2 - \alpha & \text{if} (II, II); (II, I); (I, II); (I, I) \end{cases}$

(Note that we actually only have to consider the first two cases in (12), i.e., only $\alpha = \alpha_1 \pm \alpha_2$ since the third one does not give rise to a different value for $\cos \alpha$.)

---

[8]To eliminate any ambiguity one must apply formulae (5) Sect. 2.2. rigorously which, in turn, makes it necessary to determine the sign and absolute value of he argument of $\cos^{-1} X$ to assure that $|X| \leq 1$. (See appendix this section.) In addition, it is also necessary that the selected results meet all the criteria of the quadrant orientation diagram. The latter eliminates all the guess work.

Again the application of the COS-TH. to the navigational triangle $P\hat{S}_1Z$ (see Fig. 2.4.2) yields:

(13) $\varphi = 90° - \cos^{-1}(\sin \delta_1 \cdot \sin a_1 + \cos \delta_1 \cdot \cos a_1 \cdot \cos \alpha) = 90° - \cos^{-1} X_\varphi$,
with $-90° < \varphi < 90°$

Finally, we calculate t_1 by employing the SIN-TH. and obtain:

$$\sin t_1 = \frac{\cos a_1 \cdot \sin \alpha}{\cos \varphi}, \quad -90° < \varphi < 90°, \text{ and hence:}$$

(14)

$$t_1 = \sin^{-1}\left(\frac{\cos a_1 \cdot \sin \alpha}{\cos \varphi}\right) = \cos^{-1}\left(\frac{\sin a_1 - \sin \delta_1 \cdot \sin \varphi}{\cos \delta_1 \cos \varphi}\right) = \cos^{-1} X_\lambda,$$

with $0 \leq t_1 \leq 180°$, and by employing formula (2) we find λ.

It should be noted that this procedure for determining the exact solution for the fundamental navigation problem involves only five trigonometric expressions, namely, (9)–(14) and does not require the plotting of any LOPs. It also involves the same number of trigonometric expressions that have to be evaluated if the navigator chooses to use the approximation of Celestial Navigation based on DR and/or assumed position as well as on the use of other plotting techniques.

Of course, the exact solution of the physical problem provided by the aforementioned formulae is contingent upon the lack of errors in the data used. In general, all data provided by physical instruments such as sextants, chronometers, barometers and thermometers, is bound to be erroneous to some degree. There may also be personal errors and/or errors that have entered into the ephemeral data. These errors will become obvious once the navigator employs data from observations of more than two celestial objects. However, irrespective of the magnitude of those errors (which enter ALL approximate methods of celestial navigation), the exact method presented here will always be superior to any non-exact method.

Regarding the mathematical procedures for minimizing the error in the evaluation of multi-body observations, please refer to Sect. 3.5.

Now, let's turn our attention to another method for calculating the exact solution to the physical problem. This method is based on finding the intersection of two circles of equal altitude by an iterative process.

Look again at Figs. 2.4.1 and 2.4.2 and evaluate the altitudes of the two triangles of equal altitude. The application of the COS-TH. readily yields:

(16) (a) $\sin a_1 = \sin \delta_1 \cdot \sin \varphi + \cos \delta_1 \cdot \cos \varphi \cdot \cos(\lambda + \text{GHA } S_1)$
  (b) $\sin a_2 = \sin \delta_2 \cdot \sin \varphi + \cos \delta_2 \cdot \cos \varphi \cdot \cos(\lambda + \text{GHA } S_2)$

These two equations, (16a) and (16b), constitute a transcendental system of equations for the latitude φ and the longitude λ and can be solved by numerical methods as outlined in the appendix of this chapter. Actually we can eliminate the unknown λ and obtain a single equation for φ representing the root of a nonlinear equation. In particular from (16a) we deduce under the assumption that $\cos \delta_{1,2} \cdot \cos \varphi \neq 0$ that: [44]

2.4 The Use of the Exact Equations for Finding ...

(17) $\lambda = -\text{GHA}(S_1) \pm \cos^{-1}\left(\frac{\sin a_1 - \sin \delta_1 \cdot \sin \varphi}{\cos \delta_1 \cdot \cos \varphi}\right) =$

$-\text{GHA}(S_2) \pm \cos^{-1}\left(\frac{\sin a_2 - \sin \delta_2 \cdot \sin \varphi}{\cos \delta_2 \cdot \cos \varphi}\right)$, where the plus sign stands if the object S_1 or S_2 lies West of the observer and the minus sign holds if the object is East of the observer.

If we define:

(18) $f(\varphi) = \pm \cos^{-1}\left(\frac{\sin a_1 - \sin \delta_1 \cdot \sin \varphi}{\cos \delta_1 \cdot \cos \varphi}\right) - (\text{GHA}(S_1) - \text{GHA}(S_2)) -$

$[\pm\cos^{-1}\left(\frac{\sin a_2 - \sin \delta_2 \cdot \sin \varphi}{\cos \delta_{21} \cdot \cos \varphi}\right)]$, then the latitude φ is merely a root of the equation $f(\varphi) = 0$ and can be found by employing a convergent iterative method (as found in a subsequent chapter). [18]

Another very important aspect of solving the system (16) is that it suggests an approximation which will turn out to be very similar to and perhaps superior to Sumner's two lines of position method.

Let us once more interpret Eq. (17) by defining a function of $g(\varphi)$ by the first right hand expression, and another function $h(\varphi)$ by the second right hand side expression so that (17) reads as follows:

(19) $\lambda = g(\varphi) = h(\varphi)$.

Then the equations $\lambda = g(\varphi)$ and $\lambda = h(\varphi)$ are two functional equations each representing a circle of equal altitude about S_1 and S_2, respectively. Now, if we plot the graphs of these two functions in the vicinity of the root of $f(\varphi) = g(\varphi) - h(\varphi) = 0$, these graphs represent the arcs of the two circles of equal altitude in the vicinity of the points of intersection of these circles. It follows then that if φ_1 lies sufficiently close to the actual solution of $f(\varphi) = 0$, we may expand $g(\varphi)$ and $h(\varphi)$ in a Taylor series about φ_1, and by neglecting terms of higher than first order derivatives, obtain the two linear equations:

$g(\varphi) \cong g(\varphi_1) + g'(\varphi) \cdot (\varphi - \varphi_1) + \cdots$
$h(\varphi) \cong h(\varphi_1) + h'(\varphi) \cdot (\varphi - \varphi_1) + \cdots$

However, because of the complexity in evaluating the derivatives of $g(\varphi)$ and $h(\varphi)$, we choose to approximate these derivatives by difference quotians and arrive at the two linear equations:

(20) $\lambda^{(1)} = g(\varphi_1) + \frac{g(\varphi_2) - g(\varphi_1)}{\varphi_2 - \varphi_1} \cdot (\varphi - \varphi_1) : L^{(1)}$

$\lambda^{(2)} = h(\varphi_1) + \frac{h(\varphi_2) - h(\varphi_1)}{\varphi_2 - \varphi_1} \cdot (\varphi - \varphi_1) : L^{(2)}$

The additional value of φ_2 used in formulae (20) represents another initial approximation to φ and is supposed to be very close to φ_1. Then the intersection of $L^{(1)}$ and $L^{(2)}$ defined by (20) constitutes another (and hopefully better) approximation to φ than φ_1 and φ_2. Like in the case of other approximations, the intersection of the two lines of position does not, in general, lie on any of the two circles of equal altitude and is not an exact solution.

In an upcoming section, I will specifically discuss approximate methods and will show how the Eq. (20) can be used to generate a sequence $\{\varphi_1\}$ of successive approximations that will converge to the solution of φ of Eq. (18) provided that φ_1 and φ_2 satisfy certain conditions. [18]

In the next section I shall apply the results of this section to some concrete solutions for which I will provide numerical solutions together with some conclusions. In addition, I will also restate all the necessary formulae in logical sequence.

Appendix One

Limitations
On the limitations of the applications of formulae (9)–(14)

The problem associated with the numerical evaluations of the aforementioned expressions basically consists in avoiding critical numerical values for the parameters for which a numerical solution cannot be found by means of using a specific calculator or computer—ILL CONDITIONED CASES.

In particular, the expressions $\cos^{-1} X_\alpha$, $\cos^{-1} X_\varphi$, and $\cos^{-1} X_\lambda$ require that the arguments of the multi-valued function $\cos^{-1} X$ are all bounded away from one, i.e., that $|X| < 1$, since for all $|X| > 1$, $\cos^{-1} X$ does not exist.

Should any of those arguments equate to a number greater than one, you may conclude that your input data, as for instance, altitude, latitude or any of the ephemeral data is erroneous.

Mathematically speaking, in the case of $|X| > 1$ this translates into saying, "There does not exist a spherical triangle with sides a, b, c and angle A as prescribed." To make this statement very clear, consider a spherical triangle with sides a, b, c and angle A opposite to a. Then the COS-TH. states:

$\cos a = \cos b \cdot \cos c + \sin b \cdot \sin c \cdot \cos A \ldots$

and it holds for any spherical triangle. However if we interpret the above equation as an abstract algebraic equation, then for some arbitrarily chosen sets of values a, b, c, it may not be possible to find a value for A that satisfies the equation. In order to prove this obvious statement it suffices to merely provide on example for which the above equation does not have a solution for A.

Example a = 75°, b = 15°, c = 45°. By substituting these values into the above formula, we find that: $\cos A = -2.317837252$ which does not have a solution. Therefore, in all cases where navigators use data that results in an absolute value of the argument $|X|$ of $\cos^{-1} X$ that is greater than one, they know that they are trying to apply the laws of spherical trigonometry to something that is not a spherical triangle.

2.4 The Use of the Exact Equations for Finding ...

Appendix Two

Error Analysis

Since the exact formulae of Astro Navigation (9)–(14) are analytic expressions based on analytic functions, a quantitative error analysis can be provided.[9] The latter is not available for plotting techniques still used by many navigators therefore those techniques must be classified as non-analytic methods.

As has been pointed out already, the criteria for all the above formulae are that the absolute values of the arguments of $\cos^{-1} X$ must be less than one, if not, the introduced values of the parameters are "blunders". It therefore appears that the key to the successful numerical evaluation are exactly these quantities: $X_\alpha, X_\varphi, X_\lambda$.

By considering the geometry of the triangles depicted in Figs. 2.4.5a–d. it is clear that only sets of celestial objects (S_1, S_2) should be considered that satisfy the following criteria:

(i) The azimuths should be bounded away from 180° and 360° and that they should show a clear cut separation. This also applies to the case where S_1, and S_2 are the same object and observe at distinct times.
(ii) The altitudes a_1 and a_2 should be bounded away from 90°.
(iii) The declination δ_1 and δ_2 should also be bounded away from 90°.
(iv) The latitude of the observer must also be bounded away from ±90°.

Whenever (i)–(iv) are satisfied, we can expect to obtain variable results by applying formulae (9)–(14) to obtain a "fix". However, these criteria still do not provide quantitative ones or any other quantitative assessment of the error. In order to obtain analytic results and quantitative criteria, we must consider the changes of the multi-valued function $\cos^{-1} X$ i.e., its derivative for all values of $|X|<1$ in all the relevant formulae (11), (13) and (14).

(Note that we don't have to invoke formula (10) since according to our hypothesis the ephemeral data is correct.)

Recalling from calculus that $\frac{d}{dx}\cos^{-1} u = \frac{1}{\sqrt{1-u^2}} \cdot \frac{du}{dx}$, we can readily see that all errors introduced by Δa_1 and Δa_2, i.e., by errors in the altitude, have a common amplification factor $(1-X^2)^{-1/2}$. In particular, we deduce that:

for $X_\alpha = X_\alpha(a_1, a_2)$: $\Delta E_\alpha \cong \Delta \alpha_2 = -(1-X_\alpha^2)^{-1/2} \cdot \text{grad} X_\alpha \cdot \vec{\Delta a}$, where $\vec{\Delta a} = (\Delta a_1, \Delta a_2)$ and $\text{grad} X_\alpha = \left(\frac{\delta X_\alpha}{\delta a_1} \cdot \frac{\delta X_\alpha}{\delta a_2}\right)$.

Similarly, we find that:

$\Delta E_\varphi \cong \Delta \varphi = -(1-X_\varphi^2)^{-1/2} \cdot \text{grad} X_\varphi \cdot \vec{\Delta \alpha}$, with $\vec{\Delta \alpha} = (\Delta a_1, \Delta \alpha_2)$, and

$\Delta E_\lambda \cong \Delta \lambda = -(1-X_\lambda^2)^{-1/2} \cdot \text{grad} X_\lambda \cdot \vec{\Delta \varphi}$, with $\vec{\Delta \varphi} = (\Delta a_1, \Delta \varphi)$

---

[9]This analysis is explicitly provided in Sect. 2.6.

It follows then that the coefficient of amplifications of ΔE_α is $(1 - X_\alpha^2)^{-1/2}$, the coefficient of amplification in ΔE_φ is $(1 - X_\alpha^2)^{-1/2} \cdot (1 - X_\varphi^2)^{-1/2}$, and the coefficient of amplification of ΔE_λ is $(1 - X_\alpha^2)^{-1/2} \cdot (1 - X_\varphi^2)^{-1/2} \cdot (1 - X_\lambda^2)^{-1/2}$. Therefore, the critical values in our computations are these coefficients of amplification, all of them to be bound away from ∞. Therefore, in order to minimize the resulting errors, we must assure that all $|X| \ll 1$.

Recalling that by definition—see formula (11)—

$$|X_\alpha| = \left| \frac{(\sin a_2' - \sin a_1 \cdot \cos d)}{\cos a_1' \cdot \sin d} \right| = |\cos \alpha_2|$$ so that the condition $|X_\alpha| \ll 1$ implies that $\alpha_2 \gg 0$,[10] and $\alpha_2 \ll 180°$, and therefore that the West-East orientation of S_1–S_2 should be such that Z, the observer's position, does not lie on the great-circle connecting S_1 and S_2.

According to formula (13) we have—

$$|X_\varphi| = |\sin \delta_1 \cdot \sin a_1 + \cos \delta_1 \cdot \cos a_1 \cdot \cos \alpha| = |\sin \varphi|,$$ hence $|X_\varphi| \ll 1$ implies that $\varphi \ll 90°$, i.e. that high latitude observations should be avoided, if possible.

Looking at formula (14), we deduce—

$$|X_\lambda| = \left| \frac{(\sin a_1 - \sin \delta_1 \cdot \sin \varphi)}{\cos \delta_1 \cdot \cos \varphi} \right| = |\cos t_1|$$ which implies that t_1 should satisfy $0 \ll t_1 \ll 180°$, i.e., very small and large local hour angles should be avoided.

Although we could have deduced some of the more obvious results merely by knowing the behavior of the multi-valued function $\cos^{-1} X$ in the vicinity of the point $|X| = 1$, we would not have been able to readily see that in the case of the error in φ, the product of the two amplification factors enters the magnitude of the resulting error, and in the case of $\Delta\lambda$, the product of all three amplification factors enters the magnitude of the resulting error.

In conclusion, the recommended procedure for the application of the exact formulae (9)–(14) should be to select the COs in compliance with the criteria established in this part of the appendix. Furthermore, it is always necessary to check the magnitude of all three values of X to make sure that $|X| < 1$ is indeed satisfied. If not, the user of these formulae will probably have "blunder" in his or her data. It is also imperative that the correct branch for $\cos^{-1} X$ is to be selected, thereby eliminating the remaining ambiguities (see formulae (5), Sect. 2.2).

---

[10]In all cases where GHA $S_1(T_1)$ < GHA $S_2(T_2)$, the parallactic angle α becomes the angle at S_2. By considering again all twelve relevant cases, we can again set up a box-diagram denoted by $(S_1, S_2) = (S_2, S_1)$, and thereby reducing those cases to the previous ones. Example: (II, III) = (III, II).

Appendix Three

Analytic Derivation of Formula (8)

The correction for the observed altitude (as a result of the motion of the vessel as given in this section) can be derived analytically from the exact equation:

$$\sin H_0' = \sin H_0 \cdot \cos \Delta t \cdot v/60 + \cos H_0 \cdot \sin \Delta t \cdot v/60 \cdot \cos(C - Z_n),$$

by assuming that the quantities of $\Delta t \cdot v/60$, and $\Delta H_0 = H_0' - H_0$, are sufficiently small relative to all the other components in those equations.

For by approximating $\sin \Delta H_0$ by $\Delta H_0 \cdot \sin 1'$, and $\cos \Delta H_0$ by 1; $\sin \Delta t \cdot v/60$ by $\Delta t \cdot v/60 \cdot \sin 1'$; and $\cos \Delta t \cdot v/60$ by 1, the above equation can be reduced to:
$\sin H_0' = \sin H_0 + \Delta t \cdot v/60 \cdot \cos H_0 \cdot \cos(C - Z_n) \cdot \sin 1'$.
Similarly we deduce that:
$\sin H_0' = \sin(H_0 + \Delta H_0) = \sin H_0 + \cos H_0 \cdot \Delta H_0 \cdot \sin 1'$, and therefore:
$\Delta H_0 = \Delta t \cdot v/60 \cdot \cos(C - Z_n)$, as was to be shown.

IN ORDER FOR THE READER TO USE THIS BOOK AS A MANUAL, AN ALGORITHM SPECIFICALLY FOR THIS SECTION HAS BEEN INCLUDED.

ALGORITHM A

FOR CALCULATING THE EXACT POSITION WHEN TWO ALTITUDE MEASUREMENTS ARE AVAILABLE—EITHER TWO DISTINCT COs OR ONE AT DISTINCT TIMES.

Definitions:

SHA(S_1): Sidereal Hour Angle of S_1	Z_n: Azimuth of S_2 at T_2
SHA(S_2): Sidereal Hour Angle of S_2	C: Course of vessel
T_1 : UTC of observation of S_1	v: Speed of vessel
T_2 : UTC of observation of S_2	a_1: True altitude of S_1 at $T_1 \geq T_2$
δ_1: Declination of S_1	a_2: True altitude of S_2 at T_2
δ_2: Declination of S_2	a_2': Corrected altitude of a_2
d: Interstellar distance between S_1 and S_2	φ: Latitude of vessel at T_1
	λ: Longitude of vessel at T_1

1. Select two COs or only one at distinct instants in accordance with the criteria and recommendations of Sect. 2.4. Note: S_1 always lies to the West of S_2.
2. Calculate:
   (i) $\Delta S = SHA(S_1) - SHA(S_2) + \Delta T \cdot 1.002737962 \cdot 15°, \quad \Delta T = T_1 - T_2$
   =
   (Note that SHA (S_1) $\geq$ SHA (S_2).)

3. Calculate:
   (ii) $X_d = \sin \delta_1 \cdot \sin \delta_2 + \cos \delta_1 \cdot \cos \delta_2 \cdot \cos \Delta S$.
   =
   If $|X_d| \leq 1$, proceed to step 4. If $|X_d| > 1$—STOP—Remove blunder.
4. Calculate:
   (iii) $d = \cos^{-1} X_d$, $0 \leq d \leq 180°$ by employing

   (iii') $\cos^{-1} X = \begin{cases} \cos^{-1}|X| \text{ if } 0 \leq X \leq 1 \\ 90° + \sin^{-1}|X| \text{ if } -1 \leq X \leq 0 \end{cases}$ or

   $\cos^{-1} X = \begin{cases} -\cos^{-1}|X| \text{ if } 0 \leq X \leq 1 \\ -(90° + \sin^{-1}|X|) \text{ if } -1 \leq X \leq 0 \end{cases}$
   =

   Recall that $\cos^{-1}|X|$ and $\sin^{-1}|X|$ are the main branches that you have in your calculator.
5. Calculate:
   (iv) $X_{\alpha 1} = \frac{\sin \delta_2 - \sin \delta_1 \cdot \cos d}{\cos \delta_1 \cdot \cos d}$
   =

   and check whether or not $|X_{\alpha 1}| \leq 1$ If yes, go to step 6. If not—STOP.
6. Calculate:
   (v) $\alpha_1 = \cos^{-1} X_{\alpha 1}$, $0 \leq \alpha_1 \leq 180°$ by employing formulae (iii'). Go to step 7.
   =
7. Calculate:

   (vi) $a_2' = \begin{cases} \sin^{-1}(\cos v \cdot \Delta T/60 \cdot \sin a_2 + \sin v \Delta T/60 \cdot \cos a_2 \cdot \cos(Z_n - C)) \\ a_2 + \cos(Z_n - C) v \Delta T/60, \Delta T = T_1 - T_2. \end{cases}$ or
   =

   (Note that the second expression is merely an approximation of the first. Also note that expression (vi) is evaluated by employing the formula:

   (vi') $\sin^{-1} X = \begin{cases} \sin^{-1}|X|, \text{ if } 0 \leq X \leq 1 \\ -\sin^{-1}|X|, \text{ if } -1 \leq X \leq 0 \end{cases}$ or

   $\sin^{-1} X = \begin{cases} 90° + \cos^{-1}|X|, \text{ if } 0 \leq X \leq 1 \\ -(90° + \cos^{-1}|X|), \text{ if } -1 \leq X \leq 0 \end{cases}$
   =
8. Calculate:
   (vii) $X_{\alpha_2} = \frac{\sin a_2' - \cos d \cdot \sin a_1}{\cos a_1 \cdot \sin d}$. If $|X_{\alpha_2}| \leq 1$ proceed to step 9. If not—STOP.
   =
9. Calculate:
   (viii) $\alpha_2 = \cos^{-1} X_{\alpha_2}$, $0 \leq \alpha_2 \leq 180°$ and proceed to step 10.
   =

2.4 The Use of the Exact Equations for Finding ...

10. Calculate:

$$(ix)\ \alpha = \begin{cases} \alpha_1 + \alpha_2 & \text{if}(IV, IV); (IV, I); (III, IV); (III, I) \\ \alpha_1 - \alpha_2 & \text{if}(IV, III); (IV, II); (III, III); (III, II) \\ \alpha_2 - \alpha_1 & \text{if}(II, II); (II, I); (I, II); (I, I) \end{cases}$$

=

(Note that, for instance, (IV, I) means that S_1 lies in quadrant IV and S_1 lies in quadrant I. "Z_n" denotes that zenith of the observer.)

11. Calculate:

(x) $X_\varphi = \sin \delta_1 \cdot \sin a_1 + \cos \delta_1 \cdot \cos a_1 \cdot \cos \alpha$

If $|X_\varphi| < 1$ proceed to step 12. If not—STOP—Remove blunder.

=

12. Calculate:

(xi) $\varphi' = 90° - \cos^{-1} X_\varphi, -90° \leq \varphi \leq 90°$

=

13. Calculate:

(xii) $X_\lambda = \frac{\sin a_1 - \sin \delta_1 \cdot \sin \varphi'}{\cos \delta_1 \cdot \cos \varphi'}$. If $|X_\lambda| \leq 1$ then go to step 14. If not—STOP.

=

14. Calculate:

(xiii) $t_1 = \cos^{-1} X_\lambda, 0 \leq t_1 \leq 180°$ by employing formulae (iii') Go to step 15.

=

15. Calculate:

$$(xiv)\ \lambda' = \begin{cases} t_1 - GHA(S_1), & \text{if } S_1 \text{ lies West of the observer} \\ -(t_1 + GHA(S_1)), & \text{if } S_1 \text{ lies East of the observer} \end{cases}$$

2.5 Conclusions and Numerical Examples

In the previous section, the navigator has been provided with all the relevant formulae for finding the exact solution to the most important problem of astronavigation provided, of course, that the input data for those equations is exact. However, "exact" data is seldom possible. Nevertheless, the exact formulae will always give you superior results than any of those provided by "approximate" methods. The errors introduced into the exact formulae will simply reflect the errors committed by the observer, errors in the auxiliary formulae and data, as well as possible instrument errors and, perhaps, even errors in the ephemeral data.

By employing more than two celestial objects, the navigator is able to obtain several solutions which can then be reduced by using the formulae of the Least Mean Square Approximation to a most probable position. The derivations of these individual solutions from the mean value can tell the navigator whether the data used was adequate or not.

Aside from providing the necessary exact solution, the two equations, a and b of Sect. 2.4 (16), independently solve the same problem which subsequently reduces the problem of solving the system of transcendental equations to the simpler problem of finding the roots of a single nonlinear equation (18).

Later, we will see how the iterative approximations obtained by employing the "Regula Falsi" will graphically constitute the method of finding an approximate solution by plotting two lines and determining their intersection. As opposed to other methods based on the knowledge of an approximate position, this method will actually converge to the exact solution provided that the two initial values for φ satisfy a simple condition.

It should be stressed here that there are only five trigonometric expressions in Algorithm A for the exact solution as compared to six trigonometric expressions required in the approximate LOP or Sumner's Method—namely three trigonometric formulae for each CO in addition to the problem of plotting the LOPs involved. [26, 27, 43]

Because of formulae (7) and (8), it is also necessary that the navigator measures or computes the azimuth of S_2 at T_2. Although azimuth measurements are usually difficult to perform on a moving vessel, it suffices here to use the same degree of accuracy as in determining the course C since only the difference $Z_n - C$ enters the calculations. Hopefully, the mutual errors will cancel themselves out. However, in light of other applications, such as in identifying stars and planets and finding latitude and longitude if altitude and azimuth are known, it is good practice to measure or, at least, estimate the azimuth every time altitudes are observed.[11]

Let us now consider representatives of the most relevant cases of applications as encountered in practical navigation:

The first kind of problems are encountered whenever the observer is stationary and observes two or more celestial objects. In this case, the times of the observations may be simultaneous; less than five minutes apart; more than five minutes apart; or, perhaps, hours apart.

The second kind of problems arise every time the navigator takes sights of one CO, for instance the sun or the moon or the planets, but at distinct instants that are hours apart.

The third kind of problems relate to a moving observer who observes two or more COs at distinct times T_1 and T_2 more than just a few minutes apart while the

---

[11]See Sect. 2.6 for an explicit error analysis.

2.5 Conclusions and Numerical Examples

vessel moves with a known velocity relative to the azimuth of the CO first observed.

The four numerical examples considered here are typical representations of the real problems encountered in navigation:

Example 1 Stationary observer with two stars observed at different times.[12]

Star S_1: **MARKAB**.
Time of observation $T_1 = 4^h \, 33^m \, 16^s$ UTC
Date: 11/20/08
SHA $S_1 = 13° \, 41'.7$
$\delta_1 = 15° \, 15'.4$ N
True Altitude $a_1 = 49° \, 14'.6$
Star S_2: **FOMALHAUT**
Time of observation $T_2 = 3^h \, 17^m \, 38^s$ UTC
Date: 11/20/08
SHA $S_2 = 15° \, 27'.6$
$\delta_2 = 29° \, 34'.6$
True Altitude $a_2 = 31° \, 26'.1$
The observer takes note that S_1 and S_2 both lie in quadrant (III), hence $(S_1, S_2) = $ (III, III).

> Problem: Determine the unknown position of the observer $Z = (\lambda, \varphi)$ at T_1.
> Solution: Step 2. $\Delta S = 17°.19517$ by formula (1) of Algorithm A
>
> Step 4. $d = 47°.80470729$ by formulae (2) and (3) of Algorithm A
> Step 6. $\alpha_1 = 159°.6937682$ by formulae (4) and (5) of Algorithm A
> Step 9. $\alpha_2 = 88°.48872802$ by formulae (7) and (8) of Algorithm A
> Step 10. $\alpha = 71°.20004018$ by formula (9) of Algorithm A
> Step 12. $\underline{\varphi} = 23°.718955 = 23° \, 43'.14$ N by formulae (10) and (11) of Algorithm A
> Step 13. $t_1 = 42°.4587362$ by formulae (12) and (13) of Algorithm A
> Step 13'. GHA $(S_1) = $ GHA Υ + SHA $S_1 = 141°.58333$
> Step 14. $\underline{\lambda} = -99°.12462002 = \underline{99°07'.47 \text{ W}}$

Example 2

Observer stationary. Sun is observed at two distinct times. On 03/23/08 the Sun S_2 ⊙ is observed at $T_2 = 19^h \, 22^m \, 42^s$ UTC, and a true altitude $a_2 = 67° \, 34'.12$ is determined. The GHA (S_2) and declination taken from the Nautical Almanac are GHA $(S_2) = 109°08'$ and $\delta_2 = 1°.4066$ N.
On the same day, but at $T_1 = 21^h \, 35^m \, 16^s$ UTC the sun ⊙ is again observed and $a_1 = 52° \, 11'.01$ is found. The GHA (S_1) and declination also taken from the Nautical Almanac are GHA $(S_1) = 142° \, 13'.70$ and $\delta_1 = 1° \, 26'.60$ N.

---

[12] Note that whenever SHA $S_1 <$ SHA S_2, it is possible that GHA $S_1(T_1) >$ GHA $S_2(T_2)$—check.

The observer also observes that S_2 lies in quadrant (II) and S_1 in quadrant (III), hence $(S_1, S_2) = (III, II)$.

Problem: Find the position of $Z = (\lambda, \varphi)$ of the observer at T_1.

Solution: Step 2. $\Delta S = 33°.14833$ by formula (1) of Algorithm A
Step 4. $d = 33°.13780313$ by formulae (2) and (3) of Algorithm A
Step 6. $\alpha_1 = 89°.63762935$ by formulae (4) and (5) of Algorithm A
Step 9. $\alpha_2 = 38°.35325032$ by formulae (7) and (8) of Algorithm A
Step 10. $\alpha = 51°.28437903$ by formula (9) of Algorithm A
Step 12. $\underline{\varphi = 23°.78241057 = 23°46'.94N}$ by formulae (10) and (11) of Algorithm A
Step 13. $t_1 = 31°.52017$ by formulae (12) and (13) of Algorithm A
Step 14. $\underline{\lambda = -110°.70816 = 110°42'.5W}$ by formula (14) of Algorithm A

Example 3 Observer moves between taking sights.

On 03/24/08 the navigator observes **ARCTURUS** (S_2) at $T_2 = 7^h\ 35^m\ 16^s$ UTC, and determines the true altitude at $a_2 = 61°\ 18'.72$. In addition the navigator determines the azimuth $Z_{n2} = 94°$, and course $C = 277°$. The approximate speed of the yacht is $v = 9.6$ knots. According to the Nautical Almanac, the sidereal hour angle and declination of **ARCTURUS** on this date are SHA (S_2) = $145°\ 59'.0$ and $\delta_2 = 19°\ 08'.1$. The Greenwich hour angle for Aries at T_2 is GHA (Υ) = $295°58'.1$. It has also been noted that S_2 lies in quadrant (II).

On the same day, but at $T_1 = 10^h 0^m 20^s$ UTC, the navigator observes **DENEBOLA** (S_1) and determines $a_1 = 49°\ 17'.36$. The sidereal hour angle and declination are also taken from the NA and are SHA (S_1) = $182°\ 37'.3$ and $\delta_1 = 14°\ 31'.41$ N. Furthermore, GHA (Υ) = $332°28'.1$ and S_1 lies in quadrant (III). Hence $(S_1, S_2) = (III, II)$.

Problem: Find the position of $Z' = (\lambda', \varphi')$ of the yacht at T_1.

Solution: Step 2. $\Delta S = 73°0'.3$ by formula (1) of Algorithm A
Step 7'. $\Delta T = T_1 - T_2 = 2^h.417777$, $v \cdot \Delta T/60 = 0.386832$ NM
Step 7 $a'_2 = 60°.92569232$ by formula (6) of Algorithm A
Step 4. $d = 69°.54217967$ by formulae (2) and (3) of Algorithm A
Step 6. $\alpha_1 = 74°.6466407$ by formulae (4) and (5) of Algorithm A
Step 9. $\alpha_2 = 4°.696181632$ by formulae (7) and (8) of Algorithm A
Step 10. $\alpha = 69°.950459$ by formula (9) of Algorithm A
Step 12. $\underline{\varphi' = 23°.98853014 = 23°59'.31N}$ by formulae (10) and (11) of Algorithm A.
Step 13. $t_1 = 42°.11623535$ by formulae (12) and (13) of Algorithm A
Step 14. $\underline{\lambda' = -112°.8404247 = 112°50'.43\ W}$ by formula (14) of Algorithm A.

2.5 Conclusions and Numerical Examples

Example 4 Date: 11/16/08

Time: $T_1 = T_2 = 02^h00^m00^s$UTC.
Stars Observed:
S_1: **PEACOCK**—SHA $(S_1) = 53°24'.7$, $\delta_1 = -56° \ 42'.6$, $a_1 = 44° \ 42'.6$.
S_2: **ARCHENAR**—SHA $(S_2) = 335° \ 28'.6$, $\delta_2 = -57° \ 11'.5$, $a_2 = 62° \ 59'.9$.
GHA(Υ) = $85°31'.3$. It has also been noted that S_1 lies in quadrant (III) and S_2 lies in quadrant (II). Hence $(S_1, S_2) = $ (III, II).

Problem: Find the position of the observer.

Solution: By following steps (1) through (14) we obtain the following results:
$\Delta S = 77°.935$
$d = 40°.1171$
$\alpha_1 = 124°.6819$
$\alpha_2 = 85°.1122$
$\alpha = \alpha_1 - \alpha_2 = 39°.5687$ since $(S_1, S_2) = $ (III, II).
$X_\varphi = -0.287380549$
$\underline{\varphi = -16°.7011981}$
$X_\lambda = 0.881255162$
$t_1 = 28°.20585451$ and
$\lambda = -110°.6274785$, hence,
$\varphi = 16°42'.07$ S and $\lambda = 110°37'.65$ W

In the next section we shall consider still another method for finding the exact solution to the fundamental problem of astronavigation.

2.6 An Exact Method Based on Cartesian Coordinates and Vector Representations

In the previous section we solved the exact navigational problem by solving a system of two transcendental equations (Sect. 2.4, (16)) by using equations of spherical trigonometry. Therefore, in order to avoid the problem of solving such equations by this approach, we must employ a different tactic that is not based on these formulae. Since this is indeed possible, the reader must wonder why I chose to introduce iterative methods for solving transcendental equations when simpler mathematical methods were available? The answer is simple. I wanted to show that Sumner's Method of celestial navigation was just an approximate method for solving these transcendental equations.

In the appendix to Sect. 2.4, I introduced a brief introduction to vectors. In the subsequent chapter, I showed that the fundamental equations of spherical trigonometry could easily be derived by using elements of vector algebra.

Therefore, it should be obvious to deduce that vectors could also be used to directly solve the exact fundamental problem of navigation thereby avoiding the use of transcendental equations.

Going one step further, permitting the introduction of matrices, the navigator can also solve the more general problem of finding an approximate position by observing more than two celestial objects or one CO at more than two distinct instants.

The reader will discover that the exact method of calculating the position of the observer based on vectors will prove itself superior to any approximate method because it clearly shows when NO numerical solution for a given set of parameters can be obtained. It also provides a straight forward error analysis for all exact methods. Such an error analysis is still not available for the two LOP method of celestial navigation.

Ardent users of celestial navigation frequently argue that exact methods are not superior to approximate methods because all exact methods, when applied to erroneous data, yield only approximations to the true solution anyway. However, this line of argument does not stand up to scrutiny since it does not take into consideration that ALL solutions provided by approximate methods are subject to the same errors as far as the input parameters are concerned. In addition, they are erroneous by virtue of being only approximations to the true solution in the mathematical sense.

The exact analytical methods always permit an analytic assessment of the error introduced by inaccurate parameters and show when a numerical solution for a given set of parameters does not exist. Approximate methods do not exhibit these features and therefore must be considered to be of second rank.

Let us recall the vector representation $\vec{x}$ of a point P in the Euclidean space E^3 in terms of its Cartesian coordinates x, y, z, namely:

$$\vec{x} = \begin{pmatrix} x \\ y \\ z \end{pmatrix} \text{ with its length—NORM } |\vec{x}| = (x^2 + y^2 + z^2)^{1/2}$$

Also let us recall the definition of the INNER PRODUCT of two vectors $\vec{x}_1 \cdot \vec{x}_2 = x_1 x_2 + y_1 y_2 + z_1 z_2$ where x_1, y_1, z_1 are the components of $\vec{x}_1$ and x_2, y_2, z_2 are the components of $\vec{x}_2$.

As can be shown easily (see Sect. 2.4), we then also have the expression:
$\vec{x}_1 \cdot \vec{x}_2 = \|\vec{x}_1\| \cdot \|\vec{x}_2\| \cdot \cos \alpha$, where α is the angle subtended by the vectors $\vec{x}_1, \vec{x}_2$—see Fig. 2.6.1.

By definition we have chosen the representation of all the vectors under consideration relative to the Cartesian Coordinate system (see Fig. 2.6.2). Here we have made use of the three unit vectors $\vec{e}_1, \vec{e}_2, \vec{e}_3$ which span the coordinate system

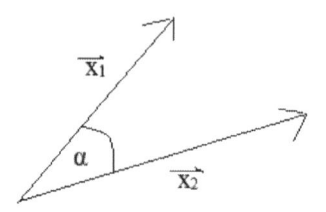

Fig. 2.6.1

2.6 An Exact Method Based on Cartesian ...

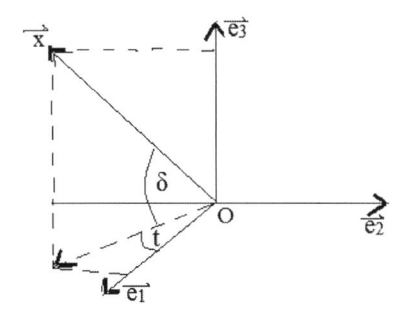

Fig. 2.6.2

and are mutually perpendicular, i.e., $\|\vec{e}_1\| = \|\vec{e}_2\| = \|\vec{e}_3\| = 1$ and $\vec{e}_1 \cdot \vec{e}_2 = \vec{e}_1 \cdot \vec{e}_3 = \vec{e}_2 \cdot \vec{e}_3 = 0$

We also have the representation:

$$\vec{e}_1 = \begin{pmatrix} 1 \\ 0 \\ 0 \end{pmatrix} \quad \vec{e}_2 = \begin{pmatrix} 0 \\ 1 \\ 0 \end{pmatrix} \quad \vec{e}_3 = \begin{pmatrix} 0 \\ 0 \\ 1 \end{pmatrix}$$

In order to relate the Cartesian Coordinates (x, y, z) to the polar coordinates, as used in astronomy and navigation, we let $\vec{e}_1$, i.e., the x-axis lie on the meridian plane passing through Greenwich and we let the z-axis point toward the celestial north pole. Finally, for the sake of convenience in the calculations, we measure the angle which $\vec{x}$, together with its projection onto the (x, y) plane, forms and the angle t which this projection forms with the vector $\vec{e}_1$, measured in the positive sense of the clock. Then this angle can be identified as the Greenwich Hour Angle (GHA). The angle that x subtends with its projection on to the (x, y) plane can now be identified as the declination of the celestial object.

In the cases where $\vec{x}$ represents the position vector of the observer, those angles become the negative longitude and latitude, respectively.

Since all the points under consideration lie on the terrestrial sphere, we may assume that all the corresponding vectors are unit vectors. As the common unit, we use the length of the radius of the standard terrestrial sphere, so that 1 min of degree corresponds to one Nautical Mile in length. We then have the following representation of the position vector $\vec{x}$ of the observer:

(1) $$\vec{x}_k = \begin{pmatrix} x_k \\ y_k \\ z_k \end{pmatrix} = \begin{pmatrix} \cos \delta_k \cdot \cos \text{GHA}(S_k) \\ -\cos \delta_k \cdot \sin \text{GHA}(S_k) \\ \sin \delta_k \end{pmatrix}, k = 1\ldots n.$$

$$\vec{x} = \begin{pmatrix} x \\ y \\ z \end{pmatrix} \begin{pmatrix} \cos \varphi \cdot \cos|\lambda| \\ -\cos \varphi \cdot \sin|\lambda| \\ \sin \varphi \end{pmatrix}$$

Since the zenith distance of a celestial object is the acute angle subtended by the position vector $\vec{x}$ and the position vector $\vec{x}_k$ of the substellar point of S_k, we conclude that:
$\vec{x}_k \cdot \vec{x} = \cos(90° - H_{0,k}) = \sin H_{0,k}, k = 1\ldots n$. Here "$H_{0,k}$" denotes the true altitude of S_k. By invoking the definition of the Inner Product, we conclude that
(2a) $x_k x + y_k y + z_k z = \sin H_{0,k}, k = 1\ldots n$, and also
(2b) $x^2 + y^2 + z^2 = 1$

Here, (2a) represents n-linear equations for three unknowns x, y, z, and (2b) is the quadratic equation that assures that the computed position vector $\vec{x}$ lies on the terrestrial sphere and not in the air or on the ocean.

Next let us consider the case which, in theory, always has a solution, i.e., the case where n = 2. Here we have only the two observation values, $H_{O,1}$, and $H_{O,2}$ to consider. According to the Eqs. (2a) and (2b), we have two linear and one quadratic equation for the three unknowns, x, y, z available, provided that:

(2') $GHA(S_1) \neq \begin{cases} GHA(S_2) \\ GHA(S_2) \pm 180° \end{cases}$

The problem defined by (2) always has a numerical solution, except in such cases where all or some of the parameters are ICPs (Incompatible Parameters—also see numerical example (3).

In order to solve the Eqs (2) explicitly, we first eliminate the term of $z_k z$ from (2a) to obtain:

(2c) $x_1 \cdot x + y_1 \cdot y = \sin H_{0,1} - z_1 \cdot z$
$x_2 \cdot x + y_2 \cdot y = \sin H_{0,2} - z_2 \cdot z$
$x^2 + y^2 + z^2 = 1$

Next we define the quantity:
(3) $D = x_1 \cdot y_2 - x_2 \cdot y_1$, that actually turns out to be the determinant of the linear system above for the components x and y in terms of z. It follows then that $D \neq O$, for
D = O implies that $\frac{x_1}{y_2} = \frac{x_2}{y_1}$.

Substituting the corresponding values for the Cartesian coordinates according to Eq. (1) into this equation, we conclude that:
tan GHA (S_1) = tan GHA (S_2), which implies that GHA (S_1) = GHA (S_2), or GHA (S_1) = GHA $(S_2) \pm 180°$, a contradiction to our hypotheses.

Therefore, it should be understood that the vanishing of the determinant D implies that the two celestial objects S_1 and S_2 lie on the same hour circle, and conversely, if the two objects lie on the same hour circle, it follows that D = 0, and the above system (2c) does not have a solution.

Almost equally important is the case when D is different from, but very close to zero. In all those cases, when the hour angles of S_1 and S_2 are close to each other—just a few degrees or less apart—the result will be that D is close to zero, and we then refer to the matrix corresponding to D to be ILL CONDITIONED. Whenever

2.6 An Exact Method Based on Cartesian ...

that occurs, the above method must fail for numerical reasons and it would be necessary to employ computers of extremely high accuracy to obtain an acceptable result. But even this approach may fail.

We also will encounter the same difficulties arising from ill conditioned matrices when we deal with the Least-Square Approximations of the cases where n > 2.

The limitations arising from numerical computations are unavoidable and not only because of round-off errors, but also due to the representation of all rational and irrational numbers by a finite number of digits as employed in all computers.

At this point in our analysis, the reader may recall the limitations of the LOP method as used in celestial navigation. Although there is no real error analysis available for said method, it is very obvious that this method must fail whenever the computed azimuths of the two COs are equal or almost equal, and also when the computed azimuths differ markedly from the true azimuths.

Next, let us define the quantities:

$h_1 = \sin H_{0,1}$

(4) $h_2 = \sin H_{0,2}$, then the unique solution of (2c) is given by:

$$x = \frac{h_1 y_2 - h_2 y_1}{D} + \frac{z_2 y_1 - z_1 y_2}{D} \cdot z \text{ and } y = \frac{h_2 x_1 - h_1 x_2}{D} + \frac{z_1 x_2 - z_2 x_1}{D} \cdot z$$

If we now define the expression:

(5) $a_1 = \dfrac{h_1 y_2 - h_2 y_1}{D}, a_2 = \dfrac{h_2 x_1 - h_1 x_2}{D}, b_1 = \dfrac{z_2 y_1 - z_1 y_2}{D}, b_2 = \dfrac{z_1 x_2 - z_2 x_1}{D},$

we may conclude that:

(6) $x = a_1 + b_1 z$, and $y = a_2 + b_2 z$.

Then in order to satisfy the third equation of (2c) we must have:

$(a_1^2 + a_2^2 - 1) + 2(a_1 b_1 + a_2 b_2)z + (b_1^2 + b_2^2 + 1)z^2 = 0.$

Again, let us define:

(7) $A = \dfrac{2(a_1 b_1 + a_2 b_2)}{b_1^2 + b_2^2 + 1}$, and $B = \dfrac{a_1^2 + a_2^2 - 1}{b_1^2 + b_2^2 + 1}$, and conclude that z must satisfy the quadratic equation: $z^2 + Az + B = 0$, which, in general has two distinct solutions, namely:

(8) $z_{1,2} = -A/2 \pm \dfrac{(A^2 - 4B)^{1/2}}{2}$. Although in theory $A^2 - 4B \geq 0$, but in praxis it can happen that the computed value of $A^2 - 4B < 0$, which implies that incorrect parameters are being used. Those incorrect values may include incorrect values for $H_{0,1}$ and $H_{0,2}$, and, perhaps, $GHA(S_1)$ and $GHA(S_2)$, as well as round-off errors which may have accumulated. Therefore, the quantity $A^2 - 4B$ serves as a filter for excluding such blunder. Hence, the proper definition for blunder should be D = 0 and $A^2 - 4B < 0$.

We still must eliminate the ambiguity due to the existence of two solutions given by (8). This can be accomplished in all cases where $(A^2 - 4B)^{1/2}$ is significantly different from zero. In the pathological case where $z_1 = z_2$ either the double root $z_{1,2} = \frac{A}{2}$ should be used or the solution to the system (2c) should be recomputed with a different set of parameters. With the exception of the pathological case, we can now obtain two distinct values for the latitude by evaluating:

(9) $\varphi_{1,2} = \sin^{-1} z_{1,2}$. Then by eliminating the latitude that is obviously wrong, we have determined the correct value for z and can then compute the corresponding values for x and y by employing Eq. (6).

Recalling that the underlying premise of our analysis is that $GHA(S_1) \neq GHA(S_2)$ and also $GHA(S_1) \neq GHA(S_2) \pm 180°$, which means that the two celestial objects S_1 and S_2 may not lie on the same hour circle, we also conceive that a sufficient separation of the 2 h circles, i.e., hour angles, will result, in general, in a sufficient separation of the two latitudes φ_1 and φ_2. Therefore, it will suffice in most practical situations that navigators need only to have a rough idea[13] of their latitude in order to be able to select the correct value for z. However, in the case when they are completely unsure of their latitude, they will have to make a rough estimate of the azimuths of the objects they are observing. It may be quite sufficient to know in which quadrant these objects can be located.

Finally, the longitude of the navigator's position is to be found by computing:

$$(10) \quad \lambda = \tan^{-1} \frac{y}{x} = \begin{cases} \pm \tan^{-1}\left|\frac{y}{x}\right| & + \text{if } 0 \leq \frac{y}{x} < \infty \\ \pm(\tan^{-1}\left|\frac{y}{x}\right| - 180°) & -\text{if } -\infty \leq \frac{y}{x} < 0, \lambda \leq 0. \end{cases}$$

This equation for λ follows directly from the polar-coordinate representation for the Cartesian coordinates as given by Eq. (1) and by the formula for the multi-valued function $\tan^{-1} x$ (see Sect. 2.4).

Next let us consider the multi-object problem where navigators have observed n distinct objects, or only one at distinct instants, or any of the above combinations. We now assume that $n > 2$. In all those situations, when the navigators have programmable calculators at their disposal, they do not have to be concerned about the amount of the calculations they have to perform and can, therefore, select any distinct pair of observations from the list of those n observations (2). As the reader can readily see, it is possible to select $n(n - 1)/2$ pairs of compatible sets of data and execute the previous program for $n = 2$ up to $n(n - 1)/2$ times thereby generation the same number of possible positions, namely $x_k = (\lambda_k \varphi_k)$, $k = 1 \ldots j = 1$, since we will eliminate-filter the incompatible sets of data in this process as explained above. By taking the Mean-Square-Value of all those possible positions, we shall arrive at the most probable position.

At this point in our discussion, it is necessary to point out that n distinct celestial objects should be observed for the purpose of pin-pointing systematic errors, as for instance, the always dominant error made in determining the atmospheric refraction R by the use of convenient formulae—(Sect. 3.1). However, it is not recommended to employ more than one CO for the purpose of assessing the personal errors which are random errors that should be evaluated by using the methods explained in the section on error analysis of Chap. 3.

Although the algorithm for finding the position of the observer has been established by executing formulae (1) through (10), it is still necessary to

---

[13] A prudent navigator always knows approximate latitude.

2.6 An Exact Method Based on Cartesian …

investigate how an error in observation or the altitudes $H_{0,1}$ and $H_{0,2}$ will effect the position vector $\vec{x}$. It also may be necessary to determine the effect of errors in determining the time of observation on the position vector.

In order to derive analytic expressions, we proceed as follows:

Firstly, let the true observation vector $\vec{L}$ be related to the observed observation vector $\vec{L}_0$ by:

(11) $\vec{L} = \vec{L}_0 + \vec{v}$, where "$\vec{v}$" denotes the error of said vector due to the errors in measurements of the observed altitudes $H_{0,1}$ and $H_{0,2}$. Since we have already solved the problem for the approximate position vector $\vec{x}_0$, defined by:

$$A\vec{x}_0 + \vec{L}_0 = 0, \text{ subject to } \|\vec{x}_0\| = 1, \text{ with } A = \begin{pmatrix} x_1 y_1 z_1 \\ x_2 y_2 z_2 \end{pmatrix}; \vec{L}_0 = \begin{pmatrix} -\sin H_{0,1} \\ -\sin H_{0,2} \end{pmatrix}.$$

We now express the true position $\vec{x}$ in terms of $\vec{x}_0$, and the error $\vec{d}$ as:

(12) $\vec{x} = \vec{x}_0 + \vec{d}$ in order to find $\vec{d}$.

Since $\vec{x}$ is the exact value, for we have assumed that the matrix A represents the exact ephemeral data, it must satisfy:

$A\vec{x} + \vec{L} = 0$, with $\|\vec{x}\| = 1$, hence $\vec{d}$ must satisfy:

(13) $A\vec{d} + \vec{v} = 0$, and $\|\vec{d}\|^2 + 2\vec{x}_0 \cdot \vec{d} = 0$, since by substituting (11) and (12) into the following equations we obtain:

$A(\vec{x}_0 + \vec{d}) + \vec{L}_0 + \vec{v} = A\vec{x}_0 + \vec{L}_0 + A\vec{d} + \vec{v} = A\vec{d} + \vec{v} = 0$, and also

$\|\vec{x}\|^2 = \|\vec{x}_0 + \vec{d}\|^2 = \|x\|^2 + 2\vec{x}_0 \cdot \vec{d} + \|d\|^2 = 1$. But $\|\vec{x}\|^2 = 1$. It follows then that: $\|d\|^2 + 2\vec{x}_0 \cdot \vec{d} = 0$.

Therefore, in order to find the error in the position vector, we must first solve the two linear equations for the two unknowns d_1 and d_2, i.e., the components of vector $\vec{d}$, and then again solve the resulting quadratic equation for the third component d_3 that is of the form:

$$d_3^2 + A_d d_3 + B_d = 0, \text{ exactly as in the case of finding } \vec{x}_0.$$

What remains to be done is to derive an explicit expression for $\vec{v} = \vec{L} - \vec{L}_0$—see expression (13') For this purpose, we define the errors introduced by employing the approximate values $H_{0,1}$ and $H_{0,2}$ in our calculations by:

$H_1 = H_{0,1} + \Delta_1$, and $H_2 = H_{0,2} + \Delta_2$ together with the corresponding error matrix:

$\Delta = \begin{pmatrix} \Delta_1 & 0 \\ 0 & \Delta_2 \end{pmatrix}$. Here "$H_1$" and "$H_2$" denote the exact altitudes and Δ_1 and Δ_2 the errors in measured altitudes. Then by definition of $\vec{L}$ we have:

(13') $\vec{L} = \begin{pmatrix} -\sin H_1 \\ -\sin H_2 \end{pmatrix} = \begin{pmatrix} -\sin(H_{0,1} + \Delta_1) \\ -\sin(H_{0,2} + \Delta_2) \end{pmatrix} = \begin{pmatrix} -\sin H_{0,1} \\ -\sin H_{0,2} \end{pmatrix} - \sin 1'' \cdot \Delta \vec{L}_0.$

Note that the errors Δ_1 and Δ_2 are measured in seconds of degrees. Henceforth we have found $\vec{v}$ to be:

(14) $\vec{v} = -\sin 1'' \, \Delta \vec{L}_0$, where $\vec{L}_0 = \begin{pmatrix} \cos H_{0,1} \\ \cos H_{0,2} \end{pmatrix}$

In the derivation of (14) we have assumed that the values Δ_1 and Δ_2 are sufficiently small as to justify the use of the following approximations:

$\cos \Delta_{1,2} \cong 1$, and $\sin \Delta_{1,2} \cong \sin 1'' \cdot \Delta_{1,2}$.

In order to facilitate the error analysis for the case n = 2, we have used the concept of matrices as used in the second part of this book. However, it should be understood that this is not absolutely necessary for the understanding and execution of the error analysis as presented in this section because everything can be done by long-hand calculations. For example, (13) reads in component form as:

$x_1 d_1 + y_1 d_2 + z_1 d_3 = -v_1$
$x_2 d_1 + y_2 d_2 + z_3 d_3 = -v_2$, and the second equation of (13) reads:

$d_1^2 + d_2^2 + d_3^2 + 2(x_0 d_1 + y_0 d_2 + z_0 d_3) = 0$, where the components of x_0, y_0, and z_0 belong to the approximate position of vector $\vec{x}_0$, calculated with $\vec{L}_0$. It is then obvious that the solution of this non-linear problem employs the same program as the solution for $\vec{x}_0$.

It should also be understood that the explicit error analysis provided in this section is valid for any exact solution (see appendix below) of the fundamental navigational problem of Astro Navigation and, in particular, for the exact solution provided in Sect. 2.4. Furthermore, the same line of arguments used in the derivation of the error, as a result of errors in the altitudes, yields the explicit error as a result of the errors in the chronometers. Note that an error in the chronometer results in an error of the ephemeral data that enters all calculations.

In one of the next sections, we shall consider the above problem for n > 2 that does not have a solution in general since the number of equations is always greater than three—the number of unknowns. Furthermore, the data of observation $\vec{L}_0$ is not exact in general. However, since the position of the observer is real, we may look for a numerical approximation to the true solution by utilizing the observation data that makes up the matrix A or our system, in context with the n + L equations of conditioning. Obviously, any substitute for the true value obtained by such a method is merely an approximation and therefore, said methods are classified as approximate methods. It is left to the navigator to decide whether or not he or she prefers to use such methods instead of using the exact methods provided in this book.

2.6 An Exact Method Based on Cartesian ...

Algorithm B

For calculating the EXACT POSITION of the observer whenever two or more than two altitude observations are taken, and a crude estimate for the latitude and/or azimuth are available.

Definitions

Z_n : True azimuth of S_2	C: Course of vessel
$H_{0,1}$: True altitude of S_1	δ_1: Declination of S_1
$H_{0,2}$: True altitude of S_2	δ_2: Declination of S_2
GHA(S_1): Greenwich Hour Angle of S_1	φ: Latitude of observer
GHA(S_2): Greenwich Hour Angle of S_2	λ: Longitude of observer
	v: Speed of vessel

1. Select two C.O.s, S_1 and S_2, such that their Greenwich Hour Angles satisfy:
$$\text{GHA}(S_1) \neq \begin{cases} \text{GHA}(S_2) \\ \text{GHA}(S_2) \pm 180°. \end{cases}$$
Also recall that GHA S_k (T_k) = GHA Υ(T_k) + SHA (S_k).

2. Calculate:
(i) $\vec{x}_k = \begin{pmatrix} x_k \\ y_k \\ z_k \end{pmatrix} = \begin{pmatrix} \cos \delta_k \cdot \cos \text{GHA}(S_k) \\ -\cos \delta_k \cdot \sin \text{GHA}(S_k) \\ \sin \delta_k \end{pmatrix}$, k = 1, 2, . . .

 =

3. Compute:
(ii) $D = x_1 y_2 - x_2 y_1 \neq 0$, and go to step 4.

 =

4. Compute:
iii).
$h_1 = \sin H_{0,1}$

 =

$h_2 = \sin H'_{0,2}$

 =

$H'_{0,2} = H_{0,2} + v \cdot \frac{(T_1 - T_2)}{60} \cdot \cos(Z_n - C)$

 =

and go to step 5.

5. Compute:
(iv)
$a_1 = \frac{h_1 y_2 - h_2 y_1}{D}$

 =

$a_2 = \frac{h_2 x_1 - h_1 x_2}{D}$

 =

$$b_1 = \frac{z_2 y_1 - z_1 y_2}{D}$$
=
$$b_2 = \frac{z_1 x_2 - z_2 x_1}{D}$$
=

and go to step 6.

6. Compute:

(v)
$$A = \frac{2(a_1 b_1 + a_2 b_2)}{b_1^2 + b_2^2 + 1}$$
=
$$B = \frac{a_1^2 + a_2^2 - 1}{b_1^2 + b_2^2 + 1}$$
=

If $A^2 - 4B \geq 0$ got to step 7. If not... stop and remove blunder.

7. Compute

(vi) $z_{1,2} = -\frac{A}{2} \pm \frac{(A^2 - 4B)^{1/2}}{2}$

(vi) =

Note the ambiguity-two solutions, which can only be removed by employing an estimate for the latitude or azimuth. Also check whether or not $|z_{1,2}| \geq 1$. If so, go to step 8. If not... STOP and remove blunder.

8. Compute:

(vii) $\varphi_{1,2} = \sin^{-1} z_{1,2}$

=

Remove the physical wrong branch to obtain:

$\varphi = \varphi_1$

=

Here we have eliminated z_2. Then go to step 9.

9. Compute:

(viii) $x = a_1 + b_1 z$

=

$y = a_2 + b_2 z$

=

with $z = z_1$

=

Then go to step 10.

10. Compute:

(ix) $\lambda = \tan^{-1} \frac{y}{x}$, by employing the formula (ix') below...

(ix') $\tan^{-1} x = \begin{cases} \pm \tan^{-1} |x| & \text{if } 0 \leq x < \infty \\ \pm(\tan^{-1} |x| - 180°) & \text{if } -\infty < x \leq 0 \end{cases}$. See Sect. 4.2.

=

2.6 An Exact Method Based on Cartesian …

Note that $\tan^{-1}|x|$ stands for the main branch of $\tan^{-1} x$ and is the function your calculator uses.

$$\sin^{-1} x = \begin{cases} \sin^{-1}|x|, \text{ if } 0 \leq x = 1 \\ -\sin^{-1}|x|, \text{ if } -1 \leq x \leq 0 \end{cases}$$

$$\sin^{-1}|x| = \begin{cases} 90° + \cos^{-1}|x|, \text{ if } 0 \leq x = 1 \\ -(90° + \cos^{-1}|x|), \text{ if } -1 \leq x \leq 0 \end{cases}$$

Appendix

The result of this error analysis can also be used to estimate the error introduced by employing any approximate method, as for instance, Sumner's LOP method of celestial navigation. If one merely replaced $\vec{L}_0$ by $\vec{L}_0^*$, defined by $A\vec{x}_0^* + \vec{L}_0^* = 0$, where "$\vec{x}_0^*$" denotes the position vector obtained by the use of the corresponding approximation, then the error matrix Δ has the elements $\Delta_{1,2}$: $H_{0_{1,2}} - H_{0_{1,2}}^*$ where $H_{0_{1,2}}^*$ are the components of $\vec{L}_0^*$, and the H_0 are the values measured by the user of the approximation method.

2.7 Numerical Examples and Conclusions

First, let's consider an example based on an observation taken from the balcony of a two-story house in Ciudad Victoria, Tamaulipas, Mexico at two distinct instants. The measured Dip-Short amounted to $2°58'.4$ and the resulting observation data obtained included:

Bodies observed: Sun (S_1 and S_2) observed twice.
Date: 02/02/13
Times of observation: $T_1 = 16^h 30^m 00^s$ UTC
$T_2 = 18^h 30^m 00^s$ UTC
True Altitudes: $H_{0,1} = 37°06'$.
$H_{0,2} = 49°23'$.
Ephemeral data: $\delta_1 = -16°37'.5$
$\delta_2 = -16°36'.1$
$GHA(S_1) = 64°03'.7$
$GHA(S_2) = 94°03'.6$

Latitude NORTH—Find the position of the observer.
Solution:

1. Calculate $\vec{x}_1, \vec{x}_2$, and $\vec{h}$ by using formulae (1) and (3) of Algorithm B:
   $x_1 = 0.41912011$, $x_2 = -0.06784071$, $h_1 = 0.60320798$
   $y_1 = -0.861673606$, $y_2 = -0.95500966$, $h_2 = 0.75908159$
   $z_1 = -0.28610648$, $z_2 = -0.28571512$.

2. Calculate $D, a_1, a_2, b_1,$ and b_2 by using formulae (2) and (4) of Algorithm B.
   $D = -0.45909757$
   $a_1 = -0.16874024 \quad b_1 = 0.05946181, \quad a_1^2 + a_2^2 - 1 = -0.35981836$
   $a_2 = -0.7824787 \quad b_2 = -0.30300336, \quad b_1^2 + b_2^2 + 1 = 1.0954134.$
3. Calculate $A, B,$ and $A^2 - 4B$, by using formula (5) of Algorithm B:
   $A = 0.41452257 \quad B = -0.32847723 \quad A^2 - 4B = 1.485737882, \quad$ O.K.
4. Calculate z_1 and z_2 by using formula (6) of Algorithm B:
   $z_1 = 0.402192954, \quad z_2 = -0.81671552$
5. Calculate φ_1 and φ_2 using formulae (7) and (9') of Algorithm B.
   $\varphi_1 = 23°43'$ NORTH, and $\varphi_2 = -54°.76$ (Close to Cape Horn.)
6. Calculate x and y using formula (8) of Algorithm B. $z = z_1$
   $x = -0.14482511$ and $y = -0.904027934$ and $\frac{y}{x} = 6.24220437.$
7. Calculate $\lambda < 0$ by using formulae (9) and (9') of Algorithm B.
   Select proper branch of $\tan^{-1} x$.
   $\underline{\lambda = \tan^{-1}|6.24220437| - 180° = -99°.101439 = 99°06'.08}$ WEST

The GPS yields: $\varphi = 23°43'.057$ NORTH and $\lambda = 99°07'.53$ WEST

This example clearly demonstrates the simplicity of this algorithm. It can be executed by anyone who can operate an inexpensive calculator with the trigonometric and inverse trigonometric functions incorporated, provided that the CO's are sufficiently separated with respect to their hour angles.

Basically, it can be done by anyone with elemental knowledge of arithmetic and algebra. It is not even necessary to know much about trigonometric functions because they can be executed by merely pressing the key with the corresponding symbol on the keyboard. Nowadays, even kids learn to push the buttons on a computer without understanding the principles behind them.

However, there is one step in the program that needs to be re-examined very carefully, namely the selection of the correct branch of the multi-valued function $\tan^{-1} x$. The navigator should note that solutions obtained by adding or subtracting 180° are also possible (see formula (9)). In those cases where we require that the correct answer to be negative for a positive value of the principle branch, we must subtract 180° from this value. Refer quickly to the case above. The principal value given by the calculator was: $\tan^{-1} 6.24220437 = 80°.89836057$. However, since λ must be negative, we had to subtract 180° from this value to arrive at the correct answer for the longitude λ.

Now, for the sake of comparison, let us consider the example that we previously executed in Sect. 2.5. Recall that in that chapter, we employed the following set of data:

S_1: ☉, $T_1 = 21^h 35^m 16^s$ UTC,
$H_{0,1} = 52°11'.01$,
$\delta_1 = 1°26'.6$,
$GHA(S_1) = 142°13'.7$
S_2: ☉, $T_2 = 19^h 22^m 42^s$ UTC,

2.7 Numerical Examples and Conclusions

$H_{0,2} = 67°34'.12$,
$\delta_2 = 1°24'.34$,
$GHA(S_2) = 109°04'.8$

Find the position of the observer.
Solution:

1. Calculate $\vec{x}_1, \vec{x}_2$, and $\vec{h}$ by using formulae (1) and (3) of Algorithm B:
   $x_1 = -0.790207212$ $x_2 = -0.32689578$ $h_1 = 0.789978475$
   $y_1 = -0.612321903$ $y_2 = -0.944778271$ $h_2 = 0.924337499$
   $z_1 = 0.025188254$ $z_2 = 0.024548486$
2. Calculate D, a_1, a_2, b_1, and b_2 by using formulae (2) and (4) of Algorithm B.
   $D = 0.546470187$ O.K
   $a_1 = -0.330049846$ $a_1^2 = 0.108932901$
   $a_2 = -0.864203458$ $a_2^2 = 0.746847617$ $a_1^2 + a_2^2 - 1 = -0.144219481$
   $b_1 = 0.016040654$ $b_1^2 = 0.000257302$
   $b_2 = 0.020435024$ $b_2^2 = 0.00041759$ $b_1^2 + b_2^2 + 1 = 1.000674892$
3. Calculate A, B, and $A^2 - 4B$, by using formula (5) of Algorithm B:
   $A = -0.045877505$ $A^2 = 0.002104745$
   $B = -0.144122214$ $A^2 - 4B = 0.57859360$ O.K.
4. Calculate z_1 and z_2 by using formula (6) of Algorithm B:
   $z_1 = 0.403265454, z_2 = -0.35738794$
5. Calculate φ_1 and φ_2 using formulae (7) and (9') of Algorithm B.
   $\varphi_1 = 23°.78247715 = 23°46'.05$ NORTH, and $\varphi_2 = -20°.93$
6. Calculate x and y using formula (8) of Algorithm B. $z = z_1$
   $x = -0.323581204$ and $y = -0.855962718$ and $\frac{y}{x} = 2.645279479$
7. Calculate $\lambda < 0$ by using formulae (9) and (9') of Algorithm B.
   $\lambda = \tan^{-1} 2.645279479 - 180°$ since $\lambda < 0$

Hence $\lambda = -110°42'.49 = \underline{110°42'.49 \text{ WEST}}$

Finally, let's consider another example previously explored in Sect. 2.5 that of two COs (stars) which were observed at distinct instants T_1 and T_2.
Date 11/20/08—
Latitude NORTH

Star S_1: **MARKAB**.	Star S_2: **FOMALHAUT**
Time of observation $T_1 = 4^h 33^m 16^s$ UTC	Time of observation $T_2 = 3^h 17^m 38^s$ UTC
$= 4^h.55444$ UTC	$= 3^h.2938$ UTC
SHA $S_1 = 13°41'.7 = 13°.695$	SHA $S_2 = 15°27'.1 = 15°.45166$
$\delta_1 = 15°15'.4$ N	$\delta_2 = -29°34'.6 = -29°.57666$
True Altitude $H_{0,1} = 49°14'.6 = 49°.24333$	True Altitude $H_{0,2} = 31°26'.1 = 31°.435$
GHA (Υ) = 127°.886666, at $T = T_1$	

Find the position of the observer.
Solution:

1. First we calculate $\Delta T = T_1 - T_2 = 1^h.26056$ UTC
   Next we calculate:
   GHA $S_1(T_1)$ = GHA Υ (T_1) + SHA (S_1) = $127°.8866 + 13°.695 = 141°.5816$,
   and
   $$\begin{aligned}\text{GHA } S_2(T_2) &= \text{GHA}\,\Upsilon(T_2) + \text{SHA}(S_2)\\ &= \text{GHA}\,\Upsilon(T_1) + \text{SHA}(S_2) - \Delta T \cdot 15° \cdot 1.002737962\\ &= 127°.88666 + 15°.45166 - 18°.96017048 = 124°.3781495.\end{aligned}$$
   Hence ... GHA S1(T1) > GHA S2(T2).

2. Calculate $\vec{x}_1, \vec{x}_2$ by using formulae (1) of Algorithm B:
   $x_1 = -0.75588169 \quad x_2 = -0.491075881$
   $y_1 = -0.599498413 \quad y_2 = -0.717785296$
   $z_1 = 0.263143358 \quad z_2 = -0.493587628$

3. Calculate by using formula (2) of Algorithm B.
   $D = 0.248351635$

4. Calculate h_1 and h_2 by using formula (3) of Algorithm B.
   $h_1 = 0.757488989$, and $h_2 = 0.521530939$

5. Calculate $a_1, a_2, b_1,$ and b_2 by using formula (4) of Algorithm B.
   $a_1 = -0.930364267 \quad a_2 = -0.089514668 \quad a_1^2 + a_2^2 - 1 = -0.126409456$
   $b_1 = 1.95199776 \quad b_2 = -2.022604789 \quad b_1^2 + b_2^2 + 1 = 8.901225388$

6. Calculate A, B, and $A^2 - 4B$, by using formula (5) of Algorithm B:
   $A = -0.367368783 \quad B = -0.014032033 \quad A^2 - 4B = 0.1910879554$ O.K

7. Calculate z_1 and z_2 by using formula (6) of Algorithm B:
   $z_1 = 0.402252432, \quad z_2 = -0.03488365$
   Hence $z = z_1$ since Latitude NORTH

8. Calculate φ using formulae (7) of Algorithm B.
   $\varphi = \sin^{-1} z = 23°.7190646 = 23°43'.14$ NORTH, and $\varphi_2 < 0$, SOUTH of Equator.

9. Calculate x and y using formula (8) of Algorithm B.
   $x = -0.14516842$ and $y = -0.903112363$ and $6.221135168 = \frac{y}{x}$

10. Calculate λ by using formulae (9) and (9') of Algorithm B.
    $\lambda = \tan^{-1} 6.221135168 - 180° = -99°.13174437 = \underline{99°07'.9 \text{ WEST}}$

In conclusion, it can be stated that the limitations of the applications of the above method are all those cases where the determinant D (see expression 2) of the corresponding matrix are either zero, or worse, nearly zero. Physically speaking, this means that the alignment of the observed COs does not permit the application of the above algorithm and/or some of the used date is ICP-blunder.

The concept of ILL-CONDITIONED matrices can be better understood if we consider a 2 × 2 matrix of a set of data representing two linear equations belonging to two LOPs that are almost parallel. Since it is virtually impossible to find the point of intersection of such lines, we would also not expect to be able to solve this problem by calculations since the determinant of the coefficient matrix of the two equations will be nearly zero depending on the uncertainty of the numerical values. Therefore, we could call those almost parallel lines ill conditioned and refer to the corresponding matrix as ill-conditioned also. In those cases where n > 2, we are actually dealing with almost parallel lines or hyper planes in the n-dimensional space E^n.

It can also be stated once more that the exact method based on vector representations also enables us to provide an explicit error analysis for any other exact method and therefore, in particular, for the exact method of Sect. 2.4.

2.8 On Approximate Solutions for Finding the Position at Sea or Air by Employing Two or More Altitude Observations

By now it should be clearly understood that an exact method for calculating the position of an observer by measuring altitudes of celestial objects COs results in the true position provided that ALL the input data such as altitudes, chronometer time, and ephemeral data are exact. Since the latter is not always possible, the resulting calculations will also be subject to those input errors. But these are not errors of methodology as in the case of all approximate methods. This can readily be seen if one chooses an exact position for an observer, obtains the ephemeral data such as GHAs and declinations (at specific instants), and then calculates the corresponding true altitudes. If one uses these parameters in our formulae of Algorithms A and B, one should get the exact position of the observer as prescribed before.

On the other hand, all approximate methods will, in general, produce only erroneous results even if all the input parameters such as altitudes, time and ephemeral data are correct.

Here I would also like to distinguish between two types of approximate methods, namely:

(a) **Non-convergent approximations.**
These comprise methods that will not generate the true solution and will, at best, produce another approximation for which no exact error analysis is available. It also compromises convergent methods that actually converge (as the name implies) but that will not necessarily converge to the exact solution. For example, the two lines of position method is one of those methods.

(b) **Convergent approximations**.
These comprise all the methods that produce a sequence of results by applying the same method-formulae successively, utilizing the result of each previous iteration and thereby approaching successively the exact solution. Such methods, if they can be devised, will also provide an error estimate at each step of iteration.

It should be noted that strictly from a mathematical point of view, it will suffice to find only one special case where a given approximate method fails to produce the exact solution to classify it as a divergent method. This, however, does not imply that such methods may be considered to be completely useless. Contrarily, such methods may prove to be useful in some special applications to celestial navigation.

Another important aspect of approximate solutions arises whenever the navigator applies these methods to situations where more than two (n > 2) COs are observed. In such cases, it is customary to apply the method of "Least Square Approximation" to arrive at one, single result. It is always assumed that the result so obtained constitutes a better approximation than any of the initial ones. It is also often assumed that by repeated application of this process to more samples the exact solution can be sufficiently closely approximated.

The reality, however, may be quite different. The Least Square method results in an approximation that has value only if all or the majority of the individual results are meaningful. In other words, the Least Square Approximation of blunder is... blunder.

In a previous section, an approximation method had been conceived strictly on grounds of mathematical arguments. This approximation could then be intrepreted geometrically, representing the approximate solution as the intersection of two staraight lines. Therefore, we are now going to look for such approximations which can be obtained by drawing straight lines and finding their intersections. We begin again by looking at the circle of equal altitudes of a celestial object, S, and the position Z of the observer. Let's concentrate on depicting the resulting navigational triangle $P\widehat{S}Z$ (see Fig. 2.8.1) together with a section of the extended arc of the circle of equal altitude in the neighborhood of Z (see Fig. 2.8.2).

Note that we have extended the arc of the great circle SZ somewhat. If it would be possible to find the parallactic angle α from one observations of S, the exact solution of φ could be obtained by a simple calculation. But we already know that this is not possible. Therefore, any type of approximation has to be based on approximating α adequately. This entails approximating the above triangle $P\widehat{S}Z'$ in such a manner that the resulting angle α' approximates α sufficiently closely.

Let us suppose that Z' represents the position of a fictitious vessel whose position λ', φ' is known. Then as long as Z' stays on the extended arc spanned by SZ, the

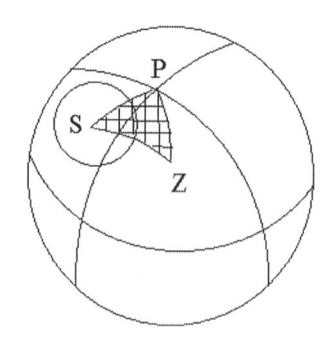

Fig. 2.8.1

2.8 On Approximate Solutions for Finding the Position …

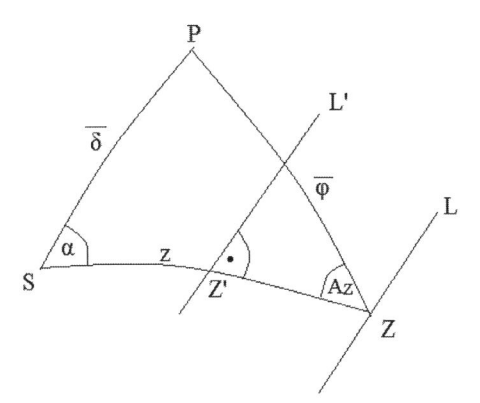

Fig. 2.8.2

exact solution can be obtained by plotting the arc ZZ′. It should be noted that the azimuth angle A′$_z$ at Z′ varies as Z′ moves along the extended arc of SZ, but the parallactic angle remains constant. Therefore (and contrary to what some proponents of the two lines of position method believe), it is not the azimuth angle of the assumed position, but the parallactic angle of said approximation that is responsible for the accuracy or even failure of said method.

In the case where α′ = α, we can now construct a line L′ perpendicular to the arc ZZ′. This line is parallel to the tangent line L to the circle of constant altitudes and passes through the position Z′ of the fictitious vessel. This means that the exact position of the observer can be obtained by moving the line L′ (the line perpendicular to the arc SZ′) by a distance equal to the arc length of arc ZZ′ parallel to itself. The intersection of this line with the arc SZ is then the exact position of the observer, and the distance ZZ′ corresponds to the difference in altitudes measured at Z and Z′ respectively.

Unfortunately, there is no way to determine the position of the fictitious vessel such that the resulting parallactic angle α′ coincides with α. Therefore, let us consider those cases where Z′ (the position of the fictitious vessel) does not lie on the extended arc SZ (see Fig. 2.8.3 highly exaggerated).

In those cases, the tangent line L$_I$[14] away from I is obtained by computing A$_{Z'}$ and hence the direction of the arc section Z′I and then by plotting the line perpendicular to this section of the arc and also passing at a distance p = z′ − z away from Z′ this section of the arc. Therefore, the tangent line L$_I$ to our circle of equal latitudes a at I does, in general, bypass the position of the observer and two such lines obtained by another CO, S′ will not intersect at the position of the observer. Moreover, it will be very difficult to develop useful error estimates for this type of approximate methods.

---

[14]A more realistic interpretation of the situation depicted in Fig. 2.8.3 results if one approximates all tangents to those circles by loxodromes that then appear as straight lines on the Mecator plotting sheets or charts.

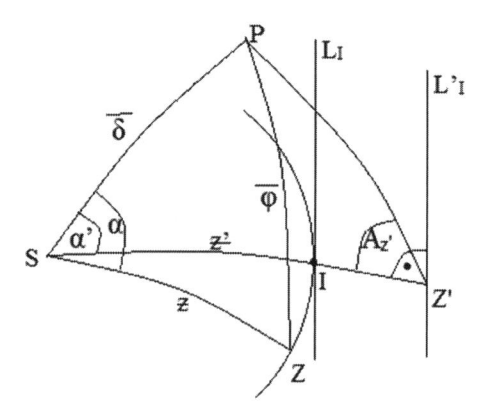

Fig. 2.8.3

It should by now be very obvious that any method based on such a fictitious vessel with known position will only produce acceptable approximations if the parallactic angle α' of the assumed position triangle $P\hat{S}Z'$ is sufficiently close to α. This implies that this method may produce satisfactory results if, for instance, the distance ZZ' is very small in comparison to z. In all those cases where the distance ZZ' is of the order of z, i.e., high altitude, and with a bad estimate of the assumed position, this method may fail to give acceptable results.

In practical navigation, the position $Z' \sim (\lambda', \varphi')$ of the fictitious vessel is merely a position close to or equal to the dead-reckoning position of the vessel and may be as far off as 50 NM in the case of sailing vessels.

The method outlined above is generally referred to as the two lines position method based on an assumed position and has been around since the times of St. Hilaires and Thomas Sumner. Obviously, this method has served navigators for almost 200 years and has given rise to many books and tables to avoid the use of elementary mathematics and is mainly justified by the need for speedy sight reductions and by the absence of suitable calculators. I, myself, learned and practiced air and sea navigation based on the instructions contained in the Air Force Manual 51–40, Vol. 1. However in the age of modern PCs and programmable calculators, the main reason for adhering to the old plotting method is simply to conform with traditional seamanship procedures. The use of sophisticated calculators, portable computers and GPS will forever change traditional navigational methods.

Those traditional methods of celestial navigation will always be a milestone in the development of modern navigation from a fine art into a science. For many of the users of traditional navigation methods it will always remain a mystery how one can obtain the position of a vessel or aircraft and only be off by not much more than a couple of nautical miles having only used approximations throughout, i.e., approximations for the altitudes, chronometer time, ephemeral data, improvised plotting sheets and a mathematical procedure that is erroneous in and of itself. An explanation for this phenomenon can perhaps be found in the Theory of Probability. [19]

2.8 On Approximate Solutions for Finding the Position ...

Because of the widespread use of the traditional assumed position method, I am going to derive all relevant formulae and provide those who want to continue to use the traditional method with a logarithm for solving their problems on an improvised plotting sheet or on the back of a high-speed, flight-computer-plotter. (It should be noted that without employing the plotting technique, the user of this approximate method will soon realize that there is no saving to be had by choosing such a method over the exact method. On the contrary, by using the above approximate method without plotting the corresponding LOPs on an improvised plotting sheet, the required calculations exceed the necessary calculation for solving the exact mathematical problem.)

In particular, if λ_A and φ_A are the coordinates of the assumed position, then:
$$t_A = \pm(GHA(S) + \lambda_A), \quad \text{where plus}(+) \text{ if S is West of the observer}$$
and minus$(-)$ if S is East of the observer.

Furthermore, the azimuth-angle A_z of our navigational triangle satisfies $0 \leq A_z \leq 180°$. Since δ, t_A and φ_A are known, A_z and the assumed altitude a_c can readily be calculated by employing formulae (1i) and (1ii) of Sect. 2.2.

Explicitly you have:

$\sin a_c = \sin \varphi_A \cdot \sin \delta + \cos \varphi_A \cdot \cos \delta \cdot \cos t_A$, i.e.

(1) $a_c = \sin^{-1}(\sin \varphi_A \cdot \sin \delta + \cos \varphi_A \cdot \cos \delta \cdot \cos t_A)$, and

$\cos a_c \cdot \cos A_z = \sin \delta \cdot \cos \varphi_A - \cos \delta \cdot \sin \varphi_A \cdot \cos t_A$, and hence...

(2) $A_z = \cos^{-1}\left(\frac{\sin \delta \cdot \cos \varphi_A - \cos \delta \cdot \sin \varphi_A \cdot \cos t_A}{\cos a_c}\right) \leq 180°$ resulting in...

(3) $Z_n = \begin{cases} A_z \text{ if S lies East of observer} \\ 360 - A_z \text{ if S lies West of observer} \end{cases}$

Because you have to apply formulae (1) and (2) twice (two lines of positions for the two observations of S_1 and S_2), you have to evaluate a total of four trigonometric expressions. If you choose to calculate the intersections of those lines, you are going to approximate the approximations (1) and (2) once more because you are going to employ an approximate transformation (see Sect. 1.3). Geometrically speaking, the tangent lines LOPs lie on the respective tangent plane and therefore, their intersection does not lie on the terrestrial sphere itself. Furthermore, calculating the intersection of the two lines of positions requires evaluating two more trigonometric expressions—see formula (5) of this section.

Also, plotting the lines of position on the same type of plotting sheets results in additional errors unless the assumed position is very close to the true position—see Sect. 1.3.

In what follows, we are going to calculate the intersection of the two lines of positions to convince the "traditionalists" that there is no advantage to be had from the computational aspect by using an approximate method.

First, let us derive the equation for any given line of positions.

From Fig. 2.8.4 we deduce that the slope of the azimuth line is:

$m_1 = \tan(90° - Z_n) = \cot Z_n$, and since L is perpendicular to the azimuth line its slope m_1 must satisfy:

$m_2 \cdot m_1 = -1$, i.e., $m_2 = -\tan Z_n$, hence... L is the line passing through (x_o, y_o) and having a slope equal to—$\tan Z_n$ which implies that the equation of L is:

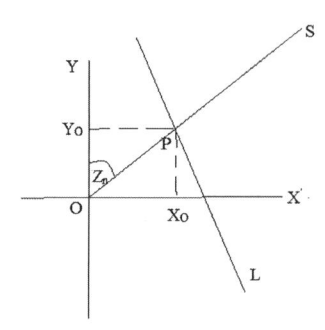

Fig. 2.8.4

$y = y_0 - \tan Z_n \cdot (x - x_0)$ But we also deduce that $x_0 = p \cdot \sin Z_n$, and $y_0 = p \cdot \cos Z_n$.
Therefore:

(4) $L : y = p \cdot \cos Z_n + \tan Z_n \cdot (p \cdot \sin Z_n - x)$

Note that in our application $p = z_c - z_n = a - a_c$, INTERCEPT

In the case of two distinct lines of positions, L and L', the point of intersection, i.e., the improved approximate solution of (x', y') must satisfy the two equations:
$L : y' = p_1 \cdot \cos Z_{n_1} + \tan Z_{n_1} \cdot (p_1 \cdot \sin Z_{n_1} - x')$
$L' : y' = p_2 \cdot \cos Z_{n_2} + \tan Z_{n_2} \cdot (p_2 \cdot \sin Z_{n_2} - x')$ Eliminating and reducing yields:

$$x' = \frac{p_2 \cdot \cos Z_{n_1} - p_1 \cdot \cos Z_{n_2}}{\cos Z_{n_1} \cdot \cos Z_{n_2} (\tan Z_{n_2} - \tan Z_{n_1})},$$

$$y' = \frac{p_1 \cdot \sin Z_{n_2} - p_2 \cdot \sin Z_{n_1}}{\cos Z_{n_1} \cdot \cos Z_{n_2} (\tan Z_{n_2} - \tan Z_{n_1})}.$$

Remember that our coordinate system x, y has been chosen in such a manner that the origin (O, O) coincides with the coordinates (λ_A, φ_A) of our assumed position and the grid corresponds to our improvised plotting sheet and is given by formula (1) in Sect. 1.3.
as:
$y' = \varphi' - \varphi_A, x' = \cos \varphi_A \cdot (\lambda' - \lambda_A)$ resulting in the coordinates (λ', φ') of our improved approximation:

$$\lambda' = \lambda_A + \frac{p_2 \cdot \cos Z_{n_1} - p_1 \cdot \cos Z_{n_2}}{\cos \varphi_A \cdot \cos Z_{n_1} \cdot \cos Z_{n_2} (\tan Z_{n_2} - \tan Z_{n_1})}, \text{ and}$$

(5) $$\varphi' = \varphi_A + \frac{p_1 \cdot \sin Z_{n_2} - p_2 \cdot \sin Z_{n_1}}{\cos Z_{n_1} \cdot \cos Z_{n_2} (\tan Z_{n_2} - \tan Z_{n_1})}$$

In conclusion, we count six trigonometric calculations as compared to five in the case of the exact solution. [43]

So far we have assumed that the two lines of positions had been established by simultaneous observations of S_1 and S_2. Now suppose S_1 is observed at $T = T_1$ and S_2

2.8 On Approximate Solutions for Finding the Position ...

is observed at $T = T_2$ with $T_1 \neq T_2$ but close to T_2. By assuming that the azimuths do not change during the interval of time $\Delta T = T_1 - T_2$, we first ask for the changes in the line L_2 that occur in this time span, and secondly, we ask for the change of altitude in a_2 in the same interval of time. According to our assumption, the change in the lines of position must occur along the line of the azimuth, i.e., in p.

The underlying mathematical problem is equivalent to asking the following question: "How does one have to changed the observed altitude in a_2 in order to compensate for a small changed, perhaps an error, in time if the above outlined method is being used?"

In order to solve this problem we adopt the following notations:

T: the exact time of observation in minutes.

$\tilde{T}$: the approximate time of observation in minutes.

a: the exact altitude at T.

a_c: the computed altitude at T.

$\tilde{a}$: the approximate (corrected) altitude at $\tilde{T}$.

$\tilde{a}_c$: computed altitude at $\tilde{T}$.

Then, because the azimuth does not change, the following equation must hold:

(i) $\tilde{a} - \tilde{a}_c = a - a_c$.

By adopting the previous definition of the intercept p, we have:

$p = a - a_c$, and by defining $\tilde{p} = a - \tilde{a}_c$; $\Delta p = \tilde{p} - p$; $\Delta T = \tilde{T} - T$ and $\Delta \lambda = \tilde{\lambda} - \lambda$, we deduce that:

(ii) $\Delta p = a_c - \tilde{a}_c$, and because of (i) we must have—

(iii) $\tilde{a} = a - \Delta p$

Next, we again adopt the improvised plotting sheet (see Sect. 1.3, formula 1) and adopt the transformation $x = (\lambda - \lambda_A) \cdot \cos \varphi_A$ and $y = \varphi - \varphi_A$—see Fig. 2.8.5— We conclude that $\alpha = -Z_n$, $\beta = Z_n - 270°$, i.e., $\cos \beta = -\sin Z_n$, and therefore:

$\Delta p = d \cdot \cos \beta = -d \cdot \sin Z_n$, and
$d = \Delta \lambda \cdot \cos \varphi_A$, i.e., $\Delta p = -\Delta \lambda \cdot \sin Z_n \cdot \cos \varphi_A$. Therefore we conclude that: $\tilde{a} = a + \Delta \lambda \cdot \sin Z_n \cdot \cos \varphi_A$, and since $\Delta \lambda = \Delta T \cdot 15'$, we find that:

(6) $\tilde{a} = a + \Delta T \cdot 15' \sin Z_n \cdot \cos \varphi_A$.

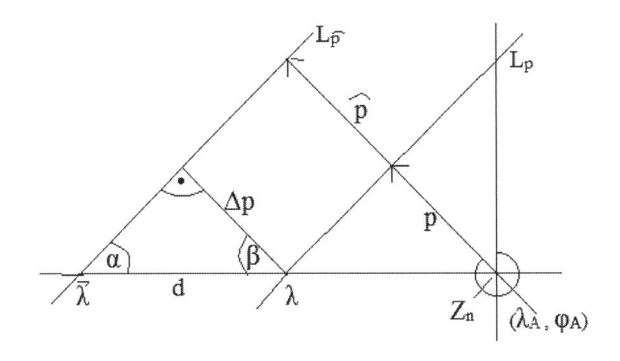

Fig. 2.8.5

Note that with $\tilde{T} > T$, $Z_n > 180°$, $\tilde{a} < a(\text{SETTING})$, and if $Z_n \leq 180°$, it follows that $\tilde{a} < a(\text{RISING})$

A disadvantage of having derived formula (6) in this manner is that we have failed to obtain a range for the validity of it. However, by employing some other reasoning and experience, one should get an estimate for the upper bound for the applicability of formula (6) in terms of the time span between T and $\tilde{T}$, i.e., in terms of ΔT. (Based more on experience than on theory a reasonable upper bound is: $|\Delta T| \leq 4$ min.)

Another way of arriving at formula (6) is based on the formulae of spherical trigonometry as used in this and in previous sections of this book. By applying those formulae to our navigational triangle, we have:

$\sin a = \sin \varphi_A \cdot \sin \delta + \cos \varphi_A \cdot \cos \delta \cdot \cos t$, differentiating yields:

$\cos a \cdot \frac{da}{dt} = -\cos \varphi_A \cdot \cos \delta \cdot \sin t$, or $\frac{da}{dt} = -\frac{\cos \varphi_A}{\cos a} \cdot \cos \delta \cdot \sin t$

and the application of the SIN-TH. yields:

$\cos \delta \cdot \sin t = \cos a \cdot \sin A_Z = \pm \cos a \cdot \sin Z_n \ldots$ hence :

$\frac{da}{dt} = \pm \sin Z_n \cdot \cos \varphi_A$ by noting that t stands for the local hour angle and therefore $dt = \pm dT \ 15'$, we deduce that: $da = dT \cdot 15' \cdot \sin Z_n \cdot \cos \varphi_A$, an infinitesimal expression with its finite equivalent:

(6') $\Delta a \cong \Delta T \cdot 15' \cdot \sin Z_n \cdot \cos \varphi_A$.

In a similar manner, by permitting both Z_n and a to depend upon the local hour angle t, by differentiating we obtain formula (2ii), and by substituting formula (2iii) from Sect. 2.2, we get:

$dZ_n = dT \cdot 15' \cdot (\sin \varphi_A - \cos Z_n \cdot \cos \varphi_A \cdot \tan a)$. Again by replacing the differential equation by the finite difference equation, we have:

(7) $\Delta Z_n = \Delta T \cdot 15' \cdot (\sin \varphi_A - \cos Z_n \cdot \cos \varphi_A \cdot \tan a)$

Aside from the fact that (6') and (7) provide approximate values for very small intervals of time only, those formulae are of very limited value in practical applications since, in general, Z_n is not known, or is approximated by Z_{An}, the azimuth of the assumed position triangle. It still remains to be shown how the lines of positions change if the vessel is underway between the sights taken (see Fig. 2.8.6).

We may deduce that:

$\Delta p = v \cdot \frac{\Delta T}{60} \cdot \sin(90° - (C - Z_n)) = v \cdot \frac{\Delta T}{60} \cos(C - Z_n)$, $v = |\vec{v}|\ldots$ hence:

(8) $\tilde{p} = p + \frac{\Delta T}{60} \cdot v \cdot \cos(C - Z_n)$, where "C" denotes course steered.

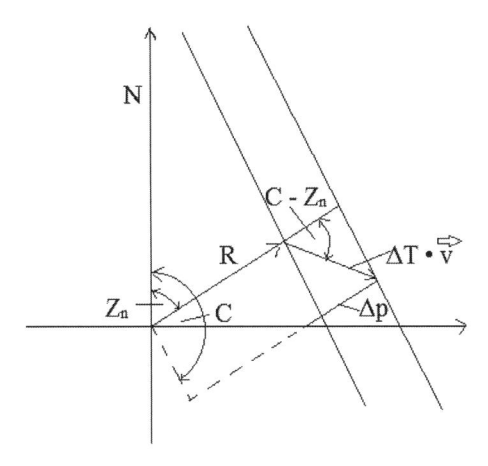

2.8 On Approximate Solutions for Finding the Position …

Fig. 2.8.6

Therefore the movement of the vessel in direction and amount $\vec{v} \cdot \Delta T$ is accounted for if we merely replace p by $\bar{p}$ as given by (8), i.e., if we move the line of position L parallel to itself by the segment Δp along $\vec{p}$.

Next let us consider the problem of multiple sights taken simultaneously or at distinct instants with the necessary adjustments made for the motion of the vessel between sights. Here we assume that n sight reductions have been made and formula (1)–(3) of this section have been computed n times, i.e., the values of a_i and $Z_n^i, i = 1, \ldots n$, and hence of p^i are known. Therefore, the n lines of positions L^i defined by (4) are also determined.

Now we want to find the coordinates λ', φ' of a point Z' so that its distance on our improvised plotting sheet from all n lines of positions is such that the sum of the squares of these distances d_i^2 is a minimum—LEAST SQUARE APPROXIMATION.

In order to do this analytically, we borrow the formula for the distance of a point x', y' from a straight line $L^i : y = A_i x + B_i$ from Analytic Geometry. Accordingly, this distance is given as:

$d_i^2 = \frac{(y' - A_i x' - B_i)}{1 + A_i^2}$. From formula (4) for the line of position we deduce that:

$A_i = -\tan Z_n^i$, and $B_i = \frac{p^i}{\cos Z_n^i}$, with $p^i = a^i - a_c^i$, i = 1, … n—and hence

$d_i^2 = \left(\cos Z_n^i y' + \sin Z_n^i x' - p^i\right)^2$ from which we deduce:

$\frac{\delta d_i^2}{\delta x'} = 2 \cdot \sin Z_n^i \cdot \cos Z_n^i y' + 2 \sin^2 Z_n^i x' - 2 \cdot p^i \cdot \sin Z_n^i$, and

$\frac{\delta d_i^2}{\delta y'} = 2 \cdot \cos^2 Z_n^i \cdot y' + 2 \sin Z_n^i \cdot \cos Z_n^i \cdot x' - 2 \cdot p^i \cdot \cos Z_n^i$.

Since we must minimize the sum of the squares of all the distances d^i, we must require that:

$$\frac{\delta P}{\delta x'} = \frac{\delta}{\delta x'} \sum_{i=1}^{n} d_i^2 = \sum_{i=1}^{n} \frac{\delta d_i^2}{\delta x'}$$
$$= 2 \cdot \sum_{i=1}^{n} \sin^2 Z_n^i \cdot x' + 2 \cdot \sum_{i=1}^{n} \sin Z_n^i \cdot \cos Z_n^i \cdot y' - 2 \cdot \sum_{i=1}^{n} p^i \sin^2 Z_n^i = 0$$

and also:

$$\frac{\delta P}{\delta y'} = \frac{\delta}{\delta y'} \sum_{i=1}^{n} d_i^2 = \sum_{i=1}^{n} \frac{\delta d_i^2}{\delta y'}$$
$$= 2 \cdot \sum_{i=1}^{n} \sin Z_n^i \cos Z_n^i \cdot x' + 2 \cdot \sum_{i=1}^{n} \cos^2 Z_n^i \cdot y' - 2 \sum_{i=1}^{n} \cdot p^i \cos Z_n^i = 0$$

To simplify the solution of this system of two linear equations, we define the parameters A, B, C, D, E, and G as follows:

$$A = \sum_{i=1}^{n} \cos^2 Z_n^i; B = \sum_{i=1}^{n} \sin Z_n^i \cos Z_n^i; C = \sum_{i=1}^{n} \sin^2 Z_n^i; D = \sum_{i=1}^{n} p^i \cos Z_n^i$$

$$E = \sum_{i=1}^{n} p^i \sin Z_n^i; \text{ and } G = AC - B^2, \text{ and deduce that } x', y' \text{ must satisfy the system:}$$

$$\begin{array}{l} C \cdot x' + B \cdot y' = E \\ B \cdot x' + A \cdot y' = D \end{array}, \text{ and resulting in}$$
$$x' = \frac{AE - BD}{G} \text{ and } y' = \frac{CD - BE}{G}$$

Recalling that the underlying transformation of our improvised plotting sheet is:

$$x = \cos \varphi_A \cdot (\lambda - \lambda_A), \text{ and } y = \varphi - \varphi_A \text{ we find that:}$$

(10) $\quad \lambda = \lambda_A + \dfrac{AE - BD}{\cos \varphi \cdot G}$ and $\varphi' = \varphi_A + \dfrac{CD - BE}{G}$ [31]

Also note that φ_A stands for the latitude of the assumed position.

For the special case n = 2, we again deduce from (9) and (10) the Eq. (5) of this section.

One word of caution: if the reader is unfamiliar with the theory of successive approximations and resorts to reiterating the Eq. (10), i.e., if the reader replaces λ_A, φ_A by λ', φ' and so on, and thereby obtains a sequence iterative approximations $\lambda_k, \varphi_k, k = 1, \ldots$ for which n + 1 iterations differs from the n-th iteration by only a minute amount, the reader may be tempted to conclude that this sequence of iterative approximations converges and perhaps converges to the true solution. This, however, is not true in general and often results in wrong conclusions. Furthermore, the procedure of reiterating Eq. (10) does not lend itself to the user who has only a simple calculator at his disposal. In order to use such a procedure, it would be necessary to have a more sophisticated, programmable calculator available, but then, why would anyone want to use such an elaborate calculation procedure for merely finding another approximate solution when a much simpler procedure is

2.8 On Approximate Solutions for Finding the Position …

available for finding the exact solution? The answer to this question can, however, not rest with the word "tradition".

After all the attempts which have been made to approximate and further simplify the underlying procedure to the extent where only tables, compass, straight edges, triangles, and improvised plotting sheets are being used, it is only fair that we also present to the reader a modern method that constitutes a true approximation to the true solution. This implies that by repeated application of this algorithm, these approximations will converge to the exact solution of Sect. 2.4, Eq. (16), provided that the initial approximations of φ_1 and φ_2 satisfy certain simple conditions.

The reader will recall that the problem of finding the exact position of the vessel consists in finding the points of intersection of the two circles of equal altitudes, i.e., in finding the values of φ and λ that satisfy the two equations in formula (16)

By adopting the definitions:

(11)
$$g(\varphi) = -\text{GHA}(S_1) \pm \cos^{-1}\left(\frac{\sin a_1 - \sin \delta_1 \cdot \sin \varphi}{\cos \delta_1 \cdot \cos \varphi}\right)$$
$$h(\varphi) = -\text{GHA}(S_2) \pm \cos^{-1}\left(\frac{\sin a_2 - \sin \delta_2 \cdot \sin \varphi}{\cos \delta_2 \cdot \cos \varphi}\right)$$
$$f(\varphi) = g(\varphi) - h(\varphi)$$

Where plus (+) if S is West, and minus (−) if S is East of observer.

As before, we conclude that our problem consists mathematically in solving the transcendental equation:

(12) $f(\varphi) = 0$ (See also Sect. 2.4, Eq. 18)

This can be accomplished by employing one or the other type of iterative methods, based on choosing suitable initial approximations $\varphi_k, k = 1, \ldots m$, and then successively computing additional approximations $\varphi_\mu, \mu \geq m$ thereby generating a sequence $\{\varphi_\mu\}$ that converges to the true solution φ of Eq. (12), i.e., $\varphi = \lim_{\mu \to \infty} \varphi_\mu$.

Here we shall use a method that employs two initial approximations, φ_1 and φ_2 respectively. The reason we are choosing this type is that the underlying problem consists in finding the intersection of two circles on the celestial sphere defined by: $\lambda = g(\varphi)$, and $\lambda = h(\varphi)$, the equivalent equations to Sect. 2.4, expressions (16) and (17). First, let's look at a small arc of on of these circles in the neighborhood of the solution λ, φ of the two equations—see Fig. 2.8.7a.

It makes perfect sense to use an approximation that uses two initial guesses φ_1 and φ_2 provided that these two values define a chore line that encloses the solution φ. If we then, at each step of our iteration φ_i choose the result of the previous iteration φ_{i-1}, together with another previous iteration $\varphi_k, 1 \leq k > i - 1$, so that the chore line defined by $(\lambda_{i-1}, \varphi_{i-1})$ and (λ_k, φ_k) encloses the actual solution, we would intuitively conclude that we would get a convergent sequence of good approximations.

On the other hand, by opting for an iterative method that is based merely on one initial guess φ_1, we would not readily see that this method converges, although it

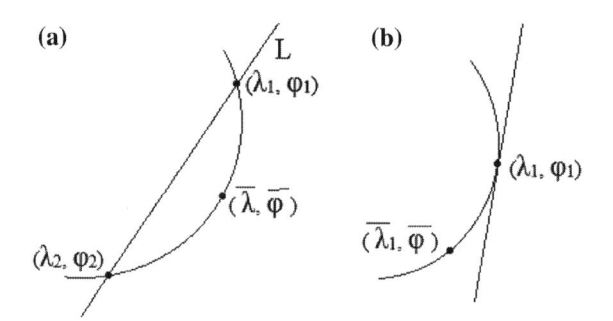

Fig. 2.8.7

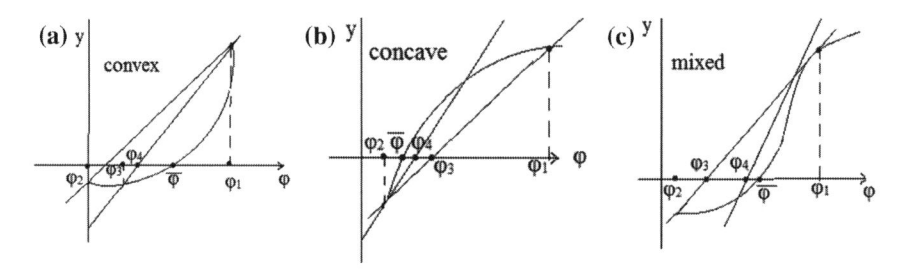

Fig. 2.8.8

may if φ_1 is sufficiently close to the solution of φ. However, such a method would entail evaluating the derivative of f (φ), i.e., f'(φ) at each step of the iteration process, something we surely would want to avoid. By looking at Fig. 2.8.7b it is clear that such a method is referred to as a tangent line method.

Let's now derive the well known numerical method know as the chore line or "Regular Falsi" method and apply it to our navigational problem. For this we picture the graph of the function y = f (φ) in the vicinity of the root $\bar{\varphi} : f(\bar{\varphi}) = 0$—see Fig. 2.8.8.

Then the equation of the secant line is given by:

$$\frac{y - f(\varphi_1)}{\varphi - \varphi_1} = \frac{f(\varphi_2) - f(\varphi_1)}{\varphi_2 - \varphi_1}, \text{ i.e., } y = f(\varphi_1) + \frac{f(\varphi_2) - f(\varphi_1)}{\varphi_2 - \varphi_1} \cdot (\varphi - \varphi_1),$$

and therefore, the first iterative approximation satisfies y = 0 resulting in:

$$\varphi_3 = \varphi_1 - \frac{(\varphi_2 - \varphi_1) \cdot f(\varphi_1)}{f(\varphi_2) - f(\varphi_1)} = \varphi_1 + \frac{(\varphi_1 - \varphi_2) \cdot y_1}{y_2 - y_1} = \frac{y_2}{y_2 - y_1} \cdot \varphi_1 + \frac{y_1}{y_1 - y_2} \cdot \varphi_2$$

with $y_i = f(\varphi_1)$. $i = 1, 2, \ldots$

From Fig. 2.8.8, we deduce that the conditions proposed for the selection of φ_1 and φ_2 are:

φ_1: Good initial approximation obtained by an independent method for approximating the latitude. There are many methods available to the navigator—see also the following section.

2.8 On Approximate Solutions for Finding the Position ...

φ_2: To be close to φ_1 but such that $f(\varphi_1) \cdot f(\varphi_2) < 0$ is satisfied in all three cases of Fig. 2.8.8.

Having found φ_3 by the previous formula, we now calculate φ_4, and so on, always checking that the condition $f(\varphi_1) \cdot f(\varphi_\mu) < 0$, for $1 \leq \mu \leq i, i \geq 2$.

If we assume that $f(\varphi)$ is convex: $\forall: \varphi_2 \leq \varphi \leq \varphi_1$, we may conclude that $f(\varphi_1) \cdot f(\varphi_i) < 0 \,\forall\, 2 \leq i$, and the sequence $\{\varphi_1\}$ is monoton increasing,[15] i.e., $\varphi_{i+1} > \varphi_i$ and thereby assuring the convergence to the exact solution, i.e., we have
$\bar{\varphi} = \lim_{i \to \infty} \varphi_i$.

In this case, the iterations of φ_i are defined by:

(13) $\varphi_{i+1} = \frac{y_i}{y_i - y_1} \cdot \varphi_1 + \frac{y_1}{y_1 - y_i} \cdot \varphi_i, i \geq 2, y_i = f(\varphi_i), f(\varphi_1)f(\varphi_2) < 0, i = 2$ and $f(\varphi_1)f(\varphi_i) < 0$ is satisfied automatically.

In cases were $f(\varphi)$ is concave, i.e. $\forall: \varphi_2 \leq \varphi \leq \varphi_1$, we may conclude that:

$f(\varphi_2)f(\varphi_i) < 0$ is satisfied automatically, provided that $f(\varphi_1)f(\varphi_2) < 0$, and the iteration formula is: [18, 35]

(14) $\varphi_{i+1} = \frac{y_2}{y_2 - y_i} \cdot \varphi_1 + \frac{y_i}{y_i - y_2} \cdot \varphi_2, i \geq 3, y_i = f(\varphi_i)$, if $i = 2$ use (13) above.

Again we obtain a monoton decreasing sequence, i.e., $\varphi_{i+1} < \varphi_i$ that is bounded from below and therefore assuring the convergence to the exact solution $\bar{\varphi}$, i.e.,

$\bar{\varphi} = \lim_{i \to \infty} \varphi_i$.

In all cases where $f(\varphi)$ is neither convex nor concave for $\varphi_1 \leq \varphi \leq \varphi_2$, we can obtain a similar formula to (13) and (14). However, since we have assumed that φ_1 and φ_2 are sufficiently close to $\bar{\varphi}$ our function $f(\varphi)$ will be necessarily either convex or concave in the vicinity of the root φ of (12). Of course we have also made use of the fact that $f(\varphi)$ is continuous on the interval specified above. Hence it suffices to use either (13) if $f(\varphi_1)f(\varphi_i) < 0$, or (14) whenever $f(\varphi_2)f(\varphi_i) < 0$.

We have therefore established the convergence of the iteration method known as "Regula Falsi" when applied to our navigational problem of finding the roots of $f(\varphi) = 0$, provided that $f(\varphi_1)f(\varphi_2) < 0$.

Next let us discuss an alternative method to the "Regula Falsi" that does away with checking for convexity or concavity all together. In this method, we utilize the last two computed points, i.e., φ_{i-1} and φ_i to generate the next point φ_{i+1} in the sequence resulting in the iterative formula:

(15) $\varphi_{i+1} = \frac{y_i}{y_i - y_{i-1}} \cdot \varphi_{i-1} + \frac{y_{i-1}}{y_{i-1} - y_i} \cdot \varphi_i, \ i \geq 2, f(\varphi_1)f(\varphi_2) < 0, y_i = f(\varphi_i)$.

This stationary iterative method is called the "Secant Method".

It can be shown that if φ_1 and φ_2 are close to the root $\bar{\varphi}$ of (12), and provided that $f'(\varphi)$ and $f''(\varphi)$, i.e., the first and second derivative of $f(\varphi)$ exist, and also

---

[15] And bounded from above.

$f'(\varphi) \geq 0$ for all φ sufficiently close to $\bar{\varphi}$, this method will converge to the solution $\bar{\varphi}$. In the case of our special function $f(\varphi)$ defined by (11) and (12), these conditions are satisfied and we can be assured to have another convergent method at hand.

Up to this point, nothing "NEW" has been revealed except for those special applications of well-known mathematical procedures to navigation. At this point of our investigations, navigators who are accustomed or compelled to plot and see LOPs will ask how these methods I've devised relate to the method of an assumed position and to the concept of position lines? Here is the answer.

In Sect. 2.4, the functions $g(\varphi)$ and $h(\varphi)$ are actually the same function F that also depends on distinct parameters, i.e., $g(\varphi) = F(\varphi, a_1, \delta_1, \text{GHA}(S_1))$, $h(\varphi) = F(\varphi, a_2, \delta_2, \text{GHA}(S_2))$, and were approximated by straight lines (See expression 20) and are restated here as:

(16) $L_1: \lambda = g(\varphi_1) + \frac{g(\varphi_1) - g(\varphi_2)}{\varphi_2 - \varphi_1} \cdot (\varphi - \varphi_1)$, and

$L_2: \lambda = h(\varphi_1) + \frac{h(\varphi_1) - h(\varphi_2)}{\varphi_2 - \varphi_1} \cdot (\varphi - \varphi_1)$, subject to $f(\varphi) = g(\varphi) - h(\varphi)$ and $f(\varphi_1)f(\varphi_2) < 0$.

Recall that φ_1 has been obtained by another independent procedure and therefore should be close to the true latitude.

For instance, φ_1 can be obtained by meridian observation, **POLARIS** observation, circumpolar stars, azimuth observations and, if everything else fails, by dead reckoning. As I said before, a prudent navigator always keeps a close track on his latitude.

If $f(\varphi_1) < 0$, then φ_2 should be close to φ_1, up to about one degree apart, if possible, and such that $f(\varphi_2) > 0$. In the case where $f(\varphi_1) > 0$, then we require that $f(\varphi_2) < 0$.

A very simple way of obtaining these values for φ is to choose a lower and upper bound for the true solution, i.e., if $\varphi_1 \leq \varphi \leq \varphi_2$ and $\varphi_2 - \varphi_1 \leq 1^0$, so that all that is actually needed is an intelligent approximation to the true latitude of the observer.

It follows then that L_1 and L_2 defined in this manner may be interpreted as lines of positions and are, by mathematical character, chord or secant lines. Their intersection $\bar{\lambda}$ and $\bar{\varphi}$ is obtained by computing $\bar{\varphi}$ from:

$g(\varphi_1) + \frac{g(\varphi_2) - g(\varphi_1)}{\varphi_2 - \varphi_1} \cdot (\bar{\varphi} - \varphi_1) \simeq h(\varphi_1) + \frac{h(\varphi_2) - h(\varphi_1)}{\varphi_2 - \varphi_1} \cdot (\bar{\varphi} - \varphi_1)$, resulting in:

(17) $\bar{\varphi} = \frac{y_2}{y_2 - y_1} \cdot \varphi_1 + \frac{y_1}{y_1 - y_2} \cdot \varphi_2 = \varphi_3$, $\bar{\lambda} = g(\bar{\varphi})$, $y_i = f(\varphi_1)$.

Therefore, the intersection for our positions line L_1 and L_2 of (16) is actually equal to the first iterative of our iterative method referred to as "Regula Falsi", (13) or (14). By replacing the set of values φ_1, φ_2 by φ_1, φ_3 or φ_2, φ_3 in our lines of positions (16) we obtain a sequence of position lines L_i, L_{i+1} whose intersections correspond exactly to the iterations of the well-known regula falsi and are not merely position lines of intelligent guesses as in other non-convergent methods.

It should be noted that at each iteration, or for each pair of positions lines defined by (16) it is only necessary to compute the values $y_i = f(\varphi_1)$ once, which implies that the same trigonometric function $F(\varphi, \ldots)$ defined by (11) has to be evaluated only twice at each step. The only exception is the first iteration that requires the evaluation of F(φ,....) four times. Furthermore, the point of intersection is found by calculating a simple algebraic formula (13) or (14) as compared to evaluating formula (5). Also note that $\lambda_\mu = g(\varphi_\mu) \cong h(\varphi_\mu)$, where "μ" denotes the index of the last iteration.

Of course, all those iterative methods require a programmable calculator in case they are intended for practical applications and not only for theoretical purposes.

2.9 An Approximate Method Based on Matrices and the Least Square Approximation

In this section let's consider the case where n > 2, i.e., the case where we have more than three equations for the three unknown components of the position vector —more than two celestial objects corresponding to n vectors x_k of the substellar points S_k given by Sect. 2.6, expression (1). The corresponding system of non-linear equations is then given by Sect. 2.6, expressions (2a) and (2b) which in matrix notations can be written as:

(1a) $A \cdot \vec{x} + \vec{L}_0 = 0$, and

(1b) $\|\vec{x}\|^2 = 1$, where "A" denotes the coefficient matrix:

$$A = \begin{pmatrix} x_1 y_1 z_1 \\ x_2 y_2 z_2 \\ \ldots\ldots\ldots \\ \ldots\ldots\ldots \\ x_n y_n z_n \end{pmatrix}, \text{ and as before } \vec{L}_0 = \begin{pmatrix} -\sin H_{0,1} \\ -\sin H_{0,2} \\ \ldots\ldots\ldots \\ \ldots\ldots\ldots \\ -\sin H_{0,n} \end{pmatrix}$$

It should be obvious that, in general, such a system as defined by (1) will not possess a solution unless $\vec{L}_0$ and A are exact which, in praxis, appears to be impossible. Then, if the mathematical problem defined by (1) does not have a solution for a given $\vec{L}_0$ and A, what are we looking for? Of course we are looking for an approximation $\vec{y}$ to the true position vector $\vec{x}$ that physically exists. In order to define such an approximation mathematically, we proceed as follows:

First we must approximate the function $f(\vec{x}) = \|\vec{x}\|^2 = 1$ by a linear function $\bar{f}(\vec{x}) = \vec{a}\vec{x} + b$. This appears the most objectionable part of this approximation method. Although we are already familiar with the problem of approximating a sphere by its tangent plane (see Sect. 1.1), we still have to determine the point $\vec{x}_0$ on the sphere where we want the tangent plane to touch the sphere. This, in turn, implies that we require an initial approximation to the position of the observer.

Although the Least-Square Approximation generates numerical values in almost all cases under consideration, even in such cases where the parameters are blunder, it remains to be seen what happens if the determinant of the matrix of the normal equation is nearly zero.

The actual problem or approximating the function $f(\vec{x})$ by a linear function $f_L(\vec{x}) = \vec{a}\vec{x} + b$ can readily be solved by applying Taylor Series expansion of functions of several variables to $f(\vec{x})$. By truncating this series after the linear part we obtain the approximation:

$$f(\vec{x}) = f(\vec{x}_0) + \frac{\partial f}{\partial x}(x - x_0) + \frac{\partial f}{\partial y}(y - y_0) + \frac{\partial f}{\partial z}(z - z_0) + \cdots \quad \text{hence:}$$

$$f(\vec{x}) = x_0^2 + y_0^2 + z_0^2 - 1 - 2(x_0^2 + y_0^2 + z_0^2) + 2(x_0 x + y_0 y + z_0 z) \text{ or}$$

$$= 2(x_0 x + y_0 y + z_0 z) - (1 + x_0^2 + y_0^2 + z_0^2) + \cdots$$

and our linear approximation $f_L(\vec{x}) = \vec{a}\vec{x} + b = 0$ becomes:

(1b) $x_0 x + y_0 y + z_0 z - \frac{1 + x_0^2 + y_0^2 + z_0^2}{2} = 0$.

By including this linear equation in our linear system (1a), we have succeeded in replacing the non-linear system (1) by the linear system:

2. $A_L \vec{x} + \vec{L}_{0,L} = 0$, where
$$A_L = \begin{pmatrix} x_1 y_1 z_1 \\ x_2 y_2 z_2 \\ \cdots\cdots \\ \cdots\cdots \\ x_n y_n z_n \\ x_0 y_0 z_0 \end{pmatrix}, \text{ and } \vec{L}_{0,L} = \begin{pmatrix} -\sin H_{0,1} \\ -\sin H_{0,2} \\ \cdots\cdots \\ \cdots\cdots \\ -\sin H_{0,n} \\ -\frac{1 + x_0^2 + y_0^2 + z_0^2}{2} \end{pmatrix}$$

Nevertheless, we still have to deal with n + 1 linear equations for the three unknown x,y,z which do not have a solution in general.

For the moment, let us assume that we have found the exact solution vector $\vec{x}$. Substituting it in (2) will result in an equation of the form:

(3) $A_L \vec{x} + \vec{L}_{0,L} = \vec{v}$, where $\vec{v}$ represents the error of the system (2).

However, we actually do not know the exact solution to our problem and therefore, cannot determine $\vec{v}$, yet... but we may interpret expression (3) as the definition of $\vec{v}$. Our approximation for solving (3) consists in computing a vector y in such a manner that the resulting error v is defined by:

(4) $\vec{v} = A_L \vec{x} + \vec{L}_{0,L}$ so as to satisfy the Least-Square requirements, namely:

$\|v_L\|^2 = \text{Min}\|v\|^2 = \text{Min}\vec{v}^T \cdot \vec{v}$

The problem of finding vector $\vec{v}_L$ is relatively easy in theory, providing that the reader is familiar with the elements of Matrix Algebra.

First, let's recall the definition of the "Transpose A^T of the matrix A". Accordingly, the transpose is obtained by merely interchanging rows and columns of A. Therefore, by taking the transpose of (4) and recalling that the transpose of a sum is the sum of the transpose, we find that:

(4') $\vec{v}^T = \vec{y}^T A_L^T + \vec{L}_{0,L}^T$. Here, we also have made use of the fact that the transpose of a product is equal the product of the transpose of the factors but in reverse order.

By recalling the definition of the inner product of vectors, we conclude that:

2.9 An Approximate Method Based on Matrices …

$\vec{v}^T \cdot \vec{v} = \|v\|^2 = \vec{v} \cdot \vec{v} = (A_L \vec{y} + \vec{L}_{0,L})^2$. By employing the well-known formula for the vector derivative of the inner product, i.e., the expression:

$\frac{\partial}{\partial x} \vec{f}(\vec{x}) \vec{g}(\vec{x}) = \vec{g}'(\vec{x})^T \cdot \vec{f}(\vec{x}) + \vec{f}'(\vec{x})^T \cdot \vec{g}(\vec{x})$ to the case where $\vec{g}(\vec{x}) = \vec{f}(\vec{x})$.

We find that $\frac{\partial}{\partial x} \|\vec{f}(\vec{x})\|^2 = 2\vec{f}'(\vec{x})^T \cdot \vec{f}(\vec{x})$, and hence for $\vec{f}(\vec{y}) = A_L \vec{y} + \vec{L}_{0,L}$ that

$\frac{\partial}{\partial y} \|v\|^2 = 2(A_L^T A_L \vec{y} + A_L^T \vec{L}_{0,L})$ which implies that the vector $\vec{y}_L$ that minimizes $v^T v = \|v\|^2$ must satisfy the equation:

(5) $A_L^T A_L \vec{y}_L + A_L^T \vec{L}_{0,L} = 0$. This equation is called the "Normal Equation" corresponding to the linear system (2)

It can be shown that if the columns of A are linearly independent, the inverse $(A_L^T A_L)^{-1}$ exist, and therefore:

(6) $\vec{y}_L = -(A_L^T A_L)^{-1} + A_L^T \vec{L}_{0,L}$. Substituting the vector $\vec{y}_L$ into Eq. (4) yields:

(7) $\vec{v}_L = -A_L (A_L^T A_L)^{-1} + A_L^T \vec{L}_{0,L} + \vec{L}_{0,L} \cdot 1$

Although the matrix $(A_L^T A_L)^{-1}$ may exist, the method explained herein may still fail if the matrix $A_L^T A_L$ is ILL-CONDITIONED, i.e., in all cases where the determinant of this matrix is nearly zero.

Therefore, it is imperative that the navigator selects the set of COs carefully in order to avoid such ill-conditioned matrices. Also note that the required approximate position of the observer is actually given by the expression (6), i.e., $\vec{x} \cong \vec{y}_L$.

In conclusion, the method of Least Squares based on matrices is by no means a fool proof method, aside from always being just an approximate method that also requires a programmable calculator or PC. Furthermore, it still remains to be seen how we can find a suitable approximation $\vec{x}_0$ in the first place since the linearization of the quadratic equation strongly depends on it. (It is my conviction that by simply employing an assumed—Dead Reckoning—position for vector $\vec{x}_0$, it may not be possible to obtain acceptable results.

A much better approach might first be to select a compatible set of two equations from the n equations and then by solving the corresponding n = 2 problem with the exact method, the navigator would obtain the desired approximation, i.e., the vector $\vec{x}_0$.

[51, 52, 53]

2.10 Sumner's Line of Assumed Position Method as Scientific Method

In this section I would like to examine the mathematical aspects of the "Line of Assumed Positions Method"—LOPs method—also referred to as Sumner's Method. This method has been classified as an empiric method until now because no analytic assessment of the errors due to various approximations employed has been available. For instance, the numerical results for the latitude and longitude depend on, among other things, the choices of the assumed position and on the deviation of the true azimuth from the calculated azimuth, just to mention a few of the critical parameters. Nevertheless, this method works well in most practical cases where the navigator selects the COs prudently and exercises great care in not including significant errors in the observation data. Therefore, it does not come as a surprise that all major Navies and Merchant Marines have, at one time or another, adopted this method as their standard method of applied arts. The fact that said method has served navigators for almost two centuries (since 1837) so well may explain why no serious efforts have been made to elevate this method to a scientific method.

From the day I started using this method for Air Navigation, it has always surprised me that I was able to obtain acceptable results despite the fact that the method itself was erroneous; the plotting sheets were all Non-Mercator sheets; the plotting instruments were all rather crude in design; and the tables for finding the calculated altitudes and azimuths lacked accuracy. At the very extreme, I was even able to plot a "fix" on the high-speed side of my aircraft computer. It also surprised me how many efforts and futile attempts have been made all over the world to avoid using elementary mathematics in the process of finding the calculated altitude and azimuth. Those efforts can only be measured in tons of paper-publications.

Heretofore, I have ignored the question of why this empirical method almost always work because later on in my navigation experience, I only used the exact methods of astronavigation. In this book, I have included three exact methods together with their algorithms and error analysis. None of these methods depend in any way on a particular calculator or PC and in an emergency such as leaked or dead batteries, Logarithmic and Trigonometric tables can be used in the place of the faulty calculators (See Sect. 2.11).

However, with the advent of the availability of sophisticated calculators that cost less than the afore mentioned tables, things have changed again. It has been brought to my attention that students of Celestial Navigation use their programmable calculators with a program for the celestial LOP method in place to approximate the true solution by successively employing the result of one iteration as the new assumed position for the next iteration. For example, one group of students on the West Coast of the United States used the coordinates corresponding to a place off the coast of New England as their first assumed position and, after several iterations, had determined their approximate position on the West Coast. In 1837, Capt. Thomas Sumner probably never conceived of a future when computational machines would turn his empiric method into a mathematical method of successive approximations.

2.10 Sumner's Line of Assumed Position Method as Scientific Method

As a mathematician, I am keenly aware that certain iterative methods for solving transcendental equations are very powerful and their convergence and rate of convergence can be determined analytically. Therefore, from the mathematical point of view, it suffices to prove that Sumner's Method, when applied successively, is equivalent to solving a well-defined mathematical problem of successive approximation by either the tangent or secant method as discussed in Sects. 2.6 and 4.1.

The problem of showing the equivalence of these two methods translates into saying that for a given problem of finding an approximate solution by using Sumner's Method, there exists an element in the corresponding sequence of iterations of the method of successive approximations as defined in Sect. 2.8. By repeated application of the LOP method to the points P_A corresponding to the points P of the convergent sequence of iterations, one can obtain the exact solution, i.e., the exact position of the observer. Therefore, Sumner's Method, when applied successively, has become a real mathematical method. Because of the great importance of this, a sketch of the proof of the statement of equivalence is given below:

On The Equivalence of Sumner's LOPs Method and the Method of Successive Linear Approximation for Solving the System of Transcendental Equations for Finding the Exact Solution of our Fundamental Problem of Navitation.

Let S_1 denote the substellar point of the first CO and S_2 the substellar point of the second CO. Furthermore, let $\lambda = g(\varphi)$ and $\lambda = h(\varphi)$ be the equations of the circles C_1 and C_2 of equal altitudes about S_1 and S_2 respectively and with $g(\varphi)$ and $h(\varphi)$ defined by the expressions (17) and (18) of Sect. 2.4.

For now, let us assume that we have found the two latitudes φ_1 and φ_2 that are close to the true latitude φ, as for instance, where φ_1 and φ_2 are lower and upper bounds respectively, i.e., $\varphi_1 \leq \varphi \leq \varphi_2$. Then φ_1 and $\lambda_1 = g(\varphi_1)$ define a point on S_1 on the circle C_1, and φ_2 and $\lambda_2 = h(\varphi_2)$ define another point S_2 on C_2. By linearization of $g(\varphi)$ and $h(\varphi)$ about S_1 and S_2 respectively, we obtain the two tangent lines L_1 and L_2 to these circles at these points. In particular, we obtain the following two linear equations:

$$L_1 : \lambda = g(\varphi_1) + g'(\varphi_1) \cdot (\varphi - \varphi_1)$$
$$L_2 : \lambda = h(\varphi_2) + h'(\varphi_2) \cdot (\varphi - \varphi_2)$$

Next, we erect the perpendicular lines L_1^N and L_2^N through S_1 and S_2 respectively. Theses lines will pass through the centers of S_1 and S_2 of the circles C_1 and C_2 respectively—(See Fig. 2.10.1) and are analytically given by:

$$L_1^N : \lambda = g(\varphi_1) - \frac{1}{g'(\varphi_1)} \cdot (\varphi - \varphi_1) \text{ and}$$
$$L_2^N : \lambda = h(\varphi_2) - \frac{1}{h'(\varphi_2)} \cdot (\varphi - \varphi_2) \quad \text{(See Fig. 2.10.1). Here } g'(\varphi_1) \neq 0,$$
$h'(\varphi_2) \neq 0$, and $g'(\varphi_1) \neq h'(\varphi_2)$.

The normals L_1^N and L_2^N to L_1 and L_2 respectively then intersect at the point P_A: (λ_A, φ_A) given by:

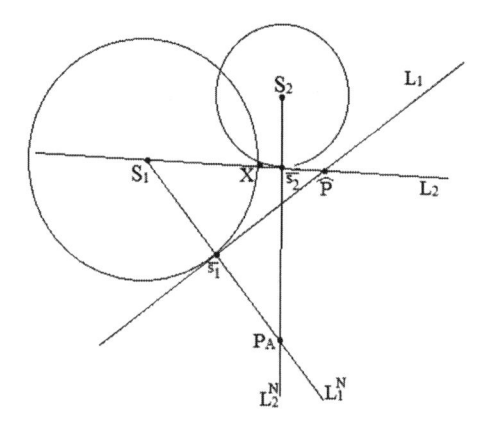

Fig. 2.10.1

$$\varphi_A = \frac{g(\varphi_1) - h(\varphi_2)}{h'(\varphi_2) - g'(\varphi_1)} \cdot g'(\varphi_1)h'(\varphi_2) + \frac{\varphi_1 \cdot h'(\varphi_2) - \varphi_2 g'(\varphi_1)}{h'(\varphi_2) - g'(\varphi_1)},$$

$$\lambda_A = g(\varphi_1) - \frac{1}{g'(\varphi_1)} \cdot (\varphi_A - \varphi_1).$$

On the other hand, the two tangent lines L_1 and L_2 intersect at a point $\tilde{P} : \left(\tilde{\lambda}, \tilde{\varphi}\right)$ that will be closer to $\vec{x} : (\lambda, \varphi)$ than the previously defined points because of the contraction mapping defined by the iterative process. Therefore $\tilde{P}$ defines the next iterative approximation.

The equivalence of said methods implies that the following results hold:

RESULT #1:
For any φ_1 and φ_2 sufficiently close to the true latitude φ the approximation for the circles of equal altitudes C_1 and C_2 by their respective tangent lines L_1 and L_2 yields a point $\tilde{P}$ defined as the point of intersection of these two lines.

The same approximation can also be found by the LOPs or Sumner's Method provided that the assumed position P_A coincides with $P_A \sim (\lambda_A \varphi_A)$ as given above.

RESULT #2:
Conversely, for any given assumed position P_A the azimuths or the corresponding navigational triangles together with the true and calculated altitudes, the two points $\overline{s_1}$ and $\overline{s_2}$ on the circle of equal altitudes and henceforth the two LOPs that coincide with the tangent lines at $\overline{s_1}$ and $\overline{s_2}$ respectively, intersect at the point $\tilde{P}$ that corresponds to the iteration defined above.

RESULT #3:
Accordingly, we may conclude that from the mathematical point of view the two methods are equivalent and hence successive application of the method of assumed positions is equivalent to the method of successive iteration of the tangent line

method that constitutes a well established mathematical method for approximating the true solution. Furthermore, the well-known error analysis for this method can now be readily applied to the LOPs method.

RESULT # 4:
For the mathematically minded only—the concept of equivalence as defined above does not, however, necessarily imply that the convergence of the tangent iterative method assures also the convergence of the corresponding LOPs method to the true solution. However, it can be shown that if the two iterative sequences $\{\varphi_i\}$ and $\{\tilde{\varphi}_i\}$ defined by the iterative tangent method and the LOPs method respectively are also contracting, i.e., if $\forall \varepsilon > 0, \ \exists: N \therefore \forall i \geq N \therefore |\varphi_i - \tilde{\varphi}_i| < \varepsilon$ then the LOPs method will also converge to the exact solution.

Proof Since $\lim_{i \to \infty}\{\varphi_i\} = \varphi \Leftrightarrow \forall \varepsilon > 0 \exists: \ N \therefore \forall i > N \therefore |\tilde{\varphi} - \varphi_i| < \frac{\varepsilon}{2}$. Because these sequences are also contracting we conclude that:

$\forall \varepsilon > 0 \exists: \tilde{N} \therefore \forall i \geq N \therefore |\varphi_i - \tilde{\varphi}_i| < \frac{\varepsilon}{2}$, hence $\forall i \geq$ max (N, N'). We have: $|\tilde{\varphi}_i - \varphi| = |(\tilde{\varphi}_i - \varphi) + (\varphi_i - \varphi)| \leq |\tilde{\varphi}_i - \varphi_i| + |\varphi_i - \varphi| < \varepsilon$ and therefore: $\lim_{i \to \infty} \tilde{\varphi}_i = \varphi$.

2.11 Numerical Example and Logarithmic Algorithm

Heretofore, I have included a section on the use and limitations of the assumed position method because of its wide-spread use and the navigators adherence to tradition. However, as an advocate of the use of calculators and computers as well as of the exact mathematical methods, I felt compelled to offer an algorithm that can also be employed in case all the computers and calculators on board fail to function (a possibility with a very low statistical probability). Therefore, I have devised an algorithm that can be executed by the use of a pocket-sized table of logarithms. Because of the wide-spread use of the assumed position method, and my own earlier exposure to it, I have based this algorithm on the assumed method formulae as discussed in previous chapters and I keep it only for emergency purposes. [50]

Let's recall the formulae corresponding to the assumed navigational triangle $P\hat{S}Z_A$ as:

(i) $\sin a_C = \sin \varphi_A \cdot \sin \delta + \cos \varphi_A \cdot \cos \delta \cdot \cos t_A$, where $t_A = \pm$ (GHA (S) + λ_A) + 360°
+ if observer is West, − if East

(ii) $\sin A_z = \dfrac{\cos \delta \cdot \sin t_A}{\cos a_C}$

(iii) $Z_A = \begin{cases} A_Z \text{ if } Z_A \leq 180° \\ 360° - A_Z \text{ if } Z_A > 180° \end{cases}$

The subscript A refers to assumed and C to computed values. The resulting algorithm is given in the form of schematic program.

ALGORITH C
FOR THE LOGARITHMIC—PRODEDURE

i.) $\sin a_C = \sin \varphi_A \bullet \sin \delta + \cos \varphi_A \bullet \cos \delta \bullet \cos t_A$, where $t_A = \pm (GHA\,(S) + \lambda_A) + 360°*$

ii.) $\sin A_Z = \dfrac{\cos \delta \bullet \sin t_A}{\cos a_C}$, where $Z_N = \begin{cases} A_Z & \text{if } Z_N \leq 180° \\ 360° - A_Z & \text{if } Z_N > 180° \end{cases}$

* + if S is West of Z; - if S is East of Z. Employ the second part of formula i.) to calculate t_A

First part:

Calculation of a_C:

I. Define

$A = \log (\pm \sin \varphi_A) = \log (\pm \sin \underline{\quad}) = \underline{\quad\quad}$ $\begin{cases} + \textit{if } \sin \underline{\quad} > 0 \\ - \textit{if } \sin \underline{\quad} < 0 \end{cases}$

$B = \log (\pm \sin \delta) = \log (\pm \sin \underline{\quad}) = \underline{\quad\quad}$

II. Calculate:

$A + B$ and $\underline{\quad\quad} = \underline{\quad\quad}$

$\log^{-1}(A+B) = \log^{-1} \underline{\quad\quad} = \underline{\quad\quad}$

III. Define:

$C = \log \cos \varphi_A = \log \cos \underline{\quad\quad} = \underline{\quad\quad}$

$D = \log \cos \delta = \log \cos \underline{\quad\quad} = \underline{\quad\quad}$

$E = \log (\pm \cos t_A) = \log (\pm \underline{\quad\quad}) = \underline{\quad\quad}$ $\begin{cases} + \text{if } \cos t_A > 0 \\ - \text{if } \cos t_A < 0 \end{cases}$

IV. Calculate

$C + D + E = \underline{\quad\quad} = \underline{\quad\quad}$

$\log^{-1}(C+D+E) = \log^{-1} \underline{\quad\quad} = \underline{\quad\quad}$

V. Calculate

2.11 Numerical Example and Logarithmic Algorithm

$F = \pm \log^{-1}(A+B) \pm \log^{-1}(C+D+E)$ use:
$\begin{cases} + \text{in first term if } \sin\varphi_A \sin\delta > 0 \\ - \text{in first term if } \sin\varphi_A \sin\delta < 0 \\ + \text{in second term if } \cos t_A > 0 \\ - \text{in second term if } \cos t_A < 0 \end{cases}$

= _____ , and

$G = \log(\pm F)$, and = _____ $\begin{cases} + \text{if } F > 0 \\ - \text{if } F < 0 \end{cases}$

$a_C = \pm (\log \sin)^{-1} G = $ _____ $\begin{cases} + \text{if } F > 0 \\ - \text{if } F < 0 \end{cases}$

Second Part: Calculation of Z_N:

VI. Define:

$H = \log(\pm \sin t_A) = \log(\pm \sin \underline{\hspace{1cm}}) = $ _____ $\begin{cases} + \text{if } \sin t_A > 0 \\ - \text{if } \sin t_A < 0 \end{cases}$

$I = \log \cos a_C = \log \cos \underline{\hspace{1cm}} = $ _____

$J = \log \cos \delta = \log \cos \underline{\hspace{1cm}} = $ _____

VII. Calculate:

$J + H - I$ = _____ , and

$A_Z = \pm (\log \sin)^{-1} (J + H - I)$ = _____ $\begin{cases} + \text{if } \sin t_A > 0 \\ - \text{if } \sin t_A < 0 \end{cases}$

$Z_N = \begin{cases} A_Z & \text{if } Z_N \leq 180° \\ 360° - A_Z & \text{if } Z_N > 180° \end{cases}$ = _____

Note that "$\log^{-1}$" denotes the anti-logarithm, i.e., $\log^{-1} x = y$ implies that $x = \log y$, and $(\log \sin)^{-1} x = y$ implies that $x = (\log \sin) y$ Also note that $(\log \sin)^{-1} x = \sin^{-1}(\log^{-1} x)$ and $\log \cos x = \log(\cos x)$.

Here's how it works. Let us consider the following example:

Example

Given the following data for a LOP of a rising CO.-S,
Assumed position $\varphi_A = -25°$, and $\lambda_A = 110°$
Ephemeral data of S: GHA (S) = 155°, $\delta = -9°$, and S rising NE of Z_A.
Find a_C and Z_A of S

Solution:
By using the second part of (i) we find that
$t_A = -(155° + 110°) + 360° = 95°$
Substituting these values in our program-algorithm C and using the logarithmic tables we find that:
$a_C = -41'$, and $Z_A = 79°44'$

Algorithm C
For the Logarithmic—Prodedure

i.) $\sin a_C = \sin \varphi_A \bullet \sin \delta + \cos \varphi_A \bullet \cos \delta \bullet \cos t_A$, where $t_A = \pm (GHA (S) + \lambda_A) + 360°*$

ii.) $\sin A_z = \dfrac{\cos \delta \bullet \sin t_A}{\cos A_C}$, where $Z_N = \begin{cases} A_z \text{ if } Z_N \leq 180° \\ 360° - A_z \text{ if } Z_N > 180° \end{cases}$

* + if S is West of Z, - if S is East of Z. Employ the second part of formula i.)

to calculate t_A

2.11 Numerical Example and Logarithmic Algorithm

First part to be employed if S is West of Z

Calculation of a_C:

I. Define

$$A = \log (\pm \sin \varphi_A) = \log (\pm \sin \text{-}25°) = \underline{9.625948 - 10} \quad \begin{cases} + \text{if sin}\underline{} > 0 \\ - \text{if sin}\underline{} < 0 \end{cases}$$

$$B = \log (\pm \sin \delta) = \log (\pm \sin \text{-}9°) = \underline{9.194332 - 10}$$

II. Calculate:

$$A + B \text{ and} \qquad = \underline{\text{-}1.17972}$$

$$\log^{-1}(A+B) = \log^{-1}\underline{\text{-}1.17972} = \underline{0.0661120}$$

III. Define:

$$C = \log \cos \varphi_A = \log \cos \underline{\text{-}25°} = \underline{9.957276 - 10}$$

$$D = \log \cos \delta = \log \cos \underline{\text{-}9°} = \underline{9.994620 - 10}$$

$$E = \log (\pm \cos t_A) = \log (\pm \underline{95}) = \underline{8.940296 - 10} \quad \begin{cases} + \text{if } \cos t_A > 0 \\ - \text{if } \cos t_A < 0 \end{cases}$$

IV. Calculate

$$C + D + E = \underline{} = \underline{\text{-}1.107809}$$

$$\log^{-1}(C+D+E) = \underline{\log^{-1}(\text{-}1.107809)} = \underline{0.0780172}$$

V. Calculate

$$F = \pm \log^{-1}(A+B) \pm \log^{-1}(C+D+E) \text{ use:} \quad \begin{cases} + \text{in first term if } \sin \varphi_A \sin \delta > 0 \\ - \text{in first term if } \sin \varphi_A \sin \delta < 0 \\ + \text{in second term if } \cos t_A > 0 \\ - \text{in second term if } \cos t_A < 0 \end{cases}$$

$$= \underline{\text{-}0.011905} \qquad \text{, and}$$

$$G = \log (\pm F), \text{ and} = \underline{\text{-}1.924271} \quad \begin{cases} + \text{if } F > 0 \\ - \text{if } F < 0 \end{cases}$$

$$a_c = \pm (\log \sin)^{-1} G = \underline{\text{-}40'.93 \cong \text{-}41'} \quad \begin{cases} + \text{if } F > 0 \\ - \text{if } F < 0 \end{cases}$$

Second Part: Calculation of Z_N:

VI. Define:

$H = \log(\pm \sin t_A) = \log(\pm \sin \underline{\;95°\;}) = \underline{\;-0.001656\;} \quad \begin{cases} + \text{ if } \sin t_A > 0 \\ - \text{ if } \sin t_A < 0 \end{cases}$

$I = \log \cos a_C = \log \cos \underline{\;-41'\;} = \underline{\;-0.000031\;}$

$J = \log \cos \delta = \log \cos \underline{\;-9°\;} = \underline{\;-0.00538\;}$

VII. Calculate:

$J + H - I = \underline{\;-0.007005\;}$, and

$A_Z = \pm (\log \sin)^{-1}(J + H - I) = \underline{\;79° \ 44'\;} \quad \begin{cases} + \text{ if } \sin t_A > 0 \\ - \text{ if } \sin t_A < 0 \end{cases}$

$Z_N = \begin{cases} A_Z \text{ if } Z_N \leq 180° \\ 360° - A_Z \text{ if } Z_N > 180° \end{cases} = \underline{\;79° \ 44'\;}$

Note that "$\log^{-1}$" denotes the anti-logarithm, i.e., $\log^{-1} x = y$ implies that $x = \log y$, and $(\log \sin)^{-1} x = y$ implies that $x = (\log \sin) y$. Also note that $(\log \sin)^{-1} x = \sin^{-1}(\log^{-1} x)$ and $\log \cos x = \log(\cos x)$.

Reviewing the calculations above, the reader will realize that the logarithmic procedure is fairly complicated by the fact that the domain of the logarithmic function log x is confined to the set of all real positive numbers. Because of this, the underlying formulae have to be altered by the introduction of the factor -1, thereby somewhat complicating the resulting algorithm. Just by mixing up a plus or a minus sign, this algorithm can produce the wrong result. Because of that, any navigator wishing to use this logarithm needs to practice with it in order to acquire the necessary efficiency. [30]

Finally, some practical advice about the acquisition of suitable logarithm-tables. There are, of course, a multitude of fine tables available, but some may cost more than the inexpensive scientific calculator that uses them. To use the above logarithm, a navigator only need a compact set of tables comparably priced to the calculator. [36]

2.12 How an Approximate Position at Sea or Air Can Be Found if an Approximate Value for the Azimuth or the Parallactic Angle Is Known in Addition to One Altitude

As has been pointed out already in some of the previous sections of this book, the main objective of any type of analysis of astronavigation should be to analyze existing methods of navigation and to promote the development of new and lesser known independent analytic methods for finding the exact or approximate position at sea or in the air by computational procedures. Although I do not advocate for the LOP method of celestial navigation as analyzed in the previous sections, I appreciate some of its merits and encourage the use of the navigational triangle of assumed position for use in other methods. In this section, we will use only one altitude observation in conjunction with an approximation of either the azimuth or the parallactic angle to determine the approximate position of the observer.

First, we will have to derive the simple formulae for finding the exact position of an observer when the altitude of one CO together with the exact azimuth or the parallactic angle is known. Secondly, we will also discover an approximate method for computing the azimuth and also the parallactic angle.

From the practical point of view, there is a difference between finding an approximate value for the azimuth and the parallactic angle. In the case of the azimuth there are basically two methods for finding an approximation to it, namely, by observations or by computation. Of course, as every navigator knows, measuring the azimuth of a CO from a moving vessel or an airplane without employing special instruments is very difficult especially if a high degree of accuracy is required.

In this context, the reader should be reminded that ancient Polynesian sea-farers hardly ever used any type of altitude observations, except for rising, setting and zenith COs, i.e., for altitudes of zero or ninety degrees. They relied chiefly on azimuth observations and on the concept of constellations.

The other method for finding the desired approximation for the azimuth rests strictly on computational methods (as presented in this section). Finding an approximate value for the parallactic angle the navigator has to rely on indirect methods only and such methods will always be based on another approximation and will involve a series of computations.

For now, let's assume that we know the elements δ, h_0, and A_Z exactly, and also let's assume that we have the exact UTC time at the instant the altitude and azimuth were measured. Then by applying the COS-TH. to our navigational triangle, we obtain:

$\sin \delta = \sin h_0 \cdot \sin \varphi + \cos h_0 \cdot \cos \varphi \cdot \cos A_z$

with $z = 90° - h_0$, $\bar{\delta} = 90° - \delta$, $\bar{\varphi} = 90° - \varphi$, and

$$t = \pm(\text{GHA}(S) + \lambda) \begin{cases} + \text{if West} \\ - \text{if East} \end{cases}$$

$$A_z = \begin{cases} Z_n & \text{if } Z_n \leq 180° \\ 360° - Z_n & \text{if } Z_n \text{ is } > 180° \end{cases}$$

Next we define the auxiliary parameters A, B, C and α as follows (Fig. 2.12.1):

$A = \sin h_0$,

$B = \cos h_0 \cdot \cos A_z$,

$C \cdot \cos \alpha = A$

$C \cdot \sin \alpha = B$

From these definitions it follows then that for all $A \neq 0$, $\alpha = \tan^{-1} \frac{B}{A}$, and

$C = (1 - \cos^2 h_0 \sin^2 A_z)^{\frac{1}{2}}$. Then the above equation for the latitude becomes:

$$\sin \delta = A \cdot \sin \varphi + B \cos \varphi = C(\cos \alpha \cdot \sin \varphi + \sin \alpha \cdot \cos \varphi)$$
$$= C \cdot \sin(\varphi + \alpha)$$
, and

(1) $\varphi = \begin{cases} \sin^{-1}\left(\dfrac{\sin \delta}{(1 - \cos^2 h_0 \cdot \sin^2 A_z)^{1/2}}\right) - \tan^{-1}\left(\dfrac{\cos A_z}{\tan h_0}\right), & \text{if } h_0 \neq 0 \\ \cos^{-1}\left(\dfrac{\sin \delta}{\sin A_z}\right), & \text{if } h_0 = 0, \text{ and } A_z \neq 90° \\ \sin^{-1}\left(\dfrac{\sin \delta}{\sin h_0}\right), & \text{if } A_z = 90°, \text{ and } h_0 \neq 0 \end{cases}$

If $h_0 = 0$ and $A_z = 90°$, then necessarily $\delta = 0$ and no value for φ can be found by the above procedure.

In order to compute the local hour angle, we employ once more the COS-TH. of spherical trigonometry and obtain the well-known formula:

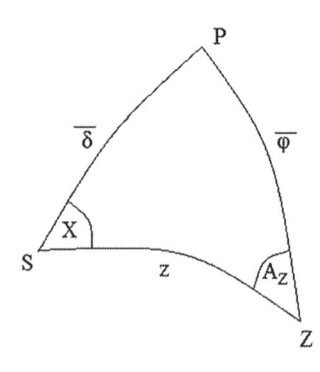

Fig. 2.12.1

2.12 How an Approximate Position at Sea or Air ...

(2) $t = \cos^{-1}\left(\frac{\sin h_0 - \sin \delta \cdot \sin \varphi}{\cos \delta \cos \varphi}\right)$, and subsequently

(3) $\lambda = -\text{GHA (S)} \pm t$, + if S is West, and − if S is East of observer.

As can readily be seen, the critical values for altitudes are the ones that are close to 90° and 0°, and azimuths close to 90° and 270°. Furthermore, all ambiguities due to the use of the multivalued functions $\sin^{-1} x, \cos^{-1} x$, and $\tan^{-1} x$ in the above formulae must be removed by using the facts that $|\varphi| < 90°$, $t > 0$, and the computed position must be in accordance with the prescribed value of Z_n.

In all critical cases where h_0 is either close to zero or ninety degrees and/or the azimuth Z_N is close to ninety or two hundred and seventy degrees, the measured or computed azimuth that enters the above equations must be determined with the highest degree of accuracy possible. In all other cases, one can expect to obtain acceptable results by approximating the true azimuth by values that differ in magnitude by not more than half a degree. Nevertheless, even in cases where the error in azimuth is in excess of half a degree, this method may still produce results that meet the actual requirements. This, in turn, suggests that it may be possible to employ approximations to the true azimuth that have been obtained by computational methods. Of course, such type of methods are always based on other approximations, as will be shown shortly. Therefore, let us devise a fairly simple method of finding an approximation to the true azimuth by computational procedures.

Since, by now, the reader should be quite familiar with the classical method of LOPs, he will appreciate that we are going to adopt the navigational triangle of the assumed position method as the base for computing an approximation to the azimuth. By doing so, we actually use only one altitude observation together with an assumed position to find the approximate position of the observer. Therefore, this approximation is bound to fail unless the assumed position is very close to the true position of the observer. (See also the numerical example.)

Although in practice it may be difficult to find an assumed position that is sufficiently close to the true position, we shall nevertheless state the well-known formulae for finding the azimuth of the assumed position triangle (Fig. 2.12.2).

Accordingly, if the assumed position Z′ has the coordinates (λ_A, φ_A), and if "A'_Z" denotes the azimuth angle of this triangle, then we conclude that:

(4) $t_A = \pm(\text{GHA(S)} + \lambda_A)$,

(5) $h_C = \sin^{-1}(\sin \delta \cdot \sin \varphi_A + \cos \delta \cdot \cos \varphi_A \cdot \cos t_A)$, and

(6) $A'_Z = \sin^{-1}\left(\frac{\cos \delta \cdot \sin t_A}{\cos h_C}\right)$, where "$h_C$" denotes the computed altitude.

Because, in general, it is virtually impossible to find an assumed position that is close to the true position, it is a good idea to use the physical method over this computational method whenever possible. (See numerical example below.)

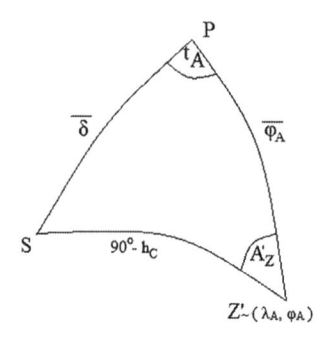

Fig. 2.12.2

Next, let's consider some numerical examples that are based on almost critical input data.

Example 1 Date: 06/13/08, UTC: $20^h\ 00^m\ 00^s$, C.O.: Sun ☉,
$h_0 = 81°29'14'' = 81°.48735571$
$Z_n = 268°38'$,
$\delta = 23°15'.8N$, (Taken from NA)
$GHA(S) = 119°56'.8$ (Taken from NA)
S observed South West of observer.
Problem: Find the position Z of the observer.
Solution:
Compute: $A_z = 360° - Z_n = 91°22' = 91°.3666...$, and
$(1 - \cos^2 h_0 \sin^2 A_z)^{\frac{1}{2}} = 0.988989521$, and
$\sin^{-1}\left(\frac{\sin \delta}{(1-\cos^2 h_0 \sin^2 A_z)^{\frac{1}{2}}}\right) = \sin^{-1} 0.399354739 = 23°.53784629$ and
$\tan^{-1}\left(\frac{\cos A_z}{\tan h_0}\right) = \tan^{-1}(-0.003569831) = -0.204535381$, and
$\varphi = 23°.74238167 = 23°44'.54N$, and
$t = \cos^{-1} 0.986941114 = 9°.269673724$, and
$\lambda = -GHA(S) + t = -110°.6769923 = 110°40'.62W$

Example 2 Same data as in the first example—except for the azimuth that is now to be approximated by $Z'_n = 269°$ (obtained by observations).

Problem: Find the approximate position Z' corresponding to this approximate azimuth Z'_n by invoking formulae (1)–(3).
Solution:
Compute: $A'_z = 360° - Z'_n = 91°$, and
$(1 - \cos^2 h_0 \sin^2 A'_z)^{\frac{1}{2}} = 0.9883865593$, and

2.12 How an Approximate Position at Sea or Air ...

$$\sin^{-1}\left(\frac{\sin \delta}{(1-\cos^2 h_0 \sin^2 A_Z')^{\frac{1}{2}}}\right) = 23°.53784629 \text{ and}$$

$$\tan^{-1}\left(\frac{\cos A_Z'}{\tan h_0}\right) = -0.204535381, \text{ and}$$

$\varphi' = \underline{23°74238167} = \underline{23°44'.54\text{N}}$, and
$t' = \cos^{-1} 0.956941114 = 9°.269673724$, and
$\lambda' = -\text{GHA}(S) + t = \underline{-110°.6769923} = \underline{110°40'.62\text{W}}$

Example 3 All relevant data as in Example 1, but here the azimuth is not known. However, an approximate, i.e., an assumed position $Z_A' \sim (\lambda_A, \varphi_A)$ is given as
$\varphi_A = 23°30'$, and $\lambda_A = -110°30'$.
Problem: Find an approximation to the azimuth by invoking formulae (4)–(6)
Solution:
Compute: $t_A = 119°.946666 - 110°.5 = 9°.44666$, and
$h_C = \sin^{-1} 0.988566146 = 81°.32741377$, and
$A_Z' = \sin^{-1} 0.999984545 = \begin{cases} 89°.68145356 \\ 90° + 0.31854644 \end{cases}$

according to Sect. 2.4, formulae (6').
This ambiguity is removed by recalling that our CO lies to the South West (quadrant III) of our observer. Hence:
$A_Z' = 90°.31854644 = \underline{90°19'.1}$

Example 4 Here, we consider the same case as in Example 3 but with a different assumed position, one that is further away from the true position, namely:
$\varphi_A = 24°$ N, and $\lambda_A = 111°$ W.
Problem: Find the approximate value for the azimuth corresponding to the above assumed position.
Solution:
Compute: $t_A = 119°.946666 - 111° = 8°.94666$, and
$h_C = \sin^{-1} 0.989706381 = 81°.77198796$, and
$A_Z' = \sin^{-1} 0.998313337 = \begin{cases} 86°.67177368 \\ 90° + 3.328226319 \end{cases}$

according to formulae (6'), Chap. 2, Sect. 2.4.
However, since S lies to the South West (quadrant III) of our observer, we conclude that:
$A_Z' = 93°.328226319 = \underline{93°19'.7}$
Therefore, this approximation is way off and not acceptable for further use.
Similar to the method of employing the azimuth angle A_z, one can also use the angle X subtended by the polar distance of S (See Fig. 2.12.3) and the zenith distance, referred to as the parallactic angle, to compute the position of the observer. From the mathematical point of view, these two methods are basically the same. However, from the point of view of the navigator, these two methods are quite different since the navigator has no physical aids available to measure, the

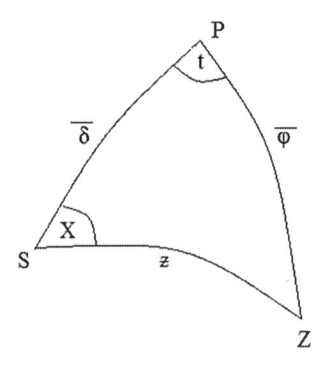

Fig. 2.12.3

parallactic angle. Therefore, by employing this angle in computational procedures, navigators have to rely on computational methods for actually finding it. This, in turn, requires the use of other approximations thereby making it more difficult.

Again, let us first derive the necessary equations for finding the position of the observer Z whenever one altitude observation together with the corresponding parallactic angle X is available.

By referring to Fig. 2.12.3, and applying the COS-TH. and SIN-TH. we find that:

(7) $\varphi = \sin^{-1}(\sin \delta \cdot \sin h_0 + \cos \delta \cdot \cos h_0 \cdot \cos X), X < 90°$, and
(8) $t = \sin^{-1}\left(\frac{\cos h_0 \sin X}{\cos \varphi}\right) > 0$, and
(9) $\lambda = \pm t - \text{GHA}(S), +\text{West}, -\text{East of observer}.$

In order to find an approximation to the parallactic angle X by computation, whenever a suitable assumed position is available, we refer to Fig. 2.12.4 and deduce that:

(10) $t_A = \pm(\text{GHA}(S) + \lambda_A) + 360°$, and
(11) $h_C = \sin^{-1}(\sin \delta \cdot \sin \varphi_A + \cos \delta \cdot \cos \varphi_A \cdot \cos t_A), < 90°$, and

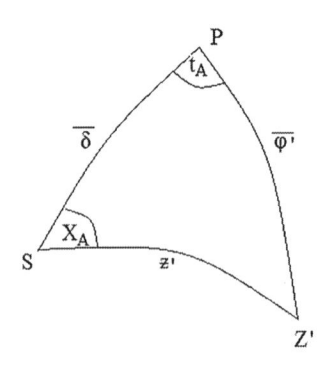

Fig. 2.12.4

2.12 How an Approximate Position at Sea or Air ...

(11′) $\sin h_C = \sin \delta \cdot \sin \varphi_A + \cos \delta \cdot \cos \varphi_A \cdot \cos t_A$, for removing ambiguities only,

(12) $X_A = \sin^{-1}\left(\dfrac{\cos \varphi_A \cdot \sin t_A}{\cos h_C}\right)$

Next let us again consider some numerical examples that are based on the same set of ephemeral data as the previous ones.

Example 5 Given the observed altitude of $h_0 = 81°.48735571$ and the parallactic angle $X = 84°.7553680501$.
Problem: Find the position of the observer.
Solution:
By employing the Eqs. (7)–(9) we deduce that:
$\varphi = \sin^{-1} 0.403037327 = 23°.76819429 = \underline{23°46'.1\text{N}}$
$t = \sin^{-1} 0.161069275 = \begin{cases} 9°.26896620 \\ 90° + 80°.7310338 \end{cases}$, and since S is visible to the West of the observer, we conclude that:
$t = 9°.26896620$, and therefore:
$\lambda = -110°.6776938 = \underline{110°40'.7\,\text{W}}$

Example 6 Same ephemeral data as in Example 5, but X is not known.
Problem: Given the coordinates of the assumed position as:
$\varphi_A = 23°42'$, and $\lambda_A = -110°42'$. Find the corresponding parallactic angle X_A by employing the Eqs. (11)–(12)
Solution:
Compute:
$t_A = 9°.24666$, and
$h_C = 81°.50933255$, and
$X_A = \underline{85°.21290697}$, and hence—
$\Delta X = X - X_A = -0°.4575\ldots = -0°.46$, that is still acceptable as an approximation.

Example 7 Same as above except that now:
$\varphi_A = 24°$, and $\lambda_A = -111°$.
Solution:
Compute:
$t_A = 8°.94666$, and
$h_C = 81°.77198796$, and
$X_A = 83°.07879612$, and hence—
$\Delta X = X - X_A = 1°.676571934$. Therefore, this approximation is not acceptable.

In conclusion, it can be said that these examples confirm the conclusions drawn in the theoretical part of this section and also demonstrate the importance of using the correct branch of the inverse trigonometric function.

2.13 On the Effect of a Change in Time on the Altitude and Azimuth

Finding a suitable estimate for the change in altitude due to a change in time is of great practical value to the navigator. Once such an estimate has been found, it will be possible to reduce two sights taken at distinct instants to two sights taken simultaneously, provided that the time elapsed between these sights is relatively small, say, less then 5 min (See Sect. 2.8).

Mathematically speaking, we would like to know how a small change in time (Δt) results in a change Δh_0 in altitude and in azimuth angle ΔA_z. To establish the desired mathematical relation, we again look at our navigational triangle $Z\hat{P}S'$ (See Fig. 2.13.1).

According to the COS-TH. of spherical trigonometry we have:
$\cos z = \sin \delta \cdot \sin \varphi + \cos \delta \cdot \cos \varphi \cdot \cos H$.

For our purposes we can consider that δ and φ are constants. Therefore, by differentiating the above expression with respect to H we have:
$\sin z \, \frac{dz}{dH} = \cos \delta \cdot \cos \varphi \cdot \sin H$

Next, by applying the SIN-TH. we find that:
$$\frac{\sin H}{\sin z} = \frac{\sin A_z}{\cos \delta}$$

Substituting $\sin H = \dfrac{\sin z \sin A_z}{\cos \delta}$ in the above expression for $\frac{dz}{dH}$ yields:

(1) $\frac{dz}{dH} = \cos \varphi \cdot \sin A_z$

With $\frac{dH}{dt} = \pm \dfrac{15.04''}{\sec}$ and $z = 90° - h_0$, we conclude that:

(1') $dh_0 = \pm 15.04 \cdot \cos \varphi \cdot \sin A_z \cdot dt$. (+ if S rising, − if S setting.)

It should be clearly understood that the expressions (1) and (1') are exact infinitesimal relations that give rise to the following finite approximation:

(2) $\Delta h_0 = \pm 15.04 \cdot \cos \varphi \cdot \sin A_z \cdot \Delta t$, with Δt in seconds and Δh_0 in arcseconds. [4, 13]

Of course the reader should be keenly aware of the limitations of this estimate for the change Δh_0 with regards to the elapsed time Δt. The application of formula

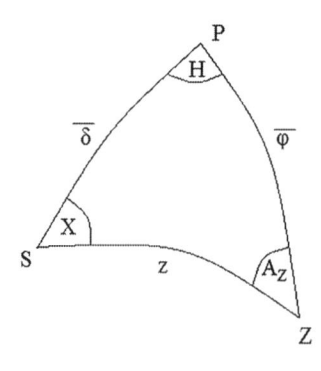

Fig. 2.13.1

2.13 On the Effect of a Change in Time on the Altitude ...

(2) to all cases where Δt is in excess of the imposed limit will certainly result in erroneous results. Furthermore, the above formula requires a good approximation or, if possible, a computation of the latitude of the observer. Once such an approximation has been found, the azimuth angle A_z can be calculated by employing formulae (4)–(6) of the previous Section (Sect. 2.12).

In all cases where the navigator uses Celestial Navigation, the required data in formulae (4)–(6) can be taken from the navigational triangle of the assumed position.

In order to improve formula (2) with respect to accuracy, the Taylor expansion of z(t), i.e., $h_0(t)$ can be readily employed. However, the latter requires an explicit expression for the rate of change of the azimuth angle, i.e., for $\frac{dA_z}{dt}$. Therefore, let us first derive this formula in terms of the known quantities.

Again, let us have another look at Fig. 2.13.1 and conclude that the application of our Analogue Formula (see Sect. 2.2, formula 2iii) to this triangle results in: $\cos\delta \cdot \cos H = \cos z \cdot \cos\varphi - \sin z \cdot \sin\varphi \cdot \cos A_z$. Then in accordance with formula (2ii) Sect. 2.2, we have:

$\sin H \cdot \cos\delta = \sin z \cdot \sin A_z$

Differentiating this expression with respect to H and recalling that z depends on H (see formula 1), we obtain:

$\cos\delta \cdot \cos H = \cos z \cdot \cos\varphi \cdot \sin^2 A_z + \sin z \cdot \cos A_z \frac{dA_z}{dH}$ Combining this expression with the first formula above, we find that:

$\frac{dA_z}{dH} = \cot z \cdot \cos\varphi \cdot \cos A_z - \sin\varphi$, and hence:

(3) $dA_z = \pm 15'.04(\cot z \cdot \cos\varphi \cdot \cos A_z - \sin\varphi)dt$, a truly infinitesimal expression for dA_z. By approximating the latter by finite quantities we obtain the desired formula:

(4) $\Delta A_z = \pm 15'.04(\cot z \cdot \cos\varphi \cdot \cos A_z - \sin\varphi)\Delta t$, $|\Delta t| < 5$ min.

We can now turn our attention to the task of improving formula (2). According to Taylor's expansion theorem, we have with $\Delta t = t - t_0$:

$\Delta z = -\Delta h_0 = \left(\frac{dz}{dt}\right)_{t_0} \cdot \Delta t + \frac{1}{2}\left(\frac{d^2z}{dt^2}\right)_{t_0} \cdot \Delta t^2 + \cdots$ and it is reasoned that the first two terms on the right hand side of this formula will suffice to achieve the desired improvement over (2).

Since we have already evaluated $\left(\frac{dz}{dt}\right)_{t_0} = \pm 15.04 \cdot \cos\varphi \cdot \sin A_z$ (see formula 1), we merely have to differentiate this formula once more to find:

$\left(\frac{d^2z}{dt^2}\right)_{t_0} = \pm 15''.04 \cdot \cos\varphi \cdot \cos A_z \frac{dA_z^r}{dt}$, where A_z^r stands for A_z in radians. Hence if A_z is given in arcseconds, then $\frac{dA_z^r}{dt} = \sin 1'' \frac{dA_z}{dt}$. Substituting the last expression in the above equation for $\frac{d^2z}{dt^2}$, the above approximation for the Taylor expansion becomes:

$$\Delta z = \pm 15''.04 \cdot \cos\varphi \cdot \sin A_z \cdot \Delta t \pm 15''.04 \cdot \cos\varphi \cdot \cos A_z \cdot \sin 1'' \frac{dA_z}{dt} \cdot \frac{\Delta t^2}{2}.$$

By employing the expression (3) for $\frac{dA_z}{dt}$ we obtain the desired result:

(5) $\Delta h_0 = \pm 15''.04 \cdot \cos\varphi \cdot \sin A_z \cdot \Delta t - \cos\varphi \cdot \cos A_z (\tan h_0 \cdot \cos\varphi \cdot \cos A_z -$

$- \sin\varphi) \cdot \left(\frac{15.04 \cdot \Delta t}{2}\right)^2 \cdot \sin 1''$, + if S Rising, − if S setting; t in seconds, and h_0 in arcseconds.

2.14 How to Determine Latitude at Sea or Air Without the Use of a Clock

Long before man invented the clock and the chronometer, sailors knew how to determine latitude at sea and sailed along circles of equal latitude. It may not have always been convenient, but it was a sure method for making a landfall. Besides this particular application, exact latitude is imperative for finding longitude and time. (See Sect. 2.21.)

Of course there are various methods available for finding latitude without the use of a time piece. However, for our purposes, we shall only look at three:

I. Determining exact latitude by observing two stars simultaneously, or within a very short interval of time.
II. By observing a circumpolar star at upper or lower culmination.
III. By observing the sun or a star at or near the passage of the local meridian.

(The case where the altitude of **POLARIS** is employed will be considered in Sect. 2.16.)

Now, let's take a look at a couple of examples:

Case #I:

This case has already been covered in Sect. 2.4. Here ΔS is either computed by formula (4) in the case of simultaneous observations, or by formula (6) where the interval $T_1 - T_2$ is determined by counting the seconds or by the use of an Hour-Glass. By entering formula (9) with the known values of the declinations for S_1 and S_2 and ΔS we find d. Next by employing formula (10) with the values for the declination we find α_1. Next by using the values for the observed altitudes a_1 and a_2, respectively, we find α_2 with the help of formula (11). By recalling the sectors in which S_1 and S_2 have been observed, we calculate α with the help of expression (12) and finally find φ by means of formula (13).

2.14 How to Determine Latitude at Sea or Air ...

Now, let us consider a numerical example of this very important application.

Example 1

On July 31, 2008, the navigator of a small sailing vessel in the Gulf of Mexico observes at $04^h\ 20^m\ 00^s$ UTC[16] **KOCHAB** and **DENEB** virtually simultaneously, and after reductions, deduces that: $a_1 = h_0^1 = 31° 31'.1$ and $a_2 = h_0^2 = 56° 27'.78$. He wants to know on which latitude he is approaching the coast.

Solution:

The application of expression (4) from Sect. 2.4 yields:
$\Delta S = SHA(S_1) - SHA(S_2) = 87°.761$. Entering expression (9) with this value, we find:
$d = \cos^{-1}(\sin 45.311 \cdot \sin 74.123 + \cos 74.123 \cdot \cos 45.311 \cdot \cos 87.761) = 46°.2645$.

Entering formula (10) with this value yields:
$$\alpha_1 = \cos^{-1}\left(\frac{\sin 45.311 - \sin 74.123 \cdot \cos 46.2645}{\cos 74.123 \cdot \sin 46.2645}\right) = 76°.5495.$$
Next by employing formula (11), we find:
$$\alpha_2 = \cos^{-1}\left(\frac{\sin 56.463 - \cos 46.2645 \cdot \sin 31.5183}{\sin 46.2645 \cdot \cos 31.5783}\right) = 39°.9594$$
Since $S_1 \in IV$ and $S_2 \in I$, (See the four sections for the azimuth.), we have $(S_1, S_2) = (IV, I)$, and in accordance with our quadrant formulae (12) deduce: $\alpha = \alpha_1 + \alpha_2 = 116°.5089$ and therefore by virtue of expression (13), we find that:
$\varphi = 90° - \cos^{-1}(\sin 74.1233 \cdot \sin 31.5783 + \cos 74.1233 \cdot \cos 31.5783 \cdot$
$\cdot \cos 116.5089) = 23°.4993 = 23°30'$ N

Next, let us consider case #II.

Case #II:

First let us recall that for a star to be a "Circumpolar Star" its declination δ must satisfy the obvious condition that $\delta > 90° - \varphi = \bar{\varphi}$ i.e., the declination of this star must be greater than the colatitude of the observer. (See Fig. 2.14.1.)

Then we readily deduce from Figs. 2.14.1 and 2.14.2 that:
$\varphi = H_0^L + P^L = H_0^L + 90° - \delta^L$, and also
$= H_0^U - P^U = H_0^U - 90° + \delta^U$, if $Z_n = 0$. Similarly we deduce that:
$\varphi = H_0^L + P^L = H_0^L + 90° - \delta^L$, and
$= 180° - (H_0^U + P^U) = 90° - (H_0^U - \delta^U)$, if $Z_n = 180°$ Hence:

$$\varphi = \begin{cases} \dfrac{H_0^L + H_0^U}{2} + \dfrac{\delta^U - \delta^L}{2} & \text{if } Z_n = 0 \\ \dfrac{H_0^L + H_0^U}{2} + \dfrac{\delta^U - \delta^L}{2} + (90° - H_0^U) & \text{if } Z_n = 180° \end{cases}$$

---

[16]The time is actually not required.

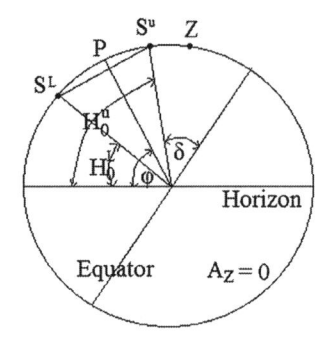

Fig. 2.14.1

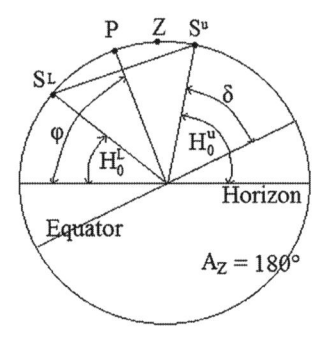

Fig. 2.14.2

Since for our particular purposes the change in declination within a period of less then 24 h is negligible, we deduce that:

(1) $\quad {}^{17}\varphi = \begin{cases} \frac{H_0^L + H_0^U}{2} \text{ if } Z_n = 0 \\ \frac{H_0^L + H_0^U}{2} + (90° - H_0^U) \text{ if } Z_n = 180° \end{cases}$ [8, 16]

Furthermore, as long as the error committed by assuming that the actual azimuth is either zero or one hundred eighty degrees respectively is negligible, expression (1) constitutes an excellent approximation for the latitude of the observer. The latter can readily be verified by carrying out an error analysis.

The major disadvantage of this method consists in not always providing a tangible result in low latitudes because the time between the two observations is

---

[17] In all cases where the vessel moves between the two culminations of the star during the period of 12^h Sid. T. = $11^h.96723467$ Sol. M. T. the altitude of the first culmination in formula (1) has to be adjusted by either adding or subtracting the latitude made good by the vessel during that interval of time. This change in latitude can be easily calculated by using the corresponding formulae of Sect. 1.2 or 1.3.

2.14 How to Determine Latitude at Sea or Air ...

exactly 12 h sidereal time and therefore will most likely render the star invisible at one of its culminations.

Let's take a look at a numerical example:

Example 2

The date is November 26, 2008 and you are on board a fishing vessel in the Sea of Labrador. Shortly after the sun sets, you find the star **DENEB** almost at the zenith and after having checked all the compasses on board, the altitude of it is determined when it passes the assumed meridian. After the necessary reductions, you determine that:

$H_0^U = 84°.19'.2$

You decide to return to the same fishing grounds the next day. $11^h\ 58^m$ later **DENEB** passes your meridian again and you take another measurement. After making the necessary adjustments, you find that:

$H_0^L = 6°19'.2$

These two values for the altitude so obtained together with expression (1) above yields:

$\varphi \cong 51°19'.2\,N$

Next, let's consider case #III. (See also Figs. 2.14.1 and 2.14.2.)

Case #III:

This case does not involve more mathematics than adding or subtracting the declination to the altitude once the declination has been found. Of course it is always assumed that a good approximation for the local meridian has been determined. However, in the case of the Sun, we require an approximation of the longitude of the observer, since no time piece is available. This is necessary since the computation of the latitude also requires the exact value of the declination at the time of transit.

Recalling that the local time of transit of the sun is listed in the Nautical Almanac or can be computed by employing one of the algorithms provided in the second part of this book, we merely need an approximate longitude in order to find the approximate UTC of transit and subsequently can extract the required declination from said publication or computation. (What has been said about the Sun also applies to the planets.)

In the case of a transit of a star, an approximation for the longitude is not required since the declination of stars does not change for the purpose of our applications within the relatively short period of 24 h.

The reader should also be acutely aware of the fact that the methods employed in Cases #II and #III require that the assumed meridian, most likely determined by compass observations, should fall within a small fraction of a degree of the true azimuth, i.e., zero or one hundred eighty degrees respectively, in order to assure that the value for the computed latitude falls within the same range with respect to its error.

In conclusion, I would like to encourage the reader to use the exact method described in Case #1 whenever possible and discourage the modern navigator from doing the "old" thing over and over again, i.e., resort to the use of approximate procedures when exact ones are available.

2.15 On Calculating the Interval Between Meridian Passage and Maximum Altitude and Finding Approximate Longitude and Latitude of a Moving Vessel, and Longitude by Equal Altitudes

For the astronomer who has a well-defined meridian at his disposal, the determination of the time of meridian transit and latitude are trivial matters. But for the navigator, only approximate methods for finding the longitude and latitude are available, notwithstanding the existence of exact mathematical methods as provided in this book.

In one of the previous sections, I have shown that the exact position of the vessel can be calculated when altitude and azimuth together with the declination and GHA of a celestial object are known. However, in general, on a ship or airplane, the exact azimuth is not known. In particular, the determination of true north by means of a compass is always subject to some errors. Therefore, in order to find a suitable approximation to the true time of meridian passage indirect methods have to be devised. One of the most obvious methods of accomplishing this task is to observe the CO continuously when it is near the meridian and to determine the time when it reaches its maximum altitude. The fact that the time of maximum altitude and meridian transit do not coincide, in general, should be obvious. In the case where the celestial object is the sun or a planet, the two times never coincide as long as their declinations change. In the case of a star, these two events will also not occur at the same time as long as the ship has a north/south component in its trajectory. Therefore, it is necessary to correlate the time of these two events, i.e., meridian and maximum altitude passage, analytically.

I will now devise a suitable formula subject to the conditions that either the latitude, course and speed of the vessel together with the approximate time of meridian transit, or the approximate time of meridian passage and the altitude of the sun, planet or other CO together with the course and speed of the vessel are known. The exact or approximate longitude can also be calculated once the navigator has determined the interval between meridian transit and maximum altitude.

First, let's depict a typical case of a moving vessel relative to the ever-moving sun. (See Fig. 2.15.1.)

Here "Z" denotes the position of the vessel when the sun ⊙ passes the meridian and "Z'" denotes the position of the vessel when the altitude of the sun ⊙ reaches its maximum altitude relative to Z'. According to Fig. 2.15.1, the local hour angle H at Z when the sun ⊙ reaches its maximum altitude relative to Z' is given by:

2.15 On Calculating the Interval Between Meridian …

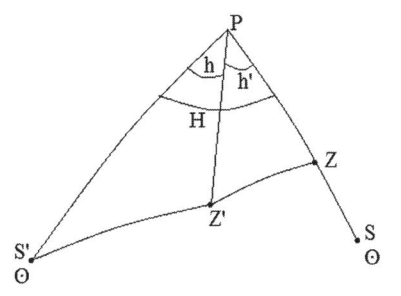

Fig. 2.15.1

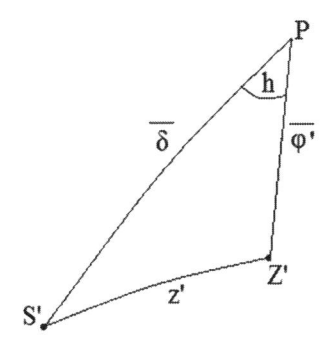

Fig. 2.15.2

(1) $H = h + h'$

Note that in reality, the angles h and h' are very small, indeed.

Since we want to find the local hour angle h of the sun at Z' at the time of maximum altitude, we need to look at the navigational triangle $P\hat{Z}'S'$ in order to establish the dependence of the zenith distance z' on the latitude φ', declination δ, and the hour angle h. (See Fig. 2.15.2.)

In Fig. 2.15.2 we have already approximated the latitude φ' of Z' by the latitude φ of Z. The application of the COS-TH. yields:

$$\cos z' = \sin \delta \cdot \sin \varphi + \cos \delta \cdot \cos \varphi \cdot \cos h.$$

Because we would like to find h so that the altitude is at maximum, we must minimize the zenith distance z', i.e., we must first find an expression for the derivative of $\frac{dz}{dt}$. The latter can be derived easily by applying the implicit differential expression provided in Sect. 4.2. If we define:

$F(z') = \cos z'$, and $G(\delta, \varphi, h) = \sin \delta \cdot \sin \varphi + \cos \delta \cdot \cos \varphi \cdot \cos h$, the application of said expression yields:

$$-\sin z' \frac{dz'}{dt} = (\sin \varphi \cdot \cos \delta - \sin \delta \cdot \cos \varphi \cdot \cos h) \cdot \frac{d\delta}{dt} + (\sin \delta \cdot \cos \varphi - \cos \delta \cdot$$
$$\cdot \sin \varphi \cdot \cos h) \cdot \frac{d\varphi}{dt} - \cos \delta \cdot \cos \varphi \cdot \sin h \cdot \frac{dh}{dt}$$

By taking into account that h is a very small angle, we may approximate cos h by one, i.e., cos h ≅ 1. Hence, the above expression reduces to:

$-\sin z' \frac{dz'}{dt} = \sin(\varphi - \delta)\left(\frac{d\delta}{dt} - \frac{d\varphi}{dt}\right) - \cos\delta \cdot \cos\varphi \cdot \sin h \cdot \frac{dh}{dt}$. Because we are looking for a minimum value of z′, the necessary condition is:

$\frac{dz'}{dt} = 0$, which reduces the above equation to:

(2) $\sin(\varphi - \delta)\frac{d}{dt}(\delta - \varphi) \cong \cos\delta \cdot \cos\varphi \cdot \sin h \cdot \frac{dh}{dt}$ [1, 13, 27]

In this equation "$\frac{dh}{dt}$" denotes the change of the local hour angle, i.e., the change of the hour angle of the sun relative to the moving vessel. If we denote the change in longitude of the sun by $d\lambda_\odot/dt$, and the change in longitude of the vessel by $d\lambda v/dt$, we have:

$\frac{dh}{dt} = d\lambda v/dt - d\lambda_\odot/dt$. Since we measure longitude negative if West and positive if East of Greenwich, and since the sum moves approximately 15°/h West, we conclude that: $d\lambda_\odot/dt = 900'$. Hence:

(3) $\frac{dh}{dt} = 900 + d\lambda v/dt$, (arc minute/hour). Furthermore, "$\frac{d}{dt}(\delta - \varphi)$" denotes the difference between the rate of change of declination and latitude that we shall denote by y, i.e.,

(4) $y = \frac{d\delta}{dt} - \frac{d\varphi}{dt}$. Then, Eq. (2) reduces to:

$\frac{(\tan\varphi - \tan\delta)}{(900 + \frac{d\lambda v}{dt})} \cdot y \cong \sin h$. But we have already taking into account that h is very small, therefore, $\sin h \cong \sin 1' \cdot h$, with $\sin 1' = \frac{\pi}{180 \cdot 60} = \frac{1}{3,437.73}$.

Substituting this approximation into the expression above, we find that:

$h = \frac{1}{\sin 1'} \cdot \frac{1}{(900 + \frac{d\lambda v}{dt})} \cdot y \cdot (\tan\varphi - \tan\delta)$. By also taking into account that $\frac{1}{900}\frac{d\lambda v}{dt} \ll 1$, we may again approximate by employing the Taylor expansion to obtain: $\frac{1}{1 + \frac{1}{900}\frac{d\lambda v}{dt}} = 1 - \frac{1}{900} \cdot \frac{d\lambda v}{dt}$, and finally resulting in the relevant equation:

(5) $h = \frac{1}{900 \cdot \sin 1'}\left(1 - \frac{1}{900}\frac{d\lambda v}{dt}\right) \cdot y \cdot (\tan\varphi - \tan\delta)$, (arc minutes).

Again, the reader is to be reminded that we measure latitude and declination in the North by assigning positive values to φ and δ, respectively; and in the Southern hemisphere, negative values to these measurements. Therefore, Eq. (5) holds for all admissible values of φ, δ, and λv. We also note that the expression (1) stands for the actual physical values of H, h, and h′, i.e., for the values obtained by our equations in arc minutes.

2.15 On Calculating the Interval Between Meridian ...

Next let us consider the ratio $\frac{h'}{H}$ that equals the ratio of change of h', i.e., $\frac{dh'}{dt}$ and the rate of change of H, i.e., $\frac{dH}{dt}$. Since "$\frac{dh'}{dt}$" stands for the change of the longitude of the vessel, we conclude that $\frac{dh'}{dt} = \frac{d\lambda v}{dt}$. The derivative "$\frac{dH}{dt}$" stands for the rate of change in longitude of the sun and therefore, we also have $\frac{dH}{dt} = -900$. Hence:

$$\frac{h'}{H} = -\frac{1}{900} = \frac{d\lambda v}{dt} \text{ or}$$

(6) $h' = -\frac{1}{900} = \frac{d\lambda v}{dt} \cdot H$, (arc minutes)

It follows then from expression (1) that $h = H - h' = H\left(1 + \frac{1}{900} \cdot \frac{d\lambda v}{dt}\right)$.

Substituting this value for h into Eq. (5) yields:

$H = 3.819722 \cdot \left(1 - \frac{1}{900}\frac{d\lambda v}{dt}\right)^2 \cdot y \cdot (\tan\varphi - \tan\delta)$. Note that we have again made use of the approximation that is based on $\frac{1}{900}\frac{d\lambda v}{dt} \ll 1$, therefore, we may also conclude that: $\left(1 - \frac{1}{900}\frac{d\lambda v}{dt}\right)^2 \cong 1 - \frac{2}{900}\frac{d\lambda v}{dt}$, and therefore we arrive at the final result:

(7) $H = 3.819722 \cdot \left(1 - \frac{2}{900}\frac{d\lambda v}{dt}\right) \cdot y \cdot (\tan\varphi - \tan\delta), y = \frac{d\delta}{dt} - \frac{d\varphi v}{dt}$.

Of course, we have also assumed that the observer has recorded the time of maximum altitude LMA. Therefore, once H has been computed, the time of meridian transit LAN has been determined and the longitude of the vessel has been found. Specifically, we may conclude that the relationship between the local apparent noon LAN and the time of maximum altitude LMA is LAN = LMA $- \frac{H}{900}$ in hours) and the longitude is given by:

(8) $\lambda = -((\text{LAN} + \text{ET}) - 12^h) \cdot 15$, where "ET" denotes the equation of time, i.e., the difference between the meridian passage of the true and mean sun, i.e.:
(9) ET = (GHA ☉ − GHA Ø)/15, ☉ true sun, Ø mean sun.

In detail, we can derive the expression (8) by noting that at the time of meridian passage of the sun:

(i) LHA ☉ = 0, but (ii) LHA ☉ = GHA ☉ + λ, hence:
(ii) GHA ☉ = −λ, and from (9) it follows that:
(iii) GHA ☉ = GHA Ø + 15 · ET. Since
(iv) GHA Ø = (LAN − 12) · 15, we conclude that:
(v) GHA ☉ = ((LAN + ET) − 12^h) · 15, and therefore, by virtue of (iii), (8) follows.

The reader may also recall that the equation of time is listed in the NA for every day twice. Furthermore, in the second part of this book, the reader will find an explicit formula for computing the equation of time for any specified instant. However, for now, it will suffice us to know that for our specific purposes the relevant values for the ET can be found in the NA. [At this point, it may be of interest to know that: $-14 \leq ET \leq 16$ min.]

So far we have assumed that the exact or approximate latitude of the vessel is known, which may not always be the case. If we now assume that the latitude is not known, we may proceed as follows:

(a) Observe the sun several minutes before the sun is even close to the assumed meridian.
(b) With the help of a well adjusted compass and pelerous, if available, estimate the actual time of meridian transit and measure the altitude of the sun.
(c) Calculate the latitude by assuming that the azimuth of the sun is exactly 180° or 0°, i.e., use the well-known formula:

(10) $\varphi = 90° - H_0 \pm \delta$. Plus (+) if ☉ is North of equator; minus (−) if ☉ is South of equator.

Hence, by virtue of the expressions (8), (9), and (10), we now have a suitable expression for the latitude of the observer.

(d) Next, by employing the latitude obtained by formula (10), we then calculate H with the help of expression (7), and after having determined the time of maximum altitude LMA, obtain the "exact" time of meridian transit LAN by virtue of the expression that precedes formula (8). If the difference between this time and the assumed time of meridian passage does not exceed four minutes, the approximate latitude so obtained will be sufficiently close to the true latitude, and further adjustments for an error in latitude are unnecessary. The latter follows directly if one employs the formula for the change in zenith distance due to a change in the time of observation as derived in Sect. 2.13. This formula is:

(11) $\Delta z' = \pm 15 \cdot \cos\varphi \cdot \sin A_z \cdot \Delta T$, ΔT in minutes, $\Delta z'$ in arc minutes.
Since $A_z = 180° \pm \Delta A_z$, ΔA_z small and less than 30' arc minutes, we conclude that $\sin(180° = \pm \Delta A_z) = \pm \sin A_z = \pm \Delta A_z \cdot \sin 1'$, but $\Delta z' = -\Delta H$ and therefore it follows that:
$\Delta H_0 = \pm \cos\varphi \cdot \Delta A_z \cdot \Delta T \cdot 0.0043633$, and
$|\Delta H_0| \leq |\cos\varphi| \cdot 30' \cdot 4 \cdot 0.0043633 \leq 0'.5236$ because $\Delta A_z \leq 30'$ and $|\Delta T| = 4$ min, $|\cos\varphi| \leq 1$.
The reader is reminded that "H_0" denotes the altitude of the sun and is not to be confused with the hour angle H of expression (7).

We therefore may conclude that the error in latitude will be less than one arc minute provided that the error in the assumed azimuth is less or equal to half a degree, and the difference between assumed and true time of transit is less or equal to 4 min. However, it should be noted that in most applications, the incurred error is considerably less than this upper bound since we have approximated the absolute value of $\cos\varphi$ by one.

2.15 On Calculating the Interval Between Meridian …

Let us now return to Eq. (7) in order to establish an important rule for the occurrence of the maximum altitude. For this purpose, it is convenient to conduct a proper case study of this equation.

Case Study

(Ia) Vessel moves West $\begin{cases} \underline{1.)\text{ vessel moves towards the sun}} \\ 2.)\text{ vessel moves away from the sun} \end{cases}$

(IIa) Vessel moves East $\begin{cases} \underline{1.)\text{ vessel moves towards the sun}} \\ 2.)\text{ vessel moves away from the sun} \end{cases}$

(a) $\tan\varphi - \tan\delta > 0$.

The remaining four cases are defined in the same manner but with the notations containing b.) signifying that:

(b) $\tan\varphi - \tan\delta < 0$.

Case: Ia. (1)

Here $\frac{d\varphi}{dt} < 0$, and $\left|\frac{d\varphi}{dt}\right| > \left|\frac{d\delta}{dt}\right|$, which implies that $y > 0$, $h > 0$, and $H > 0$. Therefore, the time of maximum altitude LMA occurs LATER than the time of meridian passage LAN. (See Fig. 2.15.3.)

Case: Ia. (2)

Here $\frac{d\varphi}{dt} > 0$, and, again, $\left|\frac{d\varphi}{dt}\right| > \left|\frac{d\delta}{dt}\right|$, which implies that $y < 0$, $h < 0$, and $H < 0$. Therefore, the time of maximum altitude LMA occurs BEFORE than the time of meridian transit. (See Fig. 2.15.4.)

Case: IIa. (1)

Here $\frac{d\varphi}{dt} < 0$, and, again, $\left|\frac{d\varphi}{dt}\right| > \left|\frac{d\delta}{dt}\right|$, resulting in $y > 0$, and also $h > 0$ as well as $H > 0$.

Therefore the time of maximum altitude LMA occurs LATER than the time of meridian transit LAN (Fig. 2.15.5).

Case IIa. (2)

Here $\frac{d\varphi}{dt} > 0$, and, again, $\left|\frac{d\varphi}{dt}\right| > \left|\frac{d\delta}{dt}\right|$, implying that $y < 0$, and also $h < 0$ as well as

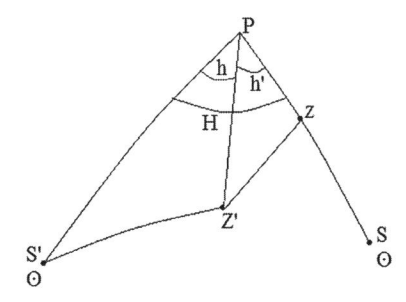

Fig. 2.15.3

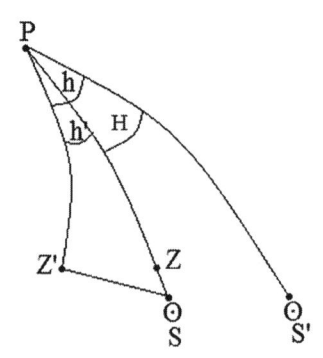

Fig. 2.15.4

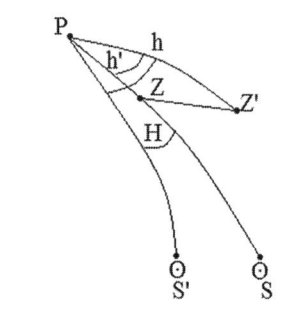

Fig. 2.15.5

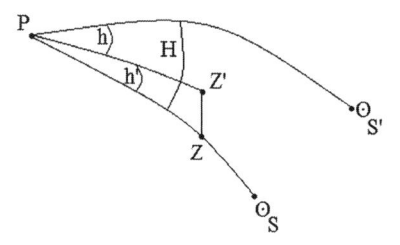

Fig. 2.15.6

H < 0. Therefore the time of maximum altitude LMA occurs BEFORE the time of meridian transit LAN (Fig. 2.15.6).

In all four cases (b), i.e., in those cases where $\tan \varphi - \tan \delta < 0$, the situation is merely reversed and therefore, we may conclude that the following rule holds:

> **RULE: Whenever the vessel move in its north/south component towards the sun H > 0, the time of maximum altitude LMA occurs AFTER the time of meridian transit, and if the vessel moves away from the sun, the time of maximum altitude LMA occurs BEFORE the time of meridian passage LAN.**

Next, let us consider some numerical examples.

2.15 On Calculating the Interval Between Meridian ...

Example #1

On March 31, 2008, a fast motor-yacht proceeds at 14-knots on latitude 49° 30′ N and longitude 51° 15′ W on a course of 227°. The time is close to the estimated LAN. The navigator would like to know the difference between the true time of meridian passage of the sun and the time of maximum altitude.

Solution:
By employing the terrestrial formulae (7) from Sect. 1.2, we find that:

(i) dep. = D · sin C = 14 · sin 227 = −10′.23895 = −0°.17065

(ii) $\frac{d\varphi}{dt} \cong \frac{\Delta\varphi}{\Delta t}$ = D/h · cos C = 14′/h · cos 227 = −9′.54798/h = −0°.15913/h,
hence:

(iii) $\overline{\varphi} = (49.5° + 49.34°)/2 = 49°.42$, and

(iv) $\frac{d\lambda v}{dt} = \frac{\Delta\lambda v}{\Delta t} = \frac{\text{dep.}}{\cos\overline{\varphi}}/h = -0°.2623/h$.

Next, consulting the NA, we find that $\frac{d\delta}{dt} = 1'$/hour, and δ = 4° 28′ = 4°.4667. Therefore, we have:

$y = \frac{d\delta}{dt} - \frac{d\varphi}{dt} = 10'.54798$, and tan φ − tan δ = 1.0927326.

Furthermore, we calculate: $1 - \frac{2}{900}\frac{d\lambda v}{dt} = 1.03497$. By applying formula (7) of this section we find that $\frac{H}{900} = 3^m\ 2^s.23$ after meridian passage.

Example #2: (Most relevant application.)

Because of the breakdown of the distance-speed indicator, a trawler has been in the Sea of Labrador for two days relying solely on the hour-meter and the RPM-indicator for dead reckoning. Therefore, no reliable DRP is available at the time the sun was approaching the local meridian. The date is May 3, 2008 and the skipper has determined that his vessel is steaming at 15.5-knots on a course of about 317°.

The navigator who has been observing the sun with his sextant notes that the time of maximum altitude LMA was $15^h\ 27^m\ 34^s$. At $15^h\ 28^m\ 30^s$, he determines that the sun was on his meridian and, after all the reductions were applied, its altitude equated to $H_0 = 56°\ 54'.6$. The skipper requires a reliable fix of the trawler's position.

Solution:
The navigator inspects the NA to determine that:
δ = 15° 54′.6, $\frac{d\delta}{dt} = 0'.7$, GHA ☉ = 52° 55′.2, and ET = $3^m\ 10'$.

He also knows that his assumed azimuth differs by not more than 30′ from 180° and therefore, he calculates the latitude by approximating Z_n by 180° to obtain φ = 90° − H_0 + δ = 49° N. As a result of this approximation, the skipper finds a good initial approximation of his actual position, namely: φ ≅ 49° N and λ ≅ 52°55′.2W.

However, since the navigator is aware of the dependence of the change in altitude on the azimuth angle and time interval (see Sect. 2.14 and formula (10) of this section), he concludes that he has already found a suitable approximation of his latitude, but his estimated longitude is still in error. Therefore, he calculates the interval between meridian passage and maximum altitude as follows:

By employing the Rhumb-Line position calculation of Sect. 1.2, he finds that: dep. $= D \cdot \sin C = 15.5 \cdot \sin 317 = -10'.5709$, and

$$\frac{d\varphi}{dt} = \frac{\Delta\varphi}{\Delta t} = D/h \cdot \cos C = 15'.5/h \cdot \cos 317 = 11'.3360/h, \text{ and}$$

$\bar{\varphi} = (49° + 49°.1889)/2 = 49°.0880$, and

$$\frac{d\lambda v}{dt} = \frac{\Delta\lambda v}{\Delta t} = \frac{\text{dep.}}{\cos\bar{\varphi}} = -16°.1413/h.$$

Hence $y = 0'.7 - 11'.3360 = -10'.6360$ and $\tan \varphi - \tan \delta = 0.8653$.

Furthermore, $1 - \frac{2}{900}\frac{d\lambda v}{dt} = 1.0359$, and therefore, by virtue of expression (7): $H = -2^m 25^s.6$.

Next, by applying H to the time of maximum altitude, the navigator concludes that the true time of meridian transit was $15^h 30^m$, and therefore a more accurate estimate for the longitude could be obtained by using formula (8) of this section, namely—

$$\lambda = -\left((15^h 30^m + 3^m 10^s) - 12^h\right) \cdot 15 = 53°17'.5 \text{ W}.$$

By noting that the difference between the estimated time of meridian transit and true time of meridian transit is merely $1^m 30^s$ and since $\cos \varphi = \cos 49 = 0.6561$, the navigator accepts that for his purpose $\varphi = 49°$ N and $\lambda = \underline{53° 17'.5 \text{ W}}$—a very nice place for a fishing trawler in the Sea of Labrador.

Longitude by Equal Altitudes

A direct and very useful application of the method described in this section is the determination of longitude by two equal altitudes. As we have seen in this particular section, the time of meridian transit LAN and the time of maximum altitude LMA coincide in all cases where $\frac{d\delta}{dt} - \frac{d\varphi}{dt} = 0$. In those cases, the curve of altitudes versus time is symmetric relative to the local meridian. Therefore, if the same altitude H_0 is observed at two distinct instants t_1 and t_2, respectively, then the time of meridian transit is given by: $t_M = \frac{t_1 + t_2}{2}$. [1]

However, in all other cases where $\frac{d\delta}{dt} - \frac{d\varphi}{dt} \neq 0$, the symmetry no longer exists and $H \neq 0$. Nevertheless, if the times t_1 and t_2 are sufficiently close to t_M, say less than 20^m, then the effect of the anti-symmetry about the local meridian can be neglected and the mean-time t_M is sufficiently close to the time of maximum altitude LMA.

Therefore, instead of clocking the time of maximum altitude by observing the sun for a short while continuously, the navigator selects a suitable time t_1 and then finds the corresponding t_2 by observing the sun for a relatively short interval of time only.

Once $t_M = $ LMA has been found, the expression (7) can be used to reduce t to the true time of meridian transit LAN. Formula (8) finally furnishes the longitude.

In the following section, I will develop analytic expressions for determining approximations to the latitude and longitude of the observer by employing **POLARIS**, as the reverend star of orientation.

2.16 To Find Latitude by Observing Polaris When Exact UTC and Longitude or an Approximation Is Available

Since **POLARIS**, the pole star, is very close (less than one degree) to the physical pole of the earth, it serves as a sky mark for finding true north and also for determining the latitude of the observer. The NA provides tables for extracting the latitude within certain limits of accuracy and also provides, as many other publications do, a formula for calculating the latitude by means of using the method of successive approximations without providing a formula that is simple enough to be readily evaluated on a calculator of the type mentioned in the previous chapters. Moreover, the tables and the stated formula have to be entered with the local hour angle of **POLARIS** or Aries which entails that the user must have knowledge of the longitude of the observer. It also implies that the user must also know the exact UTC.

The reader may recall that knowing the UTC, the altitude, the declination and longitude implies that the reader also knows the azimuth of the CO being observed. This becomes very obvious if one looks a the corresponding navigational triangle. By applying the SIN-TH. to this triangle, one readily deduces that:

(1) $\quad A_z = \sin^{-1}\left(\dfrac{\sin t \cos \delta}{\cos h}\right)$

Note that the ambiguity of this formula is removed by knowing in which quadrant S lies, and by applying formulae (3) from Sect. 2.2.

In a previous section, we have seen how the latitude φ can be readily calculated when the above data is provided. All the aforementioned hold for any celestial object and not just for **POLARIS**. Therefore, we must ask ourselves what is so special about this star when it comes to finding the exact latitude? The answer should be very obvious: by employing **POLARIS** in our navigational triangle, we note that one of the sides, namely, $p = 90° - \delta$, the polar distance is very small compared to the other two sides of our NT—unless the observer is extremely close to the north pole. Furthermore, we may also recall, as discussed in the section on finding the exact time, that an approximation for the longitude can be used for finding a suitable approximation for the sidereal hour angle Θ of the observer and hence for the actual local hour angle t. Taking these circumstances into account, we may then express:

(2) $\varphi = h_0 - \varepsilon$, where ε is a quantity of about the same magnitude as
(3) $p = 90° - \delta$, i.e., ε is a small quantity in comparison to φ and h_0.

From the NT, with the help of the COS-TH., we deduce:

(4) $\sin h_0 = \sin \varphi \cdot \cos p + \cos \varphi \cdot \sin p \cdot \cos t$.

Expanding $\sin \varphi = \sin(h_0 - \varepsilon)$, and $\cos \varphi = \cos(h_0 - \varepsilon)$ in power series of ε about h_0—Taylor series, and also expanding $\sin p$, and $\cos p$ in power series of p about $p = 0$, we find that:

(5) $\varepsilon = p \cdot \cos t - \frac{1}{2}(\varepsilon^2 - 2 \cdot \varepsilon \cdot p \cdot \cos t + p^2) \cdot \tan h_0 + E_3(p, \varepsilon)$, where E_3 (p, ε) involves powers of p and ε, and also products of those powers of degree three and higher. Taking into account that p and ε are now radial quantities and hence are of the order $\frac{\pi}{180}$, if the corresponding values in (4) are measured in degrees. In the case where those quantities are measured in seconds of degrees, they are of the order of $\sin 1'' = \frac{\pi}{180 \cdot 3600}$, hence, they are very small indeed. Because of the extremely small magnitude of p and ε, we may neglect the error term in (5) to obtain:

(6) $\varepsilon \cong p \cdot \cos t - \frac{1}{2}(\varepsilon^2 - 2 \cdot \varepsilon \cdot p \cdot \cos t + p^2) \cdot \tan h_0$.

The application of the method of successive approximations to (6), with the initial approximation:

$\varepsilon_0 = p \cdot \cos t$, yields:

$\varepsilon \cong p \cdot \cos t - \frac{p^2}{2} \cdot \sin^2 t \cdot \tan h_0$.

If we want to express ε in seconds with p also given in seconds of degrees, we find, after having cancelled out the common factor $\sin 1''$, that:

(7) $\varepsilon \cong p \cdot \cos t - \frac{p^2}{2} \sin 1'' \cdot \sin^2 t \cdot \tan h_0$.

In this expression, p is given in seconds of circular measurements and so is ε. Converting ε again to degrees and substituting in (2) yields the final explicit equation for the latitude, namely:

(8) $\varphi \cong h_0 + \frac{1}{3600}\left(\frac{p^2}{2} \cdot \sin 1'' \cdot \sin^2 t \cdot \tan h_0 - p \cdot \cos t\right)$. [13]

Again, t can be calculated by using the formula:

$t = \Theta \pm$ SHA*. The local sidereal hour angle Θ can be found either by direct computation using the equation $\Theta = \pm$ (GHA Υ + λ), provided that the longitude λ is known, or by the method employed in the section of finding time when longitude is only known approximately.

The above formula (8) should be accurate of up to 1″, provided that t also satisfies the came criterion for accuracy. Of course, it is always assumed that h_0 has been measured with the highest degree of accuracy possible.

Numerical Example Date: 06/13/08
Time $T_0 = 03^h\ 00^m\ 00^s$ UTC
CO = POLARIS
$h_0 = 23°\ 07'\ 29''$
λ = 110° 40′.66 W

Problem:
Find the latitude.

Solution:
From the NA we extract GHA Υ = 306° 48′, GHA* = 266° 43′.3, δ* = 89° 17′.9. Hence p = 90° − δ = 0°.701667 = 2,562.0012″
SHA* = 319.9216, Θ = GHA Υ + λ = 196°.12235, and

2.16 To Find Latitude by Observing Polaris When Exact UTC ...

t = Θ + SHA* − 360° = 156°.044016
Substituting these values in expression (8) yields:
φ = 23° 45′.97

Annex

Instead of finding an approximation to the solution of (4), we might as well solve this equation analytically and thereby obtain an exact solution. It will turn out that the numerical evaluation of the exact formula is not more elaborate than the evaluation of the approximate expression (8). If this is the case, why use an approximation in the first place?

In order to solve (4) for φ, we need to proceed as follows:

Define auxiliary parameters C and β by:

C · cos β = cos p,   C · sin β = sin p · cos t. Then it follows that:

β = tan^{-1}(tan p · cos t), and C^2 = cos^2 p + sin^2 p cos^2 t .   Hence,

$\frac{\sinh_0}{C}$ = cos β · sin φ + sin β cos φ = sin(φ + β),, and therefore,

(9)   φ = sin^{-1} $\left(\frac{\sinh_0}{C}\right)$ − tan^{-1}(tan p · cos t), C^2 < 1.

Again, the quantity t is to be found by computing

t = Θ ± SHA*, Θ = LHA ♈, Star * = **POLARIS**

If by chance the longitude of the observer λ is known then: Θ = ± (GHA ♈ + λ).

If not, then the method employed in the aforementioned section can be used to find an approximation to Θ.

Another method for finding t approximately is to look for a star ε that has the same, or approximately the same SHA as **POLARIS**. (See Fig. 2.16.1.) Then the

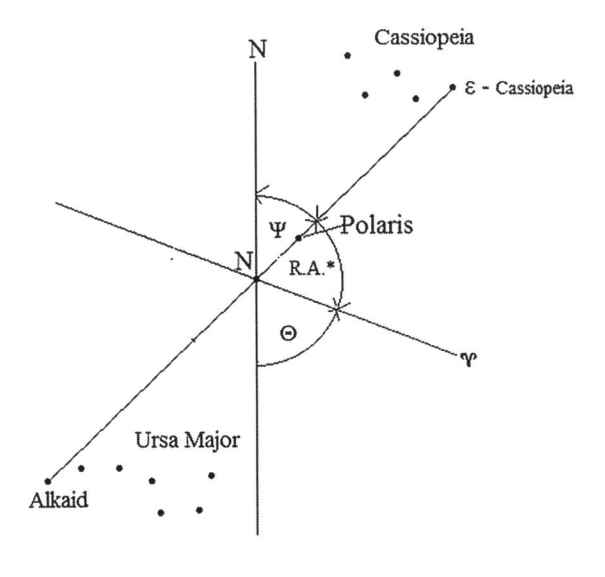

Fig. 2.16.1

three objects, namely the north-pole N, the star * ε, and **POLARIS** P lie on a straight line-meridian through N. Hence we have defined a "hand" that subtends an angle Ψ together with the local meridian.

Using the corresponding transformation: Ψ = 180° − t, it follows that cos t = −cos Ψ, and by recalling that $\tan^{-1}(-x) = -\tan x$, we deduce from (9) that:

(9') $\varphi = \sin^{-1}\left(\dfrac{\sinh_0}{C}\right) + \tan^{-1}(\tan p \cdot \cos \Psi)$. [16, 37]

Since p is a very small quantity, we may use the approximation:
$\tan^{-1} x = x - \dfrac{x^3}{3} + \dfrac{x^5}{5} - \cdots$
to obtain an approximation to φ, namely:

(10) $\varphi \cong h_0 + p \cdot \cos \Psi$, also know as "The Life-Boat" formula.

Note that the "Hand" defined by Ψ actually serves as "Sidereal Clock", since Ψ = 180° − LHA* implies that:

(11) Θ = 180° − (Ψ − SHA*) = 180° − (Ψ + RA*), or explicitly in terms of local sidereal time T:

(11') $T = \left((\Psi + RA^*) \pm 180°\right) \cdot \dfrac{1}{15}$, + if Ψ ≤ 180° − RA*; − if Ψ ≥ 180° − RA*

2.17 The Most Probable Position When Only One LOP and DRP Are Known

Although it is possible that a mariner will find him or herself in a situation where he or she has to rely solely on their Dead Reckoning Position (DRP) and only one LOP, this situation is less probable and dangerous than in the case of an aviator who may be encountering adverse meteorological conditions that do not permit the establishment of a second LOP (See annex to this section.) For this reason, the concept of the most probable position MPP, as used in aerial navigation, is less known and used by seafarers.

First, let's consider the case where at a given instant T, one LOP and one DRP are at hand. It may not even be known whether or not the two are correlated. (See Fig. 2.17.1.)

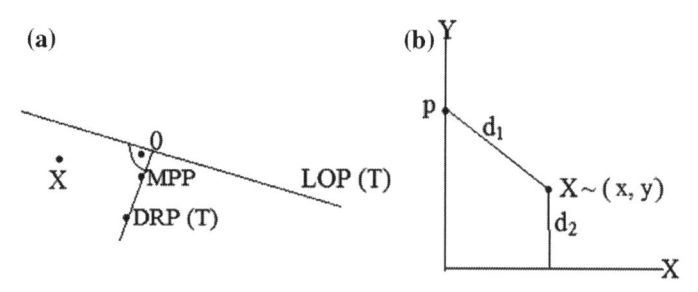

Fig. 2.17.1

2.17 The Most Probable Position When Only One ...

Even without knowing much about "least square" approximations and the theory of probability, the reader would intuitively place the most probable position of the vessel where it can be found in Fig. 2.17.1a.

If we now place a coordinate system with the origin (0, 0) at the point 0 and with the x-axis in direction of the LOP, and the positive y-axis towards the DRP with the coordinates (0, p) and the MPP with the coordinates $(0, \frac{p}{2})$, the arbitrary point X will have the coordinates (x, y). (See Fig. 2.17.1b.)

As everyone who has some knowledge of celestial navigation knows, it is highly improbable that the true position lies on the line segment DRP-0, i.e., that the coordinates of the true position are $(0, \bar{y})$, unless the DRP and the LOP are completely uncorrelated. Therefore, the true position, in general, will lie either to the right or left of the y-axis.

Next, let us define the sum of the weighted squares of the distances of 0 from the DRP and the LOP, namely: $\sum_{i=1}^{n} w_1 d_1^2 + w_2 d_2^2 = w_1(x^2 + (y-p)^2) + w_2 y^2$, where w_1 and w_2 are constants to be determined later. [19]

Since the probability density P(x, y) has the configuration: $P(x, y) = K \cdot e^{-h^2 \cdot \sum w_1 d_1^2}$, with K, h, and w_1 constants, we readily deduce that the value $\bar{X} \sim (\bar{x}, \bar{y})$, where the maximum of P(x, y) occurs, must make the weighted sum of the squares

$$\sum_{i=1}^{n} w_i d_i^2 = \sum_{i=1}^{n} w_i((x-x_i)^2 + (y-y_i)^2) \text{ a minimum. (See Fig. 2.17.2.)}$$

If we plot the above probability density as a function of x in the one dimensional case, the resulting curve, which is also called the probability curve, attains it maximum at $\bar{x}$ and its equation is simply:

$P(x) = k \cdot e^{-h^2(x-\bar{x})^2}$, with k and h constants such that $\int_{-\infty}^{+\infty} P(x) dx = 1$.

The necessary conditions for a maximum of the sum of the weighted squares of the residuals d_i are:

$\frac{\partial \Sigma}{\partial x} = 2 \cdot w_1 \cdot \bar{x} = 0$, implying that $\bar{x} = 0$, and

$\frac{\partial \Sigma}{\partial y} = 2 \cdot w_1 \cdot (\bar{y} - p) + 2 \cdot w_2 \cdot \bar{y} = 0$, and hence $\bar{y} = \frac{w_1}{w_1 + w_2} \cdot p$.

The coordinates of the most probable position MPP so found and denoted by $\bar{x}$ and $\bar{y}$, respectively, are: $\bar{x} = 0$, $\bar{y} = \gamma \cdot p$, $\gamma = \frac{w_1}{w_1 + w_2} \leq 1$, $w_1 \geq 0$, $w_2 \geq 0$.

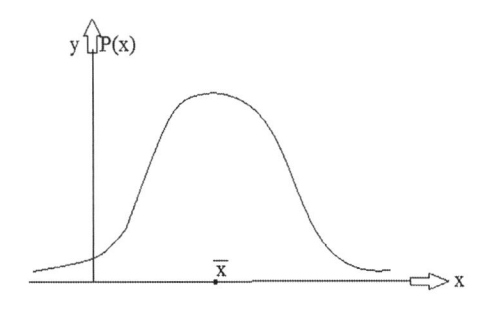

Fig. 2.17.2

Therefore, we still have two degrees of freedom left in our formulae for the MPP that we can correlate to the two objects DRP and LOP. If no additional information pertaining to the DRP and/or the LOP are available, we shall put $w_1 = w_2 = 1$ and again obtain the least probable solution to our problem.

However, if we know the time T_0 of our last fix "F", we now have two relevant parameters that will give some indications with regards to the accuracy of our approximation $(\bar{x}, \bar{y})$ to the true solution, namely, the quantity p and the elapsed time $t = T - T_0$ since the last fix. If we choose w_1 to be equal to t and w_2 to be identically to p, our probability factor γ becomes:

$\gamma = \frac{t}{t+p}$, and the coordinates of our MPP are:

(2) $\bar{x} = 0, \quad \bar{y} = \frac{t}{t+p} \cdot p$

The above formula can also be found in the Air Force Manual [2]. As has been pointed out already, these formulae still lack the necessary accuracy with regards tot he component $\bar{x}$ of the MPP. The reason for this deficiency is the lack of additional information.

Let us now assume that the position of the fix "F" at T_0 is also known in addition to the resulting track, i.e., the vector $\vec{T}$ connecting F, the T_0 fix, to the DRP$_F$ at T. (See Fig. 2.17.3.)

This additional information still does not enable us to find the x-component of the MPP that we are trying to determine. However, if we also assume that the DRP at T_0 of the assumed position on which the fix "F" is based is known, then we may advance this DRP(T_0) parallel to the track T (see Fig. 2.17.3a) to find another DRP (T) that is not directly correlated to the LOP and thereby establish the MPP for all values of $t \gg p$ as MPP $\sim$ (L, p).

If we reason then that with:

$\bar{y} = \gamma \cdot p$, we must also have: $\bar{x} = \gamma \cdot D$ with $\frac{\bar{y}}{\bar{x}} = \frac{p}{L}$, we deduce that: $D = L = 1 \cdot \cos \varepsilon$. (See Fig. 2.17.3b.)

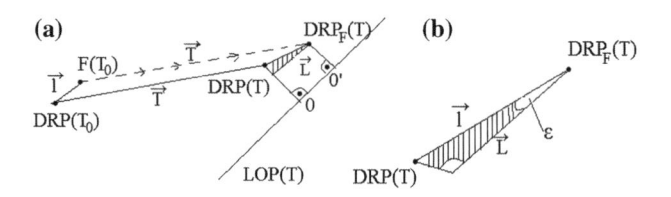

Fig. 2.17.3

2.17 The Most Probable Position When Only One …

Therefore, we have found a more realistic MPP, namely:

(3)
$$\bar{x} = \frac{t}{t+p} \cdot l \cdot \cos \varepsilon$$
$$\bar{y} = \frac{t}{t+p} p$$
$$t = T - T_0$$

Here "l" denotes the distance between DRP(T_0) and F(T_0), and "ε" denotes the angle the LOP forms with the vector $\vec{l}$ that connects the DRP(T_0) with F(T_0). (See Fig. 2.17.3a.)

To convince the reader that the case where $l = 0$, corresponding to $\bar{x} = 0$, does not constitute a realistic situation, it suffices to point out that in this case the MPP lies on the line segment between the assumed, i.e., DRP and the point 0 on the circle of equal altitude with its tangent equal to the LOP. Obviously, these points, DRP(T_0) and 0 are correlated and therefore should not be used in a statistical analysis.

In conclusion, the navigator will note that formulae (2) and (3) are definitely an improvement over mere dead-reckoning procedures, but are inferior relative to a celestial fix and therefore occupy only a third place in the praxis of navigation.

ANNEX—From the Personal Exploits of the Author—

A long time before writing this book, I believed that celestial navigation in aerial navigation had, in general, become obsolete. In particular, commercial airline pilots are no longer required to pass any kind of exam pertaining to this discipline. However, there are still some adventurous explorer pilots around the world that practice it. I also believe that some air forces still require that the navigational officer has to demonstrate a certain degree of proficiency in celestial navigation since there always exists the possibility that the enemy forces could disable the GPS that is currently in place.

As for myself, I first acquired a British Air Force bubble sextant Mark IX and a DR-computer with a high speed scale for plotting LOPs even before I bought myself a Cessna Float Plane. My object was to become a Bush Pilot. My initial training included practice take-offs and landings in farmer's backyards between apple trees, telephone and power lines as well as other obstacles. I also had to practice taking sextant sights while the plane was on autopilot. Of course, all this happened before GPS became available.

Reflecting on the more serious things that involved navigating a small float plane under adverse meteorological conditions, I vividly recall one incident that nearly cost me my plane and the lives of everyone on board. I have always considered this incident as a classic example of how accurate navigation can prevent serious accidents from happening. It is also an example of how prudent navigation and the usefulness of the MPP method.

Back in 1974, long before GPS became available, I found myself, together with my instrument-instructor who was there to certify my proficiency in instrument navigation, on an overland-sea flight from Ontario, Canada to Newfoundland. I was piloting my own float plane. I was also carrying cargo which limited by fuel carrying capacity. Nor was I equipped with an oxygen supply system.

I recall that my instructor was surprised that I had brought my bubble sextant and a high speed DR-computer along. He also noted a table with a chart that I had mounted in the cockpit that enabled me to determine the glide-slope of my plane as a function of air-speed, drift, flap-setting, and attitude of the airplane. He was also amazed by the sticker that I had attached to the bubble sextant that showed the formula for the Coriolus correction. (See Sect. 3.3.)

Everything went smoothly on the mainland including the last refueling at a remote sea-plane base in Northern Quebec. However, once we crossed the stretch of land that belonged to Labrador, ominous signs began to appear on the horizon. As a precaution, I took a shot at the sun with my Mark IX to establish in speedy succession the data for a sun-sight.

Once we reached the Strait of Belle Isle, the weather began to deteriorate. When we reached the Island of Newfoundland, the plane was engulfed in solid white clouds and the visibility was reduced to zero. I climbed the plane above the clouds. The instrument instructor suggested lowering the plane so as to find the actual ceiling. However, from personal experience, I knew that this was not a good idea. I was very familiar with Newfoundland weather patterns having owned property at the Bay of Exploits near Gander. On Newfoundland, it is not unusual for clouds to hug the ground. I asked the instructor to make contact with the airport in Gander and confirm my prognosis.

We received confirmation that the clouds were literally on the ground. Fog had brought all highway traffic to a stand-still and all flights at the airport were grounded until further notice. Making a safe landing at the sea-plane base in Gander was no longer an option because of the fog. After hearing this report, I immediately took inventory: I had only enough fuel to last me for a maximum of 2 h. Returning to Northern Quebec was also out of the question. The only option open to me was to attempt a landing in the Sea of Labrador.

At the time I was cruising above 10,000-feet. I was above the clouds, but I was at an altitude that would soon require oxygen if I did not want to impair my mental capacity. But I wasn't carrying oxygen. Time was not on my side. After rechecking my DRP and after doing some probability estimates in my head, I was able to correlate this data to the LOP that I had advanced according to my track and drift calculations.

It was then that I recalled that the U.S. Air Force was using a method referred to as "establishing the most probable position MPP". But I also recalled that the formula given in their manual could be misleading unless the pilot also took into account the actual probable error of his DRP. After having adjusted my DRP in accordance with my own probability estimates, I applied the formula provided in the Air Force Manual and arrived at my MPP that turned out to be only about two N.M. away from my LOP. (See Fig. 2.17.1.)

Based on this information and my actual altitude, I determined that I could set my plane down about three nautical miles away from shore near an inlet. My instrument instructor was not happy to see water under his feet. He had not been

2.18 How to Calculate the Time of Rising and Setting of Celestial Objects and How to Use the Measured Time of These Phenomena to Find Longitude

taught the concept of making instrument landings without instruments. While I was descending on the Sea of Labrador, I casually mentioned to him that the maneuver I was attempting was locally known as "Newfoundland-Roulette". He had never heard of it. I explained that it meant that we could get killed trying to do what I was about to do.

"Look at the chart," I said. "I will be making contact with a sea with waves about three-feet high around here." I pointed to the spot on the aeronautical chart. "We have a margin of error of about one nautical mile. We also have to worry about ice floats and ice bergs which are not marked on the chart. Our visibility will be less than a hundred feet." I could tell by his reaction that even the brave sometimes get a little scared.

My good, old Cessna did not let me down. She came in like a Mallard-Canadian Duck, made contact with the waves and came to a complete stop in less than about 50-ft away from a large ice float. However, the landing was not without incident. Something, a small chunk of ice or flotsam damaged the rudder on the starboard pontoon. Fortunately the damage was minor and we were able to find shelter in a near-by inlet. It was while we were waiting there for a break in the weather that I started contemplating the possibility of writing a book about the science of navigation fully self-contained with calculational ephemeris.

2.18 How to Calculate the Time of Rising and Setting of Celestial Objects and How to Use the Measured Time of These Phenomena to Find Longitude

Until we get to the second part of this book, we will still need to depend on our NA or some other ephemeris for most of our calculations. (That will change.) Therefore, throughout part I, it's assumed that the reader has such a publication at his or her disposal and also knows how to use it. Having made this basic assumption, we may refer to the approximate times listed for the rising and setting or the Sun and Moon as listed on the daily pages of said publication. That said, I will first provide the general equations for these phenomena as they apply to any celestial object and then add whatever is necessary to cover the case of celestial objects with finite semi-diameters such as the Sun and Moon. With the modifications so obtained, we will be able to use the measured time of sunrise and sunset to find the longitude.

Again, let's look at our navigational triangle $Z\hat{P}S$ (See Fig. 2.18.1) and deduce that according to the COS-TH.:

$\sin \delta = \sin \varphi \cdot \sin h_0 + \cos \varphi \cdot \cos h_0 \cdot \cos A_z$.

This fundamental equation reduces in case of rising-setting phenomena defined by $h_0 = 0$ to:

$\sin \delta = \cos \varphi \cdot \cos A_z$, i.e., to: $\cos A_z = \sin \delta \cdot \sec \varphi$.

Provided that $|\sin \delta| \leq |\cos \varphi|$, this equations yields:

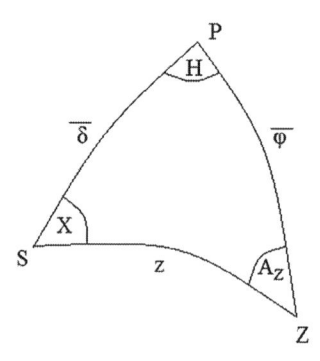

Fig. 2.18.1

(1) $A_z = \cos^{-1}(\sin\delta \cdot \sec\varphi)$.

In the case that $\delta > 90° - \varphi$ or $\varphi > 90° - \delta$, we have $\sin\delta > \cos\varphi$, and the above equation has no real solution as it is supposed to be since for all $\delta > 90° - \varphi$, the celestial object is a CIRCUMPOLAR object.

If we now apply the COS-TH. to the side $z = 90°$, corresponding to $h_0 = 0$, we find that:

$0 = \sin\delta \cdot \sin\varphi + \cos\delta \cdot \cos H \cdot \cos\varphi$, resulting in

$\cos H = -\tan\delta \cdot \tan\varphi$, an equation for H provided that $|\tan\delta| \leq |\cot\varphi|$ holds. That is to say that if the celestial object is not a circumpolar object, we have:

(2) $H = \cos^{-1}(-\tan\delta \cdot \tan\varphi)$. [1, 2, 4]

The values so obtained for A_z and H are the theoretical values for the established phenomena under consideration. Accordingly, the theoretically established value for the time of rising-setting are then given by:

(3) LMT (RIS/SET) = 12^h − ET $\pm \frac{H}{15}$ + [RA(CO) − RA☉][18] − if rising, + if setting.

Hence:

(3') LMT(SRIS) = 12^h − ET $\pm \frac{H}{15}$, Equation for the Sun.

Therefore, the local mean time of the theoretical phenomena, i.e., when the center of the celestial object touches the horizon, is known provided that the Equations of Time (ET) and the RAs are also known. The latter can readily be taken from the daily pages of the Nautical Almanac or can be calculated by an equation provided in the second part of this book.

In order to compute A_z and H, we also require the declination of the celestial object. In the case where this object is a star, no estimate for δ at the time of the event is necessary. However, in the cases of the Sun, Moon and Planets, this

---

[18] RA(CO) = [GHA♈ − GHA(CO)]/15.

additional information is required. For now, we will obtain this estimate by inspecting the NA from which we will also take the actual value of δ.

We may conclude that the prediction of the time of the theoretical rise or setting of a celestial object does not require more than the evaluation of Eqs. (2) and (3). However, since the theoretically predicted time does not coincide with the time when we actually visualize these phenomena in the case of the Sun and the Moon, on account of the necessary corrections for the altitude, we still must provide a suitable estimate for the time that elapses between theoretical and visual sunrise/set.

It should be clearly understood that a precise time for the visual sunrise/set cannot be established theoretically since the meteorological conditions are variables that are, in general, not available to the navigator.[19] As we shall see in Sect. 4.1, the determination of refraction for very low altitudes, like $h_0 = 0$, or $h_0 = -1°$, imposes a serious restriction on precise navigation—Astro and GPS. Therefore, in order to arrive at a suitable approximation for the time between theoretical sunrise/set and the corresponding visual event, it is necessary to rely more on experimental data than on theories.

It has been found that with the exception of very extreme conditions it may be assumed that the difference in altitude between these two events is approximately one full degree. This full degree accounts for the semi-diameter of the sun, the effect of refraction on the altitude, and the effect of parallax on the altitude. (A more accurate formula that shows the dependence on refraction, semi-diameter and parallax is provided in the appendix to this section.)

Based on the above assumption, we can now establish the lapse of time between theoretical and visual sunrise/set with the help of our formula (4) Sect. 2.9, namely by solving:

$60' = 15' \cos \varphi \cdot \sin A_z \cdot \Delta T$ for ΔT, thereby obtaining:

(4) $\Delta T = \dfrac{4 \sec \varphi}{\sin A_z \cdot 60}$ [hour].

The change in azimuth that occurs between the two events can be evaluated by the use of formula (4) Sect. 2.13 that yields:

(5) $\Delta A_z = -\dfrac{\tan \varphi}{\sin A_z}$ [degree].

The great importance of these formulae lies in the possibility of finding the approximate longitude by observing sunrise/set.

Now, let's assume that we have clocked the UTC of sunrise/set at a specific date and let us denote that time by GMT (SR/S)$_{vi}$. Then by applying formula (4) we find the theoretical time of sunrise/set GMT (SR/S)$_{th}$ by:

---

[19]When I practiced navigation onboard a chartered "DWOW" in the Bay of Bombay, India (now the Bay of Mumbai) back in 1974, I relied primarily on sunrise/set observations and also on low altitude observations executed with the help of an ancient Arabian "KAMEL".

(6) $GMT(SR/S)_{th} = GMT(SR/S)_{vi} \pm \frac{4 \sec \varphi}{\sin A_z \cdot 60}$, (the − sign applies at sunset).

From the definition of the Equation of Time (ET), we deduce that:
GHA ⊙ (SR/S) = GHA Ø (SR/S) + ET 15,
but GHA Ø (SR/S) = (GMT (SR/S)$_{vi}$ −12^h) 15.
(As usual ⊙ denotes the true sun and Ø denotes the mean sun.)
We also conclude that:
GHA ⊙ (SR/S) = −($\lambda \pm$ H). Recall that λ is negative if WEST and positive if EAST. The + sign holds if rising; the minus if setting.
Combining the last three expressions, we find that:
−λ = GHA ⊙ (SR/S) $\pm$ H = GHA Ø (SR/S) + ET 15 $\pm$ H = [GMT (SR/S)$_{th}$ − 12^h + ET] 15 $\pm$ H.
On the other hand, we deduce from 3'.) that:
H = $\pm$ [LMT (RS/S) − 12^h + ET] · 15. Substituting into the above equation for λ yields:
−λ = [GMT(SR/S)$_{th}$ − LMT(RS/S)$_{th}$] · 15, or

(7) $\lambda = -[GMT(SR/S)_{vi} \pm \frac{4 \sec \varphi}{\sin A_z \cdot 60} - LMT(RS/S)_{th}] \cdot 15$, − sign if setting.

Although it's very tempting to argue that by measuring the azimuth of the Sun at rising or setting, one can find the latitude by solving Eq. (1) for φ, the truth is, however, that this formula, which is equivalent to formula (1b) of Sect. 2.9, clearly shows that a slight error in the azimuth produces a significant error in the latitude, something that has already been pointed out in Sect. 2.9.

Next, let us consider some typical cases by applying the formulae or this section to actual numerical values.

Numerical Examples #1 On March 29, 2008 at latitude $\varphi = 24°.7$, and approximate longitude 97°.5 W in the Gulf of Mexico, predict the visual sunrise and azimuth of the Sun.

Solution:
From the NA we extract the approximate local time for sunrise: 5^h 56^m. Using the approximate longitude we arrive at an approximation to the GMT of sunrise to be 5^h 56^m + 6^h 30^m = 12^h 26^m UTC. Entering the NA, again, we find the declination of the Sun to be: δ (12^h 26^m) = 3°.38'.6 = 3°.6433 and ET = −4^m 39^s.
Next we evaluate formula (1) and obtain:
$A_z = \cos^{-1}\left(\frac{\sin 3.6433}{\cos 24.7}\right) = 85°59'.35$ and by evaluating formula (2) we find:
$H = \cos^{-1}(-\tan 3.6433 \tan 24.7) = 91°.6782$ and hence by formula (3')
LMT (SR)$_{th}$ = 12^h −(−04^m 39^s) − $\frac{91°.6782}{15}$ = 5^h 57^m 50^s.

This gives us the time of the theoretical sunrise. Next we employ formula (4) in order to find the time of the visual sunrise. Accordingly we have:

2.18 How to Calculate the Time of Rising and Setting ...

$$\Delta T = \frac{4}{\cos 24.7 \cdot \sin 85.9892 \cdot 60} = \frac{4.4136}{60} = 0.0736 = 4.414^m,$$ and therefore the LMT of visual sunrise is:

LMT (SR) = $5^h\ 57^m\ 50^s - 4^m\ 24^s.8 = 5^h\ 53^m\ 25^s$. The azimuth at this time is:

$$A_z = 85°59'.35 - \frac{\tan 24.7}{\sin 85.9892} = 85°59'.35 - 27'.6647 = 85°31'.7.$$

Next let us consider an example that illustrates the case where the navigator determines the longitude by suing the time of visual sunrise.

#2 The date is August 26, 2008. The latitude of the vessel in the Gulf of Mexico is φ = 23°.5 N. The navigator clocks the time of visual sunrise at $12^h\ 03^m\ 25^s$ U.T.C. The navigator has to find the longitude of the vessel.

Solution:

Entering the daily page of the current NA with this information, the navigator finds that:

ET = $01^m 44^s$, and δ = $10°10'.8 = 10°.18$. Substituting these values into formulae (1) and (2), respectively, he finds that:

H = $\cos^{-1}(-\tan 10.18 \tan 23.5) = 94°.4781$, and

$$A_z = \cos^{-1}\left(\frac{\sin 10.18}{\cos 23.5}\right) = 78°.8881.$$ By employing formula (3) he deduces that:

LMT (SR)$_{th}$ = $12^h - (-1^m\ 41^s) - \frac{94°4781}{15} = 5^h\ 43^m\ 46^s$.

Next in order to find the theoretical sunrise, we evaluate formula (4), and find:

$$\Delta T = \frac{4}{\cos 23.5 \cdot \sin 78.88 \cdot 60} = \frac{4.44}{60} = 4^m 27^s.$$

Applying this to the visual sunrise, we find that:

GMT (SR/S)$_{th}$ = $12^h\ 07^m\ 52^s$. Substituting these values into Eq. (7) yields:

$\lambda = -[\text{GMT}(\text{SR/S})_{th} - \text{LMT}(\text{RS/S})_{th}] \cdot 15 = -[12^h 7^m 52^s - 5^h 43^m 46^s] \cdot 15$
$= -96°.025 = 96°02'\text{W}.$

Appendix

In what follows, the underlying hypothesis is that an approximate value for the declination of the Sun at theoretical sunrise/set is available.

If the NA together with an approximation for the longitude is available, an approximation for the GMT can be readily found and the value for the declination can then be extracted from it. Of course in all cases where UTC is available, the declination of the Sun can be taken directly from the daily pages of the NA, or can be computed with the help of the corresponding formulae offered in the second part of this book. Furthermore, in all cases where the sunrise/set is being clocked, the time of the visual sunrise/set can be used to find the corresponding value for the declination of the Sun.

But first, let us adopt the following definitions:

T_{th} = Time of theoretical sunrise/set.

T_{vi} = Time of visual sunrise/set.
$\delta_{vi} = \delta(T_{vi})$ = Declination of the sun at visual sunrise/set.
$\tilde{\delta}$ = Approximation to the declination of the sun at theoretical sunrise/set.
Note that the symbol "$\sim$" always indicates that the corresponding value is an approximation.

By employing the formulae (1)–(3′) of this section we conclude that:

$$\tilde{H}_{th} = \cos^{-1}(-\tan\varphi \cdot \tan\tilde{\delta}), \tilde{A}_z = \cos^{-1}(\sec\varphi \cdot \sin\tilde{\delta}), \text{ or } \varphi = \cos^{-1}\left(\frac{\sin\tilde{\delta}}{\cos\tilde{A}_z}\right),$$

and

$L\tilde{M}T(SR/S_{th}) = 12^h - ET \pm \frac{\tilde{H}_{th}}{15}$, the − sign applies in case of rising.

If "D" denotes the dip, "R" the refraction, "SD" the semi-diameter, and "Π" the horizontal parallax of the sun at sunrise/set, we have:

$\cos(90° - D - R - SD + \Pi) = \cos[90° - (-\Pi + D + R + SD)] = \sin(D + R + SD - \Pi).$

The application of the COS-TH. to the corresponding navigational triangle yields:

$\sin(D + R + SD - \Pi) = \sin\varphi \cdot \sin\tilde{\delta} + \cos\varphi \cdot \cos\tilde{\delta} \cdot \cos\tilde{H}_{vi}$, and hence:

$\tilde{H}_{vi} = \cos^{-1}\left[\frac{\sin(D + R + SD - \Pi) - \sin\varphi \cdot \sin\tilde{\delta}}{\cos\varphi \cdot \cos\tilde{\delta}}\right]$. According to formula (3), we have:

$L\tilde{M}T(SR/S)_{th} = 12^h - ET \pm \frac{\tilde{H}_{th}}{15}$, and $L\tilde{M}T(SR/S)_{vi} = 12^h - E.T \pm \frac{\tilde{H}_{vi}}{15}$, and therefore:

$\Delta LMT = L\tilde{M}T_{th} - L\tilde{M}T_{vi} = \pm\left(\frac{\tilde{H}_{th} - \tilde{H}_{vi}}{15}\right)$. Here "$\Delta LMT$" denotes the time correction in hours. Hence we conclude that:

(1a) $\Delta LMT = \pm\frac{1}{15}\left\{\cos^{-1}\left[-\tan\tilde{\delta}\cdot\tan\varphi + \frac{\sin(D + R + SD - \Pi)}{\cos\varphi\cdot\cos\tilde{\delta}}\right] - \cos^{-1}\left(-\tan\tilde{\delta}\cdot\tan\varphi\right)\right\}$, − if rising, + if setting. Note that $\Pi = 8''.794 = 0'.1466$

Because we also have:
$G\tilde{M}T(SR/S)_{th} - G\tilde{M}T(SR/S)_{vi} = L\tilde{M}T(SR/S)_{th} - L\tilde{M}T(SR/S)_{vi} = \Delta LMT$
we conclude that:

(2a) $G\tilde{M}T(SR/S)_{th} = G\tilde{M}T(SR/S)_{vi} + \Delta LMT$

Formula (2a) gives the time of the theoretical sunrise/set whenever the UTC of the visual sunrise/set is known. It also enables the navigator to predict the time of the visual sunrise/set when the theoretical sunrise/set is known.

With the help of (2a), the approximate longitude $\tilde{\lambda}$ of the observer can be found in two different ways: either by using the $G\tilde{M}T(SR/S)_{th}$ to look up the LHA ⊙ and also by finding $\tilde{\delta} = \delta(GMT(SR/S)_{th})$ and then by recalculating $\tilde{H}_{th}$ with this value for δ; or by employing formula (3a) below:

(3a) $\lambda = -15\left(G\tilde{M}T_{vi} + \Delta LMT + - L\tilde{M}T_{vi}\right)$. Also note that $\Delta LMT \approx \Delta T$

The above formula follows directly from the definition of LMT, i.e., from $\lambda = -15 (GMT - LMT)$, and formula (2a).

2.19 On the Identification of Stars and Planets

The methods of recognition or identification of heavenly bodies—stars, planets, Moon, etc., can be divided into two classes. The first method is solely based on the ability of our eyes to distinguish sharply between light sources of various "brightness" and geometric patterns as they occur on the celestial sphere. The other method is based on the measurements of altitude and azimuth of a celestial body and on star charts that are based on those measurements. In addition to the two fundamental parameters employed, we still use the classifications based on magnitudes referred to as simple means of indicating the illumination of the celestial body.

One fundamental difference, between the ancient methods and the scientific methods employed today is that in our epoch we rely on measurements of the light flux per unit surface area measured by employing sensitive photo cells and by defining magnitudes according to a mathematical formula that is based on our knowledge of psychology and physiology.

The astronomers in general refer to the classification of Hipparchus and Ptolemy that was made more than two thousand years ago to classify the apparent brightness of all stars visible to the naked eye into six classes from the brightest of magnitude one to the barely visible stars of magnitude six. By having chosen the parameters in our exact method so that this original classification scale is nearly preserved, the astronomer has widened the spectrum of magnitudes to -26.8 (Sun) to 20. (Stars of magnitude greater than 20 number more than 10^9.)

In reality, classifications and pattern recognition have been known for much longer than twenty centuries. One of the first men who conceived that the Earth was a sphere and also conceived the concept of the celestial sphere was the great Archimedes. However, the Arabic literature shows that the Moors with their origin in Ethiopia had classified the visible stars also as the numerous Arabic star names testify.

Since this book is supposed to be a mathematical treatise of the subject under consideration, I shall refrain from discussing constellations and star maps as well as star finders in general and merely provide the mathematical tools, i.e., the mathematical projection for making star charts and finders. Nevertheless, it's necessary to base the formulae for computing apparent and absolute magnitude on the definitions and laws of physics and also on the underlying neurological laws of eye recognition of illumination.

Photometry

Here we adopt the following definitions as used in physics, namely:

i. Luminous Flux (F_L): measured in Lumens (lm).
ii. Luminous Intensity (I_L): measured in candles (cd) or lm/sr, i.e., $I_L = \frac{dF}{d\Omega}$.
iii. Steradian (sr): is the solid angle Ω subtended at the center of the sphere by a portion of the surface area S equal to the square of the radius of the sphere, i.e., $\frac{d\Omega}{dS} = \frac{1}{R^2}$.

iv. Illumination (E_L): defined as the light flux per unit area of the surface measured in $\frac{lm}{m^2}$, i.e., $E_L = \frac{dF_L}{dS}$.

Hence from definitions (ii) and (iv), we deduce that:

(1) $E_L = \frac{I_L}{R^2}$.

It should be clear from those definitions that the human eye or the photo cell does not quantify I_L only E_L. Therefore, the underlying psycho-physiological law of our eyes will merely involve quantities E_m of various magnitudes m. This law is based chiefly on experiments and states that the ratio of illuminations received from two different light sources designated by their magnitudes m and n, respectively, shall only depend on the difference of these two magnitudes and not in any other form of combinations of m and n. This statement translates into the equivalent mathematical statement, namely:

(2) $\frac{E_k}{E_l} = \frac{E_m}{E_n}$, if and only if $k - l = m - n$. Furthermore, if "p" denotes the magnitude of E_p then $E_p = \frac{dF_p}{dS}$. By putting $E_0 = 1$, and $E_1 = \varrho^{-1}$, $\varrho > 1$ to be determined, we find that: $\frac{E_m}{E_n} = \varrho^{-(m-n)}$. [7, 8, 41]

In order to preserve the ancient classifications closely, we require that a star of magnitude one should be one hundred times as bright as a star of magnitude six. We readily deduce from the above expression that this condition will be satisfied if $\varrho^5 = 10^2$, i.e., if $\log \varrho = 0.4$, hence we have found ϱ to be $\varrho = 10^{0.4}$.

Substituting that into the above formula, we find that:

(3) $\frac{E_m}{E_n} = 10^{-0.4(m-n)}$ For $n = 0$, we deduce that $E_m = 10^{-0.4m}$ from which we deduce the formula for the magnitude m to be:

(4) $m = -2.5 \log E_m$.

So far we have merely obtained a measure for the apparent brightness of a celestial object as it appears at the distance the observer is away from this light source. However, if we actually would like to know how strong a given light source might be, we must use another classification, one which does not depend on the distance of the celestial object is away from the observer. According to our definitions, the luminous intensity I_L is the quantity that does not depend on the distance and therefore serves as a base for defining an absolute scale of brightness, namely the magnitude M as compared to the apparent magnitude m. In order to find a mathematical formula we reduce the illumination E_m to a common distance $R = r$, and by using formulas (1) and (3), we arrive at $\frac{I_m}{I_n} = 10^{-0.4(m-n)}$. This formula hold whenever the two light sources are the same distance away from the observer.

In order to arrive at an absolute magnitude scale, we must reduce the observations of all celestial light sourced to a common distance denoted by r_0 and defined to be:

2.19 On the Identification of Stars and Planets

$r_0 = 10 \cdot \text{PARSEK} = 3.086 \cdot 10^{14}$ km (as adopted by the Astronomers International.) Then, by using M and N in lieu of m and n because of the common distance r_0 employed, the above equation becomes:

(5) $\quad \frac{I_M}{I_N} = 10^{-0.4(M-N)}$.

Furthermore, according to Eq. (1), we also have $I_0 = r_0^2$ since $E_0 = 1$ by definition. Therefore, by Eq. (5) we find that:

$I_0 = r_0^2 10^{-0.4M}$. By using Eq. (1) once more, we deduce that $E_m = \frac{I_M}{r^2}$ and therefore obtain $I_M = r^2 E_m$, i.e., $I_M = r^2 \, 10^{-0.4M}$. Combining the two equations for I_m we find that: $\left(\frac{r_0}{r}\right)^2 = 10^{-0.4(m-M)}$ that implies the equation for M, namely:

(6) $M = m + 5 \log r_0 - 5 \log r$. This equation reduces to:

$M = m + 5 + 5 \cdot \log \Pi$, where Π is the parallax of the star. [73]

Now, let's consider some special examples of celestial objects that are of importance to the navigator. First let's look at the Sun (very dangerous indeed.[20]) It has been measured that m(SUN) = −26.8, and taking into account that the Earth is about one astronomical unit AU = a = 1.495979 · 10⁸ km away from the Sun, formula (6) gives M(SUN) = 4.77 absolute magnitude. The latter implies that if we move 10 PARSEKS way from the Sun, we would barely see it with the naked eye.

Next, the MOON (variable), at full-moon has an apparent magnitude of m(MOON) = −12.5.

Next are the most brill iant Planets:
VENUS (variable) with −3.1 ≤ m ≤ −4.3.
MARS(variable)average of m(MARS) ≅ −0.2.
JUPITER (variable)average of m(JUPITER) ≅ −2.2
SATURN (variable) average of m(SATURN) ≅ 1.4

So far as the most important navigational stars are concerned, we note that the brightest ones in the sky are:

SIRIUS: m = −1.58	**CANAPUS**: m = −0.9	**RIGEL KENTAURUS**: m = −0.3
ARCTURUS: m = 0.0	**VEGA**: m = 0.0	**RIGEL**: m = 0.1
CAPELLA: m = 0.1	**PROCYON**: m = 0.4	**ARCHENAR**: m = 0.5
HADAR: m = 0.6	**ALTAIR**: m = 0.8	**BETELGEUS**—variable

The spectrum of absolute magnitudes for celestial objects varies between minus five and plus fifteen, i.e., −5 ≤ M ≤ 15, approximately. (See the back of this book for the listing of the navigational stars and their magnitudes.)

---

[20]Also see the evaluation of said danger in Chap. 3, Sect. 3.2.

Although it is certainly very helpful for the navigator to be ale to identify a star in the night sky merely by looking at the corresponding constellation, it is by no means necessary to have this ability. On the other hand, there are many people with very little formal education, like shepherds in the remote mountains of Turkey or Camel drivers in the Sahara who know the night sky very well.

For the navigator, there are basically two problems associated with star identification: the first is the problem of actually recognizing a star by looking at it, i.e., by knowing its name; and the second is the problem of selecting a star from the NA or a star map and then knowing where to find this particular star at a specific time in the night sky.

In Sect. 2.1, we have learned to associate four real numbers with any celestial object depending on how we look at that object. Those numbers represent the COORDINATES of the celestial object relative to one or the other frame of reference. In the case that we are looking at such an object from a known position on the globe, we employ the HORIZON system of coordinates which uses the two numbers referred to as altitude and azimuth, respectively. In the case that we take a star from a Celestial Globe, map, or catalogue, those two numbers refer to the declination and hour angle (Greenwich Hour Angle). Therefore, the first problem of star identification can be solved by measuring the altitude with a sextant and the azimuth with a pylorus or compass. By using formulae (2) Sect. 2.2, we can then calculate the declination and the local hour angle and subsequently the Greenwich hour angle of the star or planet.

By entering a star map, star catalogue, or the NA with declination and Greenwich hour angle or SHA, we can then extract the name of the star or planet. [11, 43]

The actual task of evaluating Eq. (2) can be eliminated if a star finder is available and only navigational stars of magnitude ≤ 3 are to be identified. This will be explained later and in more detail.

The second problem is solved in a similar manner. First we select a star or planet from the celestial globe, chart, or catalogue and extract the coordinates with respect to the celestial equator, namely: declination and Greenwich Hour Angle. Here we also have assumed that we know the approximate latitude and longitude of the observer. Therefore, we have determined the local hour angle as well. Finally, by entering formulae (1) Sect. 2.1, with the declination δ, hour angle t, and latitude φ, we find the altitude a and azimuth Z_n.

As in the case of the first type of problem, we can eliminate the need for evaluating the corresponding formulae (1) if a star finder or star globe is available provided that the star belongs to the class of navigational stars of magnitude ≤ 3.

Let's consider two examples to demonstrate the method employed herein:

Numerical Example #1 On September 1, 2008 at 00^h 45^m UT, the skipper of a sailing vessel in the Gulf of Mexico north of Tampico at $\varphi \cong 23° \, 48'$ and longitude $\lambda = -96° \, 54'$ decides to take a sight of a somewhat faint star, but he is not 100 % sure about the identity of it. Therefore, he measures its altitude and azimuth simultaneously. The approximate values he comes up with are: $h_0 = 21°.05$ and $A_z = 88°$. He then proceeds with the identification of this star.

2.19 On the Identification of Stars and Planets

Solution:

Entering formula (2i) with $\varphi = 23°.8$, $a = 21°$, $A_z = 88°$. He finds that
$\delta = \sin^{-1}(\sin 23.8 \cdot \sin 21 + \cos 23.8 \cdot \cos 21 \cdot \cos 88) = 10$
He then evaluates Eq. (2ii) and obtains:
$t = \sin^{-1}\left(\dfrac{\cos 21 \cdot \sin 88}{\cos 10}\right) = 71°.3348 = 71°.33$.
Since the longitude is WEST, i.e., $\lambda < 0$, it follows that:
$-\lambda = $ GHA* $+ t$ and hence: $-\lambda = $ GHA Υ + SHA* + t from which follows that:
SHA* = $-(\lambda + $ GHA $\Upsilon + t)$. Entering the NA with the time $00^h 45^m$ UTC, he finds that GHA $\Upsilon = 351° 48'.6$ and entering the equation for SHA* he obtains:
SHA* = $-(-90.9 + 71.33 + 351.81) = 33°.76$. By inspecting the list of navigational
stars in the NA, he identifies this star of magnitude m = 2.6 as **ENIF** (constellation Pegasus) with SHA* = $33° 50'.4$ and $\delta = 9° 55'$.

Numerical Example #2 On the same day relative to Greenwich, September 1, 2008 and at the same position except for the leeway of the sea anchor, the skipper wanted to know where he could find the navigational star **ARCHERNAR** (constellation Eridanus), m = 0.5, at about the time of local twilight, around 11^h UTC.

Solution:

The skipper extracts the SHA* and declination of this star from the daily pages of the NA to obtain: SHA* $= 335°28'.8 = 335°48$, and $\delta = 57°11'2, S = -57°.1866$.

He also finds that GHA Υ (11^h) = $145° 58'.9$ and deduces that: GHA* = $145°58'.9 + 335°28'.8 = 481°27'.7 - 360° = 121°27'.7$. Recalling that $\lambda = -96° 54'.0$, he obtains t = GHA* $+ \lambda = 24° 33'.7 = 24°.5616$. Therefore, together with $\varphi = 23°.8$, he has all the required data to compute $a = h_0$.
Formula (1i) readily yields:
$a = h_0 = \sin^{-1}(\sin 23°.8 \cdot \sin -57.1866 + \cos 23.8 \cdot \cos 57.1866 \cdot \cos 24.5616)$
$= 6°25'.15 = 6°.4191$.
Next he calculates the azimuth angle by using formula (1ii) to obtain:
$A_z = \sin^{-1}\left(\dfrac{\cos -57.1866 \cdot \sin 24.5616}{\cos 6.4191}\right) = \begin{cases} 13°.1014 \\ 90° + \cos^{-1} 0.226675 \end{cases}$
He finally verifies his result by eliminating the ambiguity with the help of formula (1) (iii) and determines that:
$Z_n = 360° - A_z = 193°.101415$

Although the analytic methods for identifying stars together with the exact methods for finding the position at sea or air (See Sect. 2.4) provide the navigator with almost unlimited choices—several hundred stars to chose from—in the selection of celestial objects for obtaining a "fix", it is still desirable that the seafarer develop a photographic memory of the night sky thereby giving him or her the necessary confidence in all matters pertaining to orientation at sea or in the air. In order to develop this type of photographic memory, it is imperative that the navigator has access to star charts and if possible to a star globe or celestial sphere and also to a suitable star finder. As far a star charts are concerned, since so many can be found, not to mention the ones in the NA, there is no need to elaborate on the use of them in this treatise. [10, 11, 31]

However, with regards to star globes, it is a little different. These devices are not as easily available as the charts. I have personally found star globes to be a wonderful aid in learning to read the night sky and in understanding the basics of celestial navigation. The reader can make his or her selection by consulting the relevant catalogues and/or looking for ones on sale on the Internet. But navigators also have the option of making their own using simple tools and materials.

Star globes are nice to have around, but they are fairly bulky and lack precision. It was for that reason that those early astronomers developed geometric methods of projecting parts of a sphere on the two dimensional Euclidean plane subject to two basic requirements: that the angles are preserved, i.e., conformity; and that all circles on the sphere are mapped either on circles or straight lines on the plane. Both of these requirements are met by the STEREOGRAPHIC PROJECTION whose mathematical mapping formulae will be provided herein.

Based on the stereographic projection, the Greek astronomer Hipparchus (ca. 150 BCE) invented the ASTROLABE, a wonderful instrument that preceded the star finder and also served as a device for solving spherical triangle calculations. One disadvantage of the stereographic projection lies in the inadequateness of mapping the entire sphere on a finite sheet of paper, since all points near the projection point of the sphere get mapped close to infinity on the plane. This shortcoming was later removed by another mathematician who invented another type of azimuthal projection, namely, the EQUIDISTANT AZIMUTHAL PROJECTION (whose mathematical mapping formula will also be derived herein.)

The reader may already know one of the most popular star finders: Star Finder #2102-D. But the reader should also know that it is basically an application of the azimuthal equidistant projection and that the azimuthal equidistant projection is an abstract mathematical projection, i.e., it can only be constructed by use of the underlying mathematical formulae. This type of projection is "Non-Conformal", i.e., it does not preserve angles, and circles on the sphere are not mapped on circles in general. Therefore the images of this projection are distorted, but distances relative to the point of projection are preserved. [39, 40]

Because of its importance in the design and construction of star charts and star finders, we are going to have a closer look at the mathematical concept of the stereographic projection and also of the non-conformal azimuthal equidistant projection. As usual, we choose the celestial sphere with a radius and equatorial plane as the plane of projection with the coordinates x', y' and the center zero.

Although in most applications to astronomy, the projecting point is either the south or north pole, we, however, will consider the more general case where the projection point can be any point on the sphere because of the aforementioned applications. Then the underlying geometry consists in connecting a given point z on the sphere with the projection point z_0 and the intersection of this straight line with the equatorial plane denoted by z' is the image point of z on this plane. (See Figs. 2.19.1 and 2.19.2.)

2.19 On the Identification of Stars and Planets

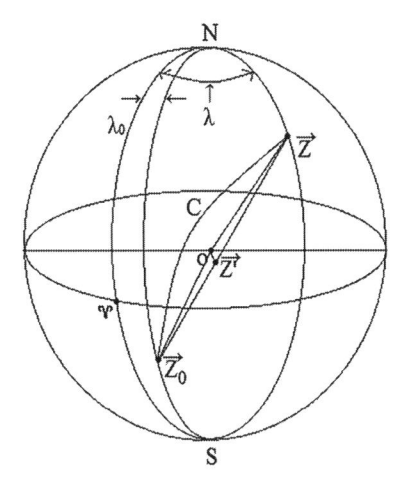

Fig. 2.19.1

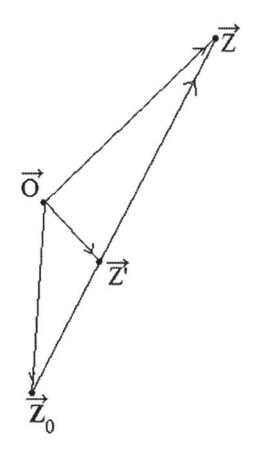

Fig. 2.19.2

Because of the underlying Euclidean geometry, the vector representation was chosen as the appropriate analytic tool. Then from Fig. 2.19.2. we readily deduct that:

$\vec{z}_0 + \alpha(\vec{z} - \vec{z}_0) = \vec{z'}$, hence:
$x' = x_0 + \alpha(x - x_0)$, and
$y' = y_0 + \alpha(y - y_0)$, and
$0 = z_0 + \alpha(z - z_0)$. Therefore:
$\alpha = \frac{z_0}{z_0 - z}$. Substituting the value for α into the above vector components yields:

(1) $x' = x_0 + \frac{z_0(x - x_0)}{z_0 - z}; \quad y' = y_0 + \frac{z_0(y - y_0)}{z_0 - z}; \quad z_0 \neq z.$

If we choose $x_0 = 0$ (Meridian), then (1) reduces to:

(2) $x' = \frac{z_0 x}{z_0 - z}$; $y' = -\frac{y_0 z_0 - y z_0}{z_0 - z}$, furthermore, if we employ the spherical coordinates φ and λ as is customary in navigation, we obtain with the help of coordinate transformation:

$x = \cos \varphi \cdot \sin(\lambda - \lambda_0)$; $y = \cos \varphi \cdot \cos(\lambda - \lambda_0)$; $z = \sin \varphi$; and $x_0 = 0$; $y_0 = \cos \varphi_0$; and $z_0 = \sin \varphi_0$ the expressions:

(3) $x' = \dfrac{1}{\sin \varphi_0 - \sin \varphi} \cos \varphi \cdot \sin \varphi_0 \cdot \sin(\lambda - \lambda_0)$ and

$y' = \dfrac{1}{\sin \varphi_0 - \sin \varphi} (\sin \varphi_0 \cdot \cos \varphi \cdot \cos(\lambda - \lambda_0) - \cos \varphi_0 \cdot \sin \varphi)$,

i.e., the polar coordinate representation of the STEREOGRAPHIC PROJECTION. In order to make the transition to the azimuthal equidistant projection, it follows that:

Since the stereographic projection is a special case of the azimuthal projection, we first transform-map the point $\vec{z}$ on to the point $\vec{z'}$ by employing a stereographic projection and then merely applying scaling factors k and k' to arrive at the equidistant projection thereby obtaining the desired projection. In particular, if x' and y' are defined by (2) and then undergo another transformation as mentioned above, we have:

$x'' = k'x'$, and $y'' = ky'$, where k and k' are to be determined subject to the requirement that the distance C on the sphere between $\vec{z}$ and $\vec{z}_0$ is preserved, i.e., that $x''^2 + y''^2 = C^2$.

Next let us put $k' = \beta \cdot k$ resulting in: $x'' = \dfrac{k \cdot \beta x z_0}{z_0 - z}$, and

$y'' = -\dfrac{k(y_0 z_0 - y z_0)}{z_0 - z}$, and determine k and β subject to the requirement that $x''^2 + y''^2 = C^2$ (C in radian) with $\cos C = \sin \varphi_0 \cdot \sin \varphi + \cos \varphi_0 \cdot \cos \varphi \cdot \cos(\lambda - \lambda_0)$. By again employing the spherical coordinates φ and λ, we find that: $\beta = \dfrac{1}{\sin \varphi_0}$, and $k = \dfrac{C}{\sin C}$, hence:

(4) $x'' = \dfrac{C}{\sin C} \cos \varphi \cdot \sin(\lambda - \lambda_0)$, and

$y'' = \dfrac{C}{\sin C} (\cos \varphi_0 \cdot \sin \varphi - \sin \varphi_0 \cdot \cos \varphi \cdot \cos(\lambda - \lambda_0))$, where $\cos C = \sin \varphi_0 \cdot \sin \varphi + \cos \varphi_0 \cdot \cos \varphi \cdot \cos(\lambda - \lambda_0)$. Formula (4) constitutes the analytic mapping of the AZIMUTHAL EQUIDISTANT PROJECTION in spherical coordinates.

In most practical applications to astronomy, the projecting point $\vec{z}_0$ is either the North or South Pole. Obviously, the stereographic projection is a much simpler projection and its main advantage lies in the simple geometry that can by handled by the use of a compass, protractor, and ruler only.

For serious sky watchers, there are now several kinds of electronic star finder-star identifiers and apps that can be loaded into you smart phones or on to your notebooks making them hand-held devices that you merely point at the sky to locate and identify thousands of celestial objects and constellations.

2.20 How to Navigate Without a Sextant

A sextant in the hands of a competent navigator is a wonderful instrument. In fact, it is indispensable. However, it is also a fairly bulky and heavy piece of equipment. In comparison to a GPS, it is huge. As a sailor and aviator, I have a collection of them. I own several marine sextants, one aircraft bubble sextant and two lifeboat drum sextants. However, I enjoy the freedom and challenge of being without a sextant relying solely on a low profile hand-bearing compass that I wear around my neck, a wrist watch-chronometer and a shirt pocket sized calculator together with a computational Ephemeris. (See second part of this book.)

Navigate without a sextant? It sounds archaic, but men have been sailing the oceans of the world long before the sextant was invented. Ancient mariners knew how to determine latitude from simply observing the altitude of the stars and measuring that altitude with a sufficient degree of accuracy. Over the millennia, man has devised a multitude of devices and methods of measuring that altitude. My favorite is the "KAMAL" because it fits in my shirt pocket and weighs nearly nothing. In fact, this device is still used by some Arab, East Indian and West African sailors. (It was most likely invented by the Moors and passed on to the Arabs. East Indians who invented the trigonometric functions claim to have used this device before the first Arab traders arrived in India, but their claim is unsubstantiated.)

Basically, a *Kamal* is a stick of wood with several notches engraved on it. It's attached to a lanyard of fixed length and worn around the neck. The purpose of the lanyard is to maintain a constant distance between the notched wood and the eye of the user. Contrary to what many westerners believe, the *Kamal* was never meant to be a measuring devise as such, but was designed to be a calibrating tool, i.e., only well defined altitudes such as 2°, 5°, 10°, and 15° could be observed. [10]

Back in 1973 when I was with the Data Institute of Fundamental Research of India in Bombay (currently Mumbai), I would charter "Dhows" (lateen rigged) on various occasions out of Mumbai (18° N, 70° E) to learn from the descendants of those ancient sailors more about the use of the *Kamal* and about low altitude navigation. Some of my findings showed that the classical methods of celestial navigation for determining refraction and dip for low altitude observations proved to be highly inadequate. For instance, I found that by assuming that the dip depends only on the height of the observer, one could end up several miles off the true position of the vessel. Later in 1984, when I was in Nigeria, I had the opportunity to sail in the Gulf of Guinea (4° N, 4.5° E) and to try and test what I had already found when I has sailed in the Arabian Sea back in 1973.

After having pointed out that there are alternatives to the use of a sextant and also that there still exist serious problems associated with the use of one, I would like to mention methods available for navigating without a sextant or substitute for it all together.

#I ZENITH STARS

The most obvious method for determining the position at sea is to observe an overhead Zenith Star and, with the help of a plumb, noting the time when it is exactly overhead. This is not always possible in practice due to the pitching and movement of the deck, but it may produce an approximation that may suffice for the purpose of navigation. Unfortunately, it is not always possible to select a suitable Zenith Star from the about 1,500–2,000 stars that can be seen under normal conditions with the naked eye. Also it may be problematic to observe the actual transit of the star from the deck of a moving vessel as well have the necessary ephemeris for proper navigation at sea.

#II OBSERVATION OF TWO OR MORE HORIZON STARS

Another more practical and accurate method is based on the exact timing of the rising or the setting of two stars that can be identified and classified in accordance with our table (see Sect. 2.4). By employing formulae (9)–(14) of Sect. 2.4, which simplify by using the values $a_1 = a_2 = 0$, we can then find the exact position of the observer. Alternately, two lines of position can be established by the approximate method described in Sect. 2.6.

#III TIMING OF SUN RISE/SETTING

Recalling the results of Sect. 2.14, one can also determine longitude by timing sunrise or sunset provided that the latitude is known. The latter may also be determined if the azimuth of the rising or setting Sun can be ascertained fairly accurately.

#IV THE STAR OF DESTINATION

Among the various methods for navigation without a sextant, I have always preferred the method of following the star of my destination. For this particular method, the navigator needs to select a bright star that passes directly over or near the destination and then use the proper ephemeral data.

For example, let us assume that we identify a Zenith Star over our destination $O' = (\varphi, \lambda)$ at Greenwich Sidereal Time—GST = T. Furthermore, our known position has the coordinates $O = (\varphi_0, \lambda_0)$. We now assume that we have only one instrument, namely as "Sidereal Clock" at our disposal (or alternatively a solar mean time clock, our never failing wristwatch chronometer) that shows GMT = UTC, and a simple calculator with the formula for converting GMT into GST (see second part of this book). Alternatively, one can look up the GHA ♈ (Aries) in the NA and extract the UTC for this particular value of GHA ♈ that is given by:

(1) $\pm\lambda = \text{GHA}\,♈ + \text{SHA}^*$. i.e., $\text{GST} = \frac{\pm\lambda - \text{SHA}^*}{15} = T$.

Then we proceed simply by following the azimuth of this star at GST = T.

About 24 h later (all sidereal time) we arrive at position $P = (\lambda_1, \varphi_1)$—highly exaggerated in Fig. 2.20.1 for clarity—which because of Loxodrome—bearing, drift, leeway, etc., does not lie on the great circle route $O \rightarrow O'$, in general.

2.20 How to Navigate Without a Sextant

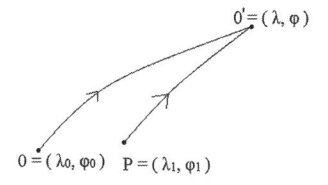

Fig. 2.20.1

Again, exactly at T, we take another bearing at the Zenith Star and then proceed on this course for another day—24-h SMT. Repeating this procedure until we arrive at our destination.

The reader should be aware that the SHA* of the star changes continuously and therefore T as defined by (1) has to be recomputed every day or, at least, at intervals of three days. It should also be obvious that on long voyages, i.e., 15 days or more, it would be necessary to choose a Zenith Star that passes over the destination after about 21^h LMT at the observer's position at O, in case this criterion is not observed, the Zenith Star may finally pass over the destination during daylight since in 15-days the star will show up over the destination approximately 1-h earlier than when the observer was at O relative to LMT. Therefore, on any long voyage, it is always recommended that one should select several Zenith Stars and employ the formulae in Sect. 1.4 to establish WAYPOINTS along the great circle route O → O′.

2.21 On Finding Time and Longitude at Sea, the Equation of Computed Time (ECT), and Being Completely Lost

Until the era when precise and reliable clocks became available to the seafarer, the plight of timekeeping at sea and restoring exact time when lost were serious and sometimes difficult problems. When chronometers and radio receivers became available, the sailors who did not have the financial resources to acquire this type of equipment still had to depend on methods that did not require any type of expensive equipment.

One of the more popular methods for correcting time at sea has been a simplified version of the Lunar Distance Method. Even among seamen who had the means to buy expensive chronometers and radios, it had become rather fashionable to be familiar with the application of this method. Although seldom recognized that this method has its own limitations, in that it requires a fairly good approximation to the exact GMT as well as a functional clock, this method became a longstanding navigational standard.

It is still possible for a navigator to lose time at sea. By that I mean exact time and under special circumstances. The possibility is there, but it is remote. With the advent of reliable quartz watches, any sailor who can afford to buy him or herself the basics for sustaining life can afford to acquire such a watch. In addition to the indispensable wristwatch, any sea-going vessel should also carry a stationary quartz watch which can also serve as a substitute for a more expensive chronometer. In addition to the above, the navigator might also consider a radio-controlled watch that is synchronized with an Atomic Clock and shows the exact UTC. In addition to all this, there are also inexpensive Single Side Band Receivers that receive time signals transmitted by a world-wide network of Atomic Time Signal transmitter stations.

As you can see, the probability of losing time completely nowadays should be of the same order as losing all other navigational equipment at the same time. The probability is low, but it is there. A solar flare; a terrorist attack; a knockdown at sea and a dead battery—any number of things may contribute to the navigator losing exact time.

In order to cover any of these remote possibilities, I have decided to include several computational procedures for determining approximate time and longitude without resorting to the use of the Lunar Distance Method (LDM). These procedures cover the circumstances where a navigator has lost time, dead reckoning position, date and, perhaps, the boat's only clock, also.

Admittedly, it is quite the challenge to address the above mentioned circumstances and the critical situation where a navigator has lost time, dead reckoning position, date and perhaps the craft's only clock. Based on what the navigator specifically has available in such a situation, we need to distinguish between the following four cases:

CASE I: Latitude and Longitude are known, but time has been lost.
CASE II: Time has been lost and only a crude estimate for the longitude and the exact latitude are known.
CASE III: Exact time and longitude are lost, but approximate time together with exact latitude are known.
CASE IV: Time, position and perhaps the only watch is lost.

We will first examine cases I and II and then treat case III and IV in detail.

In order to provide mathematical solutions to I and II, it is necessary to derive a differential formula for calculating the change in the altitude as a function of the changes in declination and hour angle. Therefore, let's look again at our navigational triangle PSZ as depicted below:

As previously explained, the application or COS-TH. to z yields:

$\cos z = \sin \delta \cdot \sin \varphi + \cos \delta \cdot \cos \varphi \cdot \cos t$. Then we ask how a change of z, i.e., dz, can be expressed in terms of the changes dδ and dt. Of course, the resulting relationship between these three differentials are only valid in the infinitesimal sense.

The above equation is basically the form $F(z) = G(\delta, t)$. We may, therefore, apply the implicit differential expression according to the formula of Sect. 4.2 and obtain:

2.21 On Finding Time and Longitude at Sea, the Equation ...

$dz = \frac{1}{F'(z)} dG$, with $dG = G_\delta \cdot d\delta + G_t \cdot dt$.

Substituting for $F(z) = \cos z$ and $G(\delta, t) = \sin \delta \cdot \sin \varphi + \cos \delta \cdot \cos \varphi \cdot \cos t$, we find that:

$dz = \frac{1}{\sin Z}[(\cos \delta \cdot \sin \varphi - \sin \delta \cdot \cos \varphi \cdot \cos t)d\delta - \cos \delta \cdot \cos \varphi \cdot \sin t \cdot dt]$ If we now ask for the infinitesimal changes in $d\delta$ and dt that result in zero changes in z, i.e., for $dz = 0$, we may conclude that:

(1) $dt = \frac{\cos \delta \cdot \sin \varphi - \sin \delta \cdot \cos \varphi \cdot \cos t}{\cos \delta \cdot \cos \varphi \cdot \sin t} d\delta = \left(\frac{\tan \varphi}{\sin t} - \frac{\tan \delta}{\tan t}\right) d\delta$

Equation (1) gives us the change in the local hour angle t in terms of the change $d\delta$ of the declination. In actual applications, these differentials will be approximated by the differences Δt and $\Delta \delta$ respectively. We can now turn our attention to the solution of case I.

CASE I: Latitude and longitude are known but time has been lost.

Solution:

By using a Sun sight at maximum altitude-Meridian passage. Suppose that the Sun is EAST of the observer. We can then extract the time of LAN from the ephemeris—NA—and denote the corresponding GMT or LAN by T_0. Our watch now shows the time T_0. Then we readily deduce that:

(2) $\Delta t \cong \frac{(\tan \varphi - \tan \delta)}{\sin t} \Delta \delta \cong \frac{\beta \cdot (\tan \varphi - \tan \delta)}{\sin t} \Delta T$: in arc minutes.

Here we have made use of the fact that t is a very small angle and therefore $\cos t = 1$. Furthermore, we also have made use of the definition of the change of δ, namely the relation $\Delta \delta = \beta \cdot \Delta T$. Note that β is also given in arc min/h. From (2) we may also deduce that:

$\sin t = \beta \cdot (\tan \varphi - \tan \delta) \frac{\Delta T}{\Delta t} = \beta \cdot \left(\frac{\tan \varphi - \tan \delta}{900}\right)$, since $\frac{\Delta t}{\Delta T} = 15°/h = 900'/h$.

However, since t is very small and of the order of arc minutes, we may use the approximation $\sin t = t \cdot \sin 1'$ and conclude that:

$\Delta t° = \beta \cdot \left(\frac{\tan \varphi - \tan \delta}{900 \cdot \sin 1'}\right) = 3.819722 \cdot \beta \cdot (\tan \varphi - \tan \delta)$.

Because $\Delta t°$ is actually the change of the local hour angle of the Sun at the time $T = T_{MA}$, we may conclude that:

$\Delta t° = 15 \cdot \Delta T \cdot 60 = 15 \cdot (T_{MA} - LAN) \cdot 60$ and therefore:

$\Delta T_{MA} = 3.819722 \cdot \beta \cdot \frac{(\tan \varphi - \tan \delta)}{15} \cdot 60 = 0.254648133 \cdot \beta \cdot \frac{(\tan \varphi - \tan \delta)}{60}$, and hence:

(3) $T_{MA} = LAN + 0.254648133 \cdot \beta \cdot (\tan \varphi - \tan \delta)/60$. [1]

This formula clearly shows that if the latitude and longitude and therefore the LAN are known, then the time of maximum altitude is also known by virtue of (3) and hence the navigator's watch can be reset.

On the other hand, if an accurate watch is available, the LAN and subsequently the longitude can be found by using formula (8) Sect. 2.15.

Now, let's consider a numerical example.

Numerical Example
Date: 03/23/08
CO: Sun
Latitude: 23°.48′.94 = 23°.7824107 N
Longitude: −110°42′.5 = 110°.70816 W
Problem: Find the accurate time for resetting your chronometer.
Solution:
By inspecting the NA, we find LAN (Greenwich) = $12^h\ 06^m$ and hence LAN = 12^h.1 + 110.708161/15 = 19.48054407. i.e., LAN = $19^h\ 28^m\ 50^s$.
From the NA, we also extract the declination δ = 1°.4083 and β = 1′.
Substituting these values in Eq. (3) yields:
T_{MA} = 19.48054407 +0.254648133 (tan 23.7824105711 − tan 1.4083)/60 = 19^h.48231006 = $19^h\ 28^m\ 56^s$.

In the case where the position of the vessel is known with a high degree of accuracy, the navigator observes the passage of the Sun over the imaginary meridian (for on a vessel, the true meridian is never known exactly), and when his or her sextant registers maximum altitude, resets the chronometer to T_{MA}.

An alternative method for finding the exact time is based on measuring the altitude of the Sun at time $\widetilde{T}_1$ when it is close to the fictitious meridian and to the EAST of the observer. The altitude of the Sun is noted at this instant and then the Sun is observed again when it is WEST of the observer. At the instant $\widetilde{T}_2$ when the altitude of the Sun has reached the same value as noted at $\widetilde{T}_1$, the instant is recorded. With the help of those two values, $\widetilde{T}_1$ and $\widetilde{T}_2$, and the formula for the noon, the exact time can be ascertained and the incorrect clock time T can be adjusted. Of course, this method only works if a functional watch that shows the incorrect time is available.

Next, let's develop the "FORMULA FOR NOON" in terms of the formula for T_{MA} above.

By employing the relation (2) again we readily derive an expression for the rate of change for the local hour angle, namely:

$$\frac{\Delta t}{\Delta T} = \frac{\beta \cdot (\tan \varphi - \tan \delta)}{\sin t}, \ t \neq 0, \ t \neq 180.$$

Since "t" stands for the local hour angle, we can no longer approximate it as in the case of the formulae for the time of maximum altitude. We therefore define: $\Delta T_m = \frac{T_2 - T_1}{2}$, and $t_m = 15\, \Delta T_m$. Next we put $t = t_M$ and deduce from the above expression that:

2.21 On Finding Time and Longitude at Sea, the Equation ...

$$\Delta t = \frac{\beta \cdot (\tan \varphi - \tan \delta)}{900 \cdot \sin t'_M} \cdot t'_M.$$ Converting this expression into time units yields the Equation of Noon, i.e.:

(4) $\Delta T_{EN} = 0.254648133 \cdot \beta \cdot (\tan \varphi - \tan \delta) \dfrac{t_M}{\sin t_M} \cdot \sin 1'.$

By making use of the expression (3) for the correction of the time for the maximum altitude T_{MA}, we deduce:

(5) $\Delta T_{EN} = \Delta T_{MA} \cdot \sin 1' \cdot \dfrac{60 \cdot t_M}{\sin t_M}$

It follows then from (5) that for small values of t_M, $\Delta T_{EN} = \Delta T_{MA}$.

Once we have computed ΔT_{EN}, we find the $\widetilde{LAN}$ that is based on the incorrect times $\widetilde{T_1}$ and $\widetilde{T_2}$ by the formula:

(6) $\widetilde{LAN} = \dfrac{\widetilde{T_1} + \widetilde{T_2}}{2} - \Delta T_{EN}.$ [1]

Next we calculate the exact LAN by employing the exact longitude and the values given in the NA to determine the error of our clock":

(7) $\Delta T = \widetilde{LAN} - LAN$

Next let's consider those cases where no exact longitude is known, including the extreme case in which no clock is available at all.

CASE II:

Here, let's consider the problem where the navigator has lost time completely, maybe even the date of the month, but does have a crude estimate of the longitude at hand. Let's consider two possible solutions.

The first solution is an approximate method for its validity depends again on parameters that are only true in the approximate sense and therefore, the final quantities like computed longitude constitute, at best, a sensible approximation provided that the initial estimate is close to the true longitude.

The second method is based on the exact formulae for the longitude when the exact latitude is known and is basically an iterative method that converges, but not necessarily, to the true solution unless all the data being used is correct.

Both methods employ a Moon-Sight and require highly accurate ephemeral data for the Moon and also an accurate latitude. The second method is the same method as outlined in CASE III.

First Method:

A method for determining longitude when time is lost but an estimate for longitude is available.

Firstly, you may want to determine the date of the month by using the last entry in your log and then measuring the distance of the Moon from a Planet, Star or the

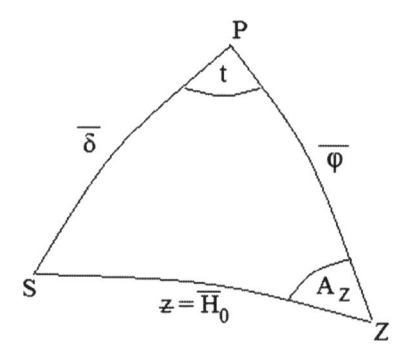

Fig. 2.21.1

Sun without applying all the necessary corrections, as, for instance, parallax, semi-diameter, etc., i.e., obtaining merely a good estimate for the true distance. Then by calculating the distances for the consecutive days by using your ephemeris, you should be able to arrive at the correct date.

Another way of determining the calendar day consists in noting the phase of the Moon and measuring the meridian altitude of the Sun and then extracting the date from the NA.

Secondly, you get your only timepiece going again. (Replace the batteries or rewind, etc.) Again, we are discussing the case of a proud vessel owner who goes to sea with only one timepiece… not a good idea.

Next by selecting two suitable COs, you take two sights simultaneously and calculate the latitude as explained in Sect. 2.14. You now have the choice of setting your timepiece to local mean time LMT either by observing the local Sun rise/set, or the meridian passage of the Sun. The LMTs for these events are listed in the NA. Suppose you choose the LMT of meridian passage of the Sun, then your timepiece shows the correct local time from then on.

Next you extract from the NA the Greenwich Hour Angle of Aries denoted by $\omega = $ GHA (12^h UTC). Since you are $\pm \lambda$ degrees in longitude away from Greenwich, the Greenwich Hour Angle or Aries at 12^h LMT is given as:

GHAΥ(12^hLMT) = GHAΥ($12^h - \lambda/15$UTC)

= GHAΥ(12^hUTC) $- \lambda/15 \cdot 1.002737916 \cdot 15$ and therefore, the local hour angle of Aries at 12^h LMT is given by:

LHAΥ(12^hLMT) = GHAΥ($12^h - \lambda/15$) $+ \lambda = \dfrac{-1.002737916}{15} \cdot \lambda + \omega + \lambda.$

Here we have used the facts that λ is negative WEST and that the Earth rotates with an angular velocity of $15° \cdot 1.002737916$/h. By using the same symbol for the local hour angle as before, we deduce that:

$\Theta(12^h\text{LMT}) = -0.002737916 \cdot \lambda + \omega.$

It follows directly that an error of $\Delta\lambda$ results in an error in the local hour angle of Aries that is merely equal to $0.002737916 \cdot \Delta\lambda$. This implies that we can safely replace λ by λ' in the above expression.

2.21 On Finding Time and Longitude at Sea, the Equation ...

Next we observe the Moon at a convenient local time $T_0\ 12^h$. Then the local hour angle of Aries is: $\Theta = \Theta(T_0) = \Theta(12^h\text{LMT}) + (T_0 - 12) \cdot 15 \cdot 1.002737916$, i.e.,

(8) $\quad \Theta = \omega - 0.002737916 \cdot \lambda' + (T_0 - 12) \cdot 15 \cdot 1.002737916$.

We have now procured a very good approximation to the local hour angle of Aries and therefore will consider that the value provided by expression (8) represents the true value. In addition, we have also procured a Moon-Sight at T_0 (LMT). We now proceed to determine the error $\Delta\lambda$ in our longitude by employing the rate of change β of the declination and also the rate of change γ of the right ascension of the Moon. However, first we need to derive an estimate for the Greenwich Mean Time of our observation by using the relation GHA Υ $(T_1) = -\lambda' + \Theta$. Here "$\lambda'$" stands again for our initial estimate of our longitude and T_1 is now reckoned in UTC. Then we extract T_1 from the NA. If "a" denotes the RA of the Moon and "t" the local hour angle at the exact time T, we have $a \pm t = \Theta$.

But since we still don't have the exact time T(UTC) of the event, we have to approximate the local hour angle by using $\delta(T_1)$ instead of $\delta(T)$, resulting in the value t_1, found by the expression (9):

(9) $\quad t = \cos^{-1}\left(\dfrac{\sin H - \sin \delta \cdot \sin \varphi}{\cos \delta \cdot \cos \varphi}\right)$

This formula follows directly from the first equation of this section corresponding to Fig. 2.21.1.

In order to maintain equilibrium in the above relation, we define a value a_1, an approximation to the RA, in degrees by: $a_1 \pm t_1 = \Theta$, minus (−) if East of the observer, plus (+) if WEST.

By computing a_1 and then consulting the NA, we extract a value T_2 corresponding to a_1. We also extract GHA Υ (T_2) from the same publication. Next we define a new and better approximation to λ which we denote by:

(10) $\quad \lambda'' = -(\text{GHA } \Upsilon \ (T_2) - \Theta)$, West of Greenwich.

By definition we have: $\Delta\lambda = \lambda - \lambda'$, $\Delta\lambda' = \lambda - \lambda''$ and we deduce that:
$\Delta\lambda = \text{GHA}\gamma(T_1) - \text{GHA}\Upsilon(T)$, and
$\Delta\lambda' = \text{GHA}\gamma(T_2) - \text{GHA}\gamma(T) = -15 \cdot (\text{GST}(T) - \text{GST}(T_2)) = -15\Delta T$. Here "GST" denotes Greenwich Sidereal Time. Next, by defining Δt and Δa according to: $(a_1 + \Delta a) \pm (t_1 + \Delta t) = \Theta$, we conclude that we must have $\Delta a = \pm \Delta t$, where Δt has been found previously by formula (1) of this section. We therefore may write: $\Delta t = \alpha \cdot \Delta\delta$ with "α" defined by:

(11) $\quad \alpha = \dfrac{\tan\varphi}{\sin t_1} - \dfrac{\tan\delta_1}{\tan t_1}$.

The value for $\delta_1 = \delta(T_1)$ is taken from the NA and t_1 is computed by using formula (9).

If we now apply the hourly rates of changes β and γ, obtainable by interpolation from the NA, we can also use the definitions: $\Delta a = \gamma \cdot \Delta T$, and $\Delta\delta = \beta \cdot \Delta\lambda/15$.

Hence we deduce that $\Delta T = \pm \dfrac{\alpha \cdot \beta \cdot \Delta \lambda}{\gamma \cdot 15}$ with plus (+) if WEST and minus (−) if EAST. Substituting this expression in the above expression for $\Delta \lambda'$ yields: $\Delta \lambda' = \pm \dfrac{\alpha \cdot \beta \cdot \Delta \lambda}{\gamma}$. Next we use the equation of equilibrium: $\lambda = \lambda' + \Delta \lambda = \lambda' + \Delta \lambda = \lambda'' \pm \dfrac{\alpha \cdot \beta}{\gamma} \Delta \lambda$ and deduce that:

(12) $\quad \Delta \lambda = \dfrac{\lambda'' - \lambda'}{1 \pm b}$ with $b = \dfrac{\alpha \beta}{\gamma}$, plus (+) if WEST and minus (−) if EAST.

It should be clearly understood that Eq. (9) is subject to an initial error of $0.002737916 \cdot \Delta \lambda$ and can be valid only if $\Delta \lambda$ is relatively small since it is based on the infinitesimal expression (1). In order to provide an approximate range for the validity of (12), λ' has to lie in a range of $|\lambda - \lambda'| \leq 30'$. Therefore if the navigator knows that their longitude falls between two values one full degree apart, i.e., $\lambda_0 \leq \lambda \leq \lambda_0 + 1$, they can take the midpoint as an approximate value for their longitude λ thereby assuring that they are no more than 30′ off the true longitude, i.e., $\Delta \lambda \leq 30$.

Numerical Example On 06/19/08 at $T_0 = 22^h\ 36^m$ LMT, the Moon was observed from a location on latitude $\varphi = 23°\ 44'.73\ N = 23°.7455$ and its altitude was cleared to $H_0 = 54°\ 03'.59 = 54°.05979243$. The longitude was estimated to be $\lambda' = -99°\ 30' = 99°.5$ W. The Greenwich Hour Angle of Aries at 12^h GMT was $\omega = \text{GHA}\ \Upsilon\ (12\text{UTC}) = 88°\ 05'.1 = 88°.085$.

Find a better approximation to the true longitude and reset your watch to nearly the exact UTC.

Solution:
Step #1: Calculate $\Theta\ (T_0)$ by using formula (8) to obtain: $\Theta\ (T_0) = 247°.7927573$.
Step #2: Calculate $\text{GHA}\gamma(T_0) = -\lambda' + \Theta = 347°.2927573$
Step #3: Extract T_1 (UTC) from the NA by using linear interpolation to obtain:
$\quad T_1 = 5^h\ 14^m\ 0^s.4 = 5^h.233445171$ on 06/20/08.
Step #4: Extract $\delta_1 = \delta\ (T_1)$ from the NA to find: $\delta_1 = 25°\ 23'.6 = 25°.3933$.
Step #5: Calculate the hour angle t_1 by employing formula (9) to obtain:
$\quad t_1 = 39°.62088448$.
Step #6: Calculate $a_1 = \Theta + t_1$, plus (+) since Moon is North East of observer to get: $a_1 = 287°.4136358$.
Step #7: By interpolating the NA data, extract the value T_2 that corresponds
$\quad$ to the RA a_1 resulting in $T_2 = 5^h\ 12^m\ 9^s = 5^h.2024928$.
Step #8: Extract GHA $\Upsilon\ (T_2) = 346°.82833$ and calculate:
$\quad \lambda'' = \text{GHA}\gamma(T_2) - \Theta = -99°.03558203$.
Step #9: Extract the rate of change β in the declination of the Moon from the
$\quad$ NA to find: $\beta = -5'.8/\text{hour} = -0.096666/\text{h}$.
Step #10: Extract the rate of change in RA from the NA to find $\gamma = 0°.7491/\text{h}$.
Step #11: Calculate α using formula (11) to obtain: $\alpha = 0.11646599$, and
$\quad$ calculate b employing formula (12), i.e., $b = \dfrac{\alpha \cdot \beta}{\gamma} = -0.015029103$.

2.21 On Finding Time and Longitude at Sea, the Equation ...

Step #12: Calculate $\Delta\lambda$ using formula (12) to find: $\Delta\lambda = 0°.457541531$ and hence: $\lambda = \lambda' + \Delta\lambda = -99°.04245847 = 99°\ 02'.55$ W and also the UTC at the instant of observation: $T = T_0 - \lambda/15 = 5^h 12^m 12^s$ on 06/20/08.

Again, the user of this method should always be aware of what has already been said, namely that the accuracy of this method is limited by the narrow range for the choice of an estimate λ' for the longitude that appears to be $\Delta\lambda \leq \pm 30'$. This, however, is to be expected since the underlying mathematics is based on a set of infinitesimal quantities than cannot be extrapolated to a set of finite values without jeopardizing the accuracy.

Next let us consider another method (my own) that has a much wider scope of applicability.

<u>CASE III</u> (and the SECOND METHOD)

A method to determine time when only an estimate for time and exact latitude are available.

Similarly, as in the previous cases, this method is also based on observing the Moon and a star simultaneously. Here, it is also assumed that the exact latitude and ephemeral data are known. Now, we must investigate the possibilities for minimizing the effect of erroneous data on the numerical results and provide a method for checking the validity of the approximation so obtained.

If we assume that all ephemeral data, the observed altitudes and given latitude, are correct, then the exact time T of the observation and exact longitude must satisfy Eq. (16) Sect. 2.1, namely:

(13) $\lambda_D(T) = -GHA_D(T) \pm \cos^{-1}\left(\dfrac{\sin H_D(T) - \sin \delta_D(T) \cdot \sin \varphi}{\cos \delta_D(T) \cos \varphi}\right)$, plus (+) if WEST

$\lambda_*(T) = -GHA^*(T) \pm \cos^{-1}\left(\dfrac{\sin H_*(T) - \sin \delta_*(T) \cdot \sin \varphi}{\cos \delta_*(T) \cos \varphi}\right)$, minus (−) if EAST.

$\lambda_D(T) = \lambda_*(T)$. * Star, D Moon. [44]

Resulting in the Equation of Computed Time (ECT):

(14) $F(T) = (GHA*(T) - GHA_D(T)) \pm (t_D(T) - t_*(T)) = 0$, with t_D and t_* representing the second terms of Eq. (13) above.

Therefore, the abstract mathematical problem of finding the exact time consists in solving the transcendental Eq. (14) with an exact set of parameters.

The continuity and convexity or concavity of the function F(T) in the vicinity of a root T assures that the iterative process defined by the equations of Sect. 2.6 will converge to the unique solution proved that the initial approximation satisfies the criteria. Hence, we may conclude that $T = \lim_{k \to \infty} T_k$.

Although in theory everything is straight forward, in actual practice, i.e., application to navigation, the solution of the ECT poses some serious problems for the following reasons:

(i) The exact values for altitudes H_D and H_* are seldom known.
(ii) The exact ephemeral data as for instance $\delta(T)$ and GHA(T) may not be available.
(iii) The functions $\delta(T)$ and GHA(T) are not given analytically, but only numerically for discrete values of T and therefore F(T) is not given analytically as a function of T, but only approximately by a set of pairs $\{T_k, F_k\}$ of numerical values.
(iv) The limitations in the accuracy of evaluating (14) on a calculator also limits the accuracy of the final numerical results.

All of this amounts to saying that in actual applications, we don't try to solve (14) but rather a similar equation, i.e., the same relation as (14) but with a different set of parameters, i.e., $\widetilde{F}(\widetilde{T}) = 0$. Therefore, if $\widetilde{T}$ is a solution of this equation it follows that in general $\widetilde{T}$ is not a solution of (14), i.e., $F(\widetilde{T}) \neq 0$. It also follows then that in practice where it is not possible to determine the exact altitudes, it is also not possible to find the exact time of observation by computation but only a approximation. However, we may reason that whenever it is possible to minimize the value of $|\Delta F| = |\widetilde{F}(\widetilde{T}) - F(\widetilde{T})|$, so that it is more or less of the same magnitude as the error in the numerical solution of Eq. (14), the root of $\widetilde{F}(\widetilde{T}) = 0$ will constitute an acceptable and probably a good approximation to T, i.e., to the solution of (14).

These conclusions are in agreement with conclusions drawn in other applications where an exact formula is available but where the input data is subject to errors as in the case of the application of the exact equations to the problem of finding your position at sea or in the air. (Review Sect. 2.4.) Therefore, the results of this section also have to be viewed as approximations and not as exact solutions. This makes it necessary to use additional information and/or calculations to assure that those approximations are sufficiently close to the true solutions. (See the Robinson Crusoe Example #4 in CASE IV.)

Next let us look at conditions that will optimize the error in the numerical calculations. According to (i)–(iv), we must assure that the following requirements are met:

(a) The values of H_D and H_* must be highly accurate with an error of less than plus or minus one arc minute. This implies that the Moon is high enough to assure that the error in parallax is minimal.
(b) If and whenever necessary interpolate the ephemeral data.
(c) Choose the approximate time of observation so that it falls in a range of time for which the change in declination of the Moon is maximal. Also choose a star so that its azimuth is close to the azimuth of the Moon or 180° apart. Furthermore, if possible, choose the time of observation so that the Moon is on or near the prime meridian since according to expression (6) Sect. 2.8: $dt = \dfrac{dH}{15 \cdot \cos \varphi \cdot \sin Z_n}$

2.21 On Finding Time and Longitude at Sea, the Equation ...

(d) By evaluating the expressions for t_D and t_*, i.e., by computing $\cos^{-1}\left(\dfrac{\sin H - \sin \delta \cdot \sin \varphi}{\cos \varphi \cdot \cos \delta}\right)$ use the highest degree of precision possible.

Now we can develop an algorithm for calculating approximations to the Equation of Computed Time (ECT). In Sect. 2.6, we had derived a procedure that is based on the mathematical method of the "Regual Falsi" (RF) and applied it to the same Eq. (14) but with the only difference that we interpreted the function F as a function of the latitude φ, i.e., we actually solved $F(\varphi) = O$ by using the RF or secant method to generate a sequence $\{\varphi_k\} \Rightarrow \varphi_*$. [18, 35]

Accordingly we may now use the same formulae for solving $F(T) = 0$, thereby establishing the existence of a solution of this equation once we have found two values, T_0 and T_1, satisfying $\widetilde{F}(T_0) \cdot \widetilde{F}(T_1) < 0$, and therefore by the continuity of F the Mean-Value Th. of calculus guaranties the existence of a solution t between T_0 and T_1. However, the efficiency of this method is questionable since it basically requires two initial values and also entails checking for the convexity or concavity of F. Because of these shortcomings, we are going to develop another iterative method that turns out to be even more suitable for solving the problem under consideration.

In order to be able to treat CASE III and CASE IV by the same iterative procedure, we must reduce CASE IV to CASE III by showing how an initial approximation can be found. The additional requirement consists of having obtained a good approximation to the azimuth of the Moon at the time of observation. We will therefore assume that Z_n as used in our method does not deviate from the true azimuth by more than half a degree.

By using the approximate value for the azimuth of the Moon, the navigator will then be able to calculate the parallax of the Moon (see Sect. 3.3) and subsequently establish the observed altitude H_D of the Moon. Then by employing the formula...

(15) $\delta = 90° - \cos^{-1}(\sin \varphi \cdot \sin H + \cos \varphi \cdot \cos H \cos A_z)$.

Here $A_z = \begin{cases} Z_n \text{ if } Z_1 \leq 180° \text{ or} \\ Z_n - 189° \text{ if } Z_n > 180° \end{cases}$

(This formula follows directly from the application of the COS_TH. to our navigational triangle—see Fig. 2.21.1.)

...the navigator finds $\widetilde{\delta}_D$, an approximation for the declination of the Moon. Entering the daily pages of the NA with $\widetilde{\delta}_D$, the navigator will find the corresponding time T_0, i.e., the required initial approximation for T. At this point, the navigator should verify that the initial approximation for the parallax of the Moon still holds and if necessary adjusts the H_D.

As has been previously mentioned, the key to an efficient iterative method is that the initial values have to be sufficiently close to the solution in order to assure the convergence of this process. Evaluating F(T) at several points numerically may prove to be a very tedious process for finding the two required values T_0 and T_1 sufficiently close to the true solution and also satisfying the additional condition that

$F(T_0) \cdot F(T_1) < 0$. This is particularly true when the given initial value is too far away from the true solution—also see Examples #3 and #4 of this section.

For the aforementioned reasons, I am going to develop a method for finding a value T_0 that may come sufficiently close to the solution. I have also reasoned that any efficient iterative method that also generates the required initial values for the secant method may also turn out to be another independent iterative method for actually finding a suitable approximation to the true solution.

In order to develop the proposed algorithm let "T_0" denote the given or computed initial value. We can then calculate the corresponding values $\lambda_D(T_0)$ and $\lambda_*(T_0)$ by using formula (14) and thereby obtaining the value $\Delta\lambda_0 = \lambda_D(T_0) - \lambda_*(T_0)$. It follows then that we have actually computed $F(T_0) = \Delta\lambda_0$, i.e., the value of the ECT. The latter follows directly from the identity:

(16) $\quad F(T) = \Delta\lambda(T) = \lambda_D(T) - \lambda_*(T)$.

Next we will try to find another value T_1 to T_0 by means of an increment ΔT_0 (to still be determined) satisfying $T_1 = T_0 + \Delta T_0$. Again we may then compute the corresponding longitudes $\lambda_D(T_1)$ and $\lambda_*(T_1)$, thereby evaluating $F(T_1) = \Delta\lambda_0$ and subsequently, ΔT_1 and $T_2 = T_1 + \Delta T_1$. Repeating this process over and over again, we obtain the following sequence of real numbers:

(17) (i) $\{T_k\}: T_{k+1} = T_k + \Delta T_k, k = 0, 1, \ldots$
      (ii) $\{\lambda_D(T_k)\}: \{\lambda_*(T_k)\}$ and $F(T_k) = \Delta\lambda_k = \lambda_D(T_k) - \lambda_*(T_k)$.

It still remains to be shown how the quantities ΔT_k are to be related to the values $\Delta\lambda_k = F(T_k)$ in order to have an algorithm.

By employing definition (9) and formulae (14), we deduce that:

$\Delta\lambda_k = GHA^*(T_k) - GHA_D(T_k) + \Delta t_k = SHA^*(T_k) - SHA_D(T_k) + \Delta t_k$, where $\Delta t_k = \pm(t_D(T_k) - t^*(T_k))$. Expressing the sidereal hour angles by their right ascensions (RAs), we can now write the above expression as a function of time T and obtain the relation:

$\Delta\lambda(T) = RA_D(T) - RA^*(T) + \Delta t$

Differentiating this equation with regard to T and taking in consideration that: $\dfrac{dRA^*}{dt} = 0$, and $\dfrac{dt^*}{dT} = 0$ since t_* does not depend on T, we obtain the expression: $\dfrac{d\Delta\lambda}{dT} = \dfrac{dRA_D(T)}{dT} \pm \dfrac{dt_D(T)}{dT}$.

Denoting the hourly change of the right ascension of the Moon by "γ", to be extracted from the NA, and denoting the hourly change of the declination of the Moon by "β", we deduce from Eq. (1) that: $\dfrac{dt_D(T)}{dT} = \alpha(T) \cdot \dfrac{d\delta}{dT} = \kappa(T)$.

With (11) $\alpha(T) = \dfrac{\tan\varphi}{\sin t_D(T)} - \dfrac{\tan\delta_D(T)}{\tan t_D(T)}$ and hence $F'(T) = \dfrac{d\Delta\lambda}{dT} = \gamma \pm \alpha \cdot \beta = \kappa(T)$. This relations suggest that the linear relationship between $\Delta\lambda_k$ and ΔT_k, that we are about to establish, should be: $-\Delta\lambda = (\gamma \pm \alpha \cdot \beta) \cdot \Delta T_k = \kappa(T) \cdot \Delta T_k$ and hence:

2.21 On Finding Time and Longitude at Sea, the Equation ...

(18) $\Delta T_k = -\frac{\Delta\lambda_k}{\kappa(T_k)}$: $\kappa(T_k) = \gamma \pm \alpha(T_k) \cdot \beta$, k = 0, 1, ...

plus (+) if Moon is WEST of observer; minus (−) if Moon is EAST of observer.

Therefore, we may conclude that our iterative process (17) and (18) is now well defined and it merely remains to be shown subject to in which conditions it will converge to the solution of (14).

First we note that by expressing the quantities occurring in Eq. (18) in terms of the ECT and its derivative, Eq. (18) assumes the form:

$\Delta T_k = -\frac{F(T_k)}{F'(T_k)}$ or $T_{k+1} = T_k - \frac{F(T_k)}{F'(T_k)} = T_k + \Delta T_k$.

It therefore turns out that our iterative procedure is equivalent to the well-know NEWTON-RAPHSON iterative method [35]. This method converges to the solution provided that the initial value of T_0 is sufficiently close to the true solution T and that the first and second derivatives of F(T) satisfy additional conditions. (See Sect. 4.2.) Since the probability that a navigator will have that information at hand is minimal, we need to rely on other, more readily available material to hand.

With the information we have on hand, namely the longitudes $\lambda_D(T_k)$ and $\lambda_*(T_k)$, we can then compute $\Delta\lambda_k$ at each step of our iteration. In addition, we also know from the general theory that the ECT function is continuous and either convex or concave in the neighborhood of a root of F(T) = 0. Therefore, we may establish the necessary and sufficient conditions for the convergence of this iterative process.

It should be obvious that the necessary conditions for the convergence of the sequence $\{\Delta\lambda_k\}$ is that $\lim_{k\to\infty}|\Delta\lambda_k| = 0$. However, this is by no means sufficient for the convergence of the sequences $\{\lambda_{Dk}\}$ and $\{\lambda_{*k}\}$ to λ. Without going too much into the abstract aspects of mathematics, those conditions are namely:

(19)
(i) $|\Delta\lambda_{k+1}| \leq |\Delta\lambda_k|, \Delta\lambda k = \lambda_D(T_k) - \lambda_*(T_k)$ for all $k \geq m \geq 0$
(ii) $\lambda_{k+1} \geq \lambda_k$ if CO approaches from the WEST
$\lambda_{k+1} \geq \lambda_k$ if CO approches from the EAST

To avoid an misinterpretation of the possible results, the reader should be aware that (19) does not imply that ALL members of the corresponding sequences have to satisfy conditions (i) and (ii), but only those members of these sequences for which the indexes satisfy $k \geq m$. However, we don't know before hand what "m" might be. Therefore, we must always look for the sequences for which m = 0 and if it turns out that $m \geq 2$, we must abort the iterative process and start over again with a new initial value. (See Example #4, below.)

It is also important that the readers know that they cannot expect to find a value T_n so that $\Delta\lambda_k = F(T_n) = 0$, only by accident when this might happen. This is because of the finite numeric involved that only permits that approximate solutions are obtainable. Accordingly, in order to terminate the iterative process a realistic error bound, i.e. $|\Delta\lambda_N| \leq 10^{-M'}, M \geq 0$ should be used.

Next, let's look at some relevant numerical examples and finally solve the "ROBINSON CARUSOE" problem, i.e., completely lost... no latitude nor longitude... no clock, watch, chronometer, etc., and no DRP.

Numerical Problems:

Example #1 Date: 06/20/08 Latitude φ: 23° 44′.73 N = 23°.7455
Celestial Objects observed: Moon H_D = 27°.72076631 SE
VEGA H_* = 59.3678133 NE
Approximate time of observation: T_0 = 05^h 00^m 00^s UTC
Approximate azimuth of Moon Z_n = 139°.5

Find the time of observation and longitude.

Solution:

Step #1. From the NA extract β = 5′.8/h = 0°.09666/h.
extract γ = 0°.7491/h.
Since the moon is EAST of the observer, we must use the (−) sign in formula
(18), i.e., $κ = γ - α · β$.
Furthermore, from the NA we take:
$GHA_D(T_0) = 56°.5233$
$GHA^*(T_0) = 64°.4683$
$δ_D(T_0) = -25°.37$
$δ_*(T_0) = 38°.79$
$SHA^* = 80°685$

Step #2. Using formula (9) calculate:
$t_D(T_0) = 39°.5535079$
$t_*(T_0) = 31°.52043343$
Using formula (11) calculate:
$α(T_0) = 1.264974986$
$κ = 0.62682757$
Using formula (13) calculate
$λ_D(T_0) = -96°.07680791$
$λ_*(T_0) = -95°.98873343$
Evaluate $Δλ_0 = -0.08807448$, and hence $|Δλ_0| > 10^{-2}$.

Step #3. Employing formula (18) calculate T_1 to obtain $ΔT_0 = 00^h · 08^m · 26^s$.
Therefore: $T_1 = T_0 + ΔT_0 = 5^h · 08^m · 26^s$.

Step #4. Repeat steps #1 to #3 by substituting T_1 for T_0 and note that it is no longer necessary to calculate $t_*(T)$ at each step since $δ_*$ = constant. Also note that
whenever $α(T)$ varies only slightly, the change in κ is insignificant and one
can use $α(T_0)$ throughout the process of iterations.

2.21 On Finding Time and Longitude at Sea, the Equation ...   163

We now have the following data:
$GHA_D(T_1) = 58°.5600$ $GHA^*(T_1) = 66°.58166$ $\delta_D(T_1) = -25°.3566$
$t_D(T_1) = 39°.57045228$ $t_*(T_1) = t_*(T_0)$ $\lambda_D(T_1) = -98°.13045228$
$\lambda_*(T_1) = -98°.10209943$ $\Delta\lambda = -0.02835285$, and hence:
$\Delta T_1 = 2^m\ 43^s$ and $T_2 = 5^h\ 11^m\ 9^s$.
Note that: $|\Delta\lambda_1| > 10^{-2}, |\Delta\lambda_1| < |\Delta\lambda_0|, \lambda_*(T_1) < \lambda_*(T_0), \lambda_D(T_1) < \lambda_D(T_0)$
Step #5. Since $|\Delta\lambda_0| > 10^{-2}$ continue. (You may choose your own error bound 10^{-n}.)
You now have:
$GHA_D(T_2) = 59°.2183$ $GHA^*(T_2) = 67°.2633$ $\delta_D(T_2) = -25°.3576$
$t_*(T_2) = t_*(T_0)$ $\lambda_D(T_2) = -98°.79507158$
$\lambda_*(T_2) = -98°.78373343$ $\Delta\lambda_2 = -0.01133815$
Note that $|\Delta\lambda_2| > 10^{-2}$ but $|\Delta\lambda_2| < |\Delta\lambda_1| < |\Delta\lambda_0|$, and $\lambda_*(T_2) < \lambda_*(T_1), \lambda_D(T_2) < \lambda_D(T_1)$.
Since $|\Delta\lambda_2| > 10^{-2}$ we calculate $\Delta T_2 = 1^m\ 5^s$ and $T_3 = 5^h\ 12^m\ 14^s$.
Because the criteria for accuracy still has not been reached, go back to steps #1 through #5 and arrive at:
$GHA_D(T_3) = 59°.4783$ $GHA^*(T_3) = 67°.5350$ $\delta_D(T_3) = -25°.35$
$t_D(T_3) = 39°.57879339$ $t_*(T_3) = t_*(T_0)$ $\lambda_D(T_3) = -99°.05709339$
$\lambda_*(T_3) = -99°.05543343$ $\Delta\lambda_3 = -0.00165996$
Note that $|\Delta\lambda_3| < 10^{-2}$ $|\Delta\lambda_3| < |\Delta\lambda_2| < |\Delta\lambda_1| < |\Delta\lambda_0|$, and $\lambda_*(T_3) < \lambda_*(T_2), \lambda_D(T_3) < \lambda_D(T_2)$. Hence the criteria for accuracy and convergence have been reached and we compute $\Delta T_3 = 10^s$ and arrive at the approximate time of observation: $\underline{T = 5^h\ 12^m\ 24^s}$, and $\lambda = -99°\ 03'.32 = \underline{99°\ 3'.32\ W}$.
Our chronometer registered $T = 5^h\ 12^m\ 30^s$ UTC.

Example #2 Same data as in Example #1, but here, the initial approximation for the time of observation is: $T_0 = 5^h\ 24^m\ 30^s$.

Find the exact time of observation.
Solution:
Again by following the iterative procedure outlined in the previous example and computing steps #1 through #5, we find:
$\lambda_D(T_0) = -102°.0473509$ $\lambda_*(T_0) = -102°.1304334$
$\Delta\lambda_0 = 0.08308253$ $\Delta T_0 = -7^m\ 57^s$ and $T_1 = 5^h\ 16^m\ 33^s$.
Subsequently:
$\lambda_D(T_1) = -100°.1072578$ $\lambda_*(T_1) = -100°.1370334$ $\Delta\lambda_1 = 0.02977563$
$|\Delta\lambda_1| > 10^{-2} |\Delta\lambda_1| < |\Delta\lambda_0|$ and $\lambda_*(T_1) > \lambda_*(T_0), \lambda_D(T_1) > \lambda_D(T_0)$
$\Delta T_1 = -2^m\ 51^s$ and $T_2 = 5^h\ 13^m\ 48^s$.
Because $|\Delta\lambda_1| > 10^{-2}$, we must continue with the iterations.
$\lambda_D(T_2) = -99°.43754136$ $\lambda_*(T_2) = -99°.44876343$ $\Delta\lambda_2 = 0.01122207$
$|\Delta\lambda_2| > 10^{-2}$, but $|\Delta\lambda_2| < |\Delta\lambda_1| < |\Delta\lambda_0|$ and $\lambda_*(T_2) > \lambda_*(T_1), \lambda_D(T_2) > \lambda_D(T_1)$
$\Delta T_2 = -1^m\ 5^s$ and $T_3 = 5^h\ 12^m\ 43^s$.
Again because $|\Delta\lambda_2| > 10^{-2}$ we must continue the iteration process.
$\lambda_D(T_3) = -99°.17379339$ $\lambda_*(T_3) = -99°.17703343$

$\Delta\lambda_3 = 0.00324004$ and hence $|\Delta\lambda_3| < 10^{-2}$.
$\Delta T_3 = -18^s$ and therefore our exact time of observation has been found to be: $\underline{T = 5^h\ 12^m\ 25^s}$.

The results of these examples encourages us to tackle the ultimate problem never before solved, namely to determine our position at sea or air without prior knowledge of an approximate position and time, maybe even without having a watch.

It should be pointed out that the proponents of the "Time Lost At Sea" problem always assume that they have a vague idea of where they are and also imply that they still have an approximation of the true time at hand. In particular, they imply that the still have a functional time piece, even if only a defective alarm clock, in their possession. Perhaps they should refer to their case as "Exact Time Lost At Sea" in order to avoid misunderstandings.

CASE IV:

Now let's consider the real problem of time lost, namely, TIME COMPLETELY LOST which means that no approximate time, nor approximate longitude is known and perhaps there is no watch at hand. However, it is always safe to assume that the latitude of the observer is known since it can be found without knowing the exact time or even the need of having a clock (Sect. 2.10). The following example will illustrate how to proceed in such an extreme case.

Example #3 Same data as in Examples #1 and #2, but now, no estimate for the time of observation is available. Instead, we make use of the measured azimuth of the moon: $Z_n = 139°.5$.

Solution:

The initial problem consists in narrowing the interval of time $0^h < T < 24^h$ to an interval of about 1 h. This can be done if the measured azimuth has an error of less than $0.5°$. Then, by assuming a value for the HP of the Moon, taking into account that it may changed by no more than $0'.3$ on the day of observation, the observed altitude can be determined.

With the date now available, an approximation of the declination δ_D of the Moon at the time of observation can readily be computed by employing formula (15). The result of said calculation is: $\delta_D \cong -25°.39370758 = -25°\ 23'.62$.

Next by entering the corresponding pages of the NA, we extract a crude estimate (here we have not interpolated) for the time that corresponds to δ_D. (The reader is advised to interpolate in order to find a refined T_0.) Here we choose:
$T_0 = 4^h\ 12^m\ 30^s$, 06/20/08.
Following the procedure previously outlined in the preceding examples, we find:
$GHA_D(T_0) = 45°.0500$ $GHA_*(T_0) = 52°.6100$ $\delta_D(T_0) = -25°.445$
$\delta_*\ (T_0) = 38°.79$ $t_D(T_0) = 45°.0301551$ $t_*(T_0) = 31°.52043343$ $\lambda_D(T_0) = -90°.0810551$ $\lambda_*(T_0) = -84°.13043334$ $\Delta\lambda_0 = -5°.95062167$
$\alpha(T_0) = 1.097081608$ $\beta = 0°.0966$ $\gamma = 0.7491$
$\kappa = \gamma - \alpha \cdot \beta = 0.643056091$ since the Moon is east of the observer. Finally,

2.21 On Finding Time and Longitude at Sea, the Equation ...

$\Delta(T_0) = 9^h.2536588837$, $T_1 = 13^h\ 27^m\ 43^s$.
Reiterating with T_1 yields:
$GHA_D(T_1) = 179°.1983$ $GHA*(T_1) = 191°.7950$ $\delta_D(T_1) = -24°.4833$
$t_D(T_1) = 40°.34846080$ $t_*(T_1) = t_*(T_0)$ $\lambda_*(T_1) = -223°.3154334$ $\lambda_D(T_1) = -219°.8467608$ $\Delta\lambda_1 = 3°.46867253$, hence:
$|\Delta\lambda_1| < |\Delta\lambda_2|$; $\lambda_D(T_1) < \lambda_D(T_0)$; $\lambda_*(T_1) < \lambda_*(T_0)$; $\Delta T_1 = -5^h23^m39^s$. Note: we have approximated α_1 by α_0. However, the reader would be advised to recompute $\alpha_1(T_1) = \alpha_1$.
We now have: $T_2 = 8^h\ 04^m\ 04^s$. Because our requirement for accuracy has not been met, we must again iterate to obtain:

$GHA_D(T_2) = 100°.9933$ $GHA*(T_2) = 110°.6616$ $\delta_D(T_2) = -25°.0633$
$t_D(T_2) = 39°.9382$ $t_*(T_2) = t_*(T_0)$ $\lambda_D(T_2) = -140°.9315$
$\lambda_*(T_2) = -142°.1820334$, $\Delta\lambda_2 = 1°.2505334$, $\Delta T_2 = -1^h56^m41^s$ and hence $T_3 = 6^h\ 07^m\ 23^s$. Note that; $|\Delta\lambda_2| < |\Delta\lambda_1| < |\Delta\lambda_0|$; $\lambda_D(T_2) > \lambda_D(T_1)$; $\lambda_*(T_2) > \lambda_*(T_1)$; And again because $|\Delta\lambda_2| > 10^{-2}$, we must iterate once more obtaining:

$GHA_D(T_3) = 72°.8016$ $GHA*(T_3) = 81°.3616$ $\delta_D(T_3) = -25°.26$
$t_D(T_3) = 39°.69223239$ $t_*(T_3) = t_*(T_0)$ $\lambda_D(T_3) = -112°.4938324$
$\lambda_*(T_3) = -112°.8820334$ $\Delta\lambda_3 = 0°.38820103$ $\Delta T_3 = -36^m13^s$ and hence $T_4 = 5^h\ 31^m\ 23^s$. Note that $|\Delta\lambda_3| < |\Delta\lambda_2| < |\Delta\lambda_1| < |\Delta\lambda_0|$; $\lambda_D(T_3) > \lambda_D(T_2) > \lambda_D(T_1) > \lambda_D(T_0)$; $\lambda_*(T_3) > \lambda_*(T_2) > \lambda_*(T_1) > \lambda_*(T_0)$ But still $|\Delta\lambda_3| > 10^{-2}$.
Therefore we must iterate again arriving at:
$GHA_D(T_4) = 64°.105$ $GHA*(T_4) = 72°.3350$ $\delta_D(T_4) = -25°.32$
$t_D(T_4) = 39°.61666912$ $t_*(T_4) = t_*(T_0)$ $\lambda_D(T_4) = -103°.7216691$
$\lambda_*(T_4) = -103°.8854334$ $\Delta\lambda_4 = 0°.1337643$ Note again that:
$|\Delta\lambda_4| < |\Delta\lambda_3| < |\Delta\lambda_2| < |\Delta\lambda_1| < |\Delta\lambda_0|$;
$\lambda_D(T_4) > \lambda_D(T_3) > \lambda_D(T_2) > \lambda_D(T_1) > \lambda_D(T_0)$;
and $\lambda_*(T_4) > \lambda_*(T_3) > \lambda_*(T_2) > \lambda_*(T_1) > \lambda_*(T_0)$ But again $|\Delta\lambda_4| > 10^{-2}$.
Therefore we must continue with our iterative process, calculating $\Delta T_4 = -12^m\ 12^s$ having used the approximation $\alpha(T_4) \cong \alpha(T_3)$ and hence $T_5 = 5^h\ 18^m\ 34^s$.

The next iteration yields:
$GHA_D(T_5) = 61°.0083$ $GHA*(T_5) = 69°.1233$ $\delta_D(T_5) = -25°.34$
$t_D(T_5) = 39°.59142562$, $t_*(T_5) = t_*(T_0)$, $\lambda_D(T_5) = -100°.5997256$
$\lambda_*(T_5) = -100°.6437334$ $\Delta\lambda_4 = 0°.04400783$ $\Delta T_5 = -4^m\ 6^s$.
$T_6 = 5^h\ 14^m\ 28^s$. Note again that: $|\Delta\lambda_5| < |\Delta\lambda_4| < |\Delta\lambda_3| < |\Delta\lambda_2| < |\Delta\lambda_1| < |\Delta\lambda_0|$;
and $\lambda_D(T_5) > \lambda_D(T_4) > \lambda_D(T_3) > \lambda_D(T_2) > \lambda_D(T_1) > \lambda_D(T_0)$;
and $\lambda_*(T_5) > \lambda_*(T_4) > \lambda_*(T_3) > \lambda_*(T_2) > \lambda_*(T_1) > \lambda_*(T_0)$ But again $|\Delta\lambda_5| > 10^{-2}$.
Therefore we must iterate again.
$GHA_D(T_6) = 60°.0233$ $GHA*(T_6) = 68°.095$ $\delta_D(T_6) = -25°.3466$
$t_D(T_6) = 39°.58308913$ $t_*(T_6) = t_*(T_0)$ $\lambda_D(T_6) = -99°.60638913$
$\lambda_*(T_6) = -99°.61543343$ $\Delta\lambda_6 = 0°.009044296$ $\Delta T_6 = -52^s$. Note again that:

$|\Delta\lambda_6| < |\Delta\lambda_5| < |\Delta\lambda_4| < |\Delta\lambda_3| < |\Delta\lambda_2| < |\Delta\lambda_1| < |\Delta\lambda_0|$; and $\lambda_D(T_6) > \lambda_D(T_5) > \lambda_D(T_4) > \lambda_D(T_3) > \lambda_D(T_2) > \lambda_D(T_1) > \lambda_D(T_0)$;
and $\lambda_*(T_6) > \lambda_*(T_5) > \lambda_*(T_4) > \lambda_*(T_3) > \lambda_*(T_2) > \lambda_*(T_1) > \lambda_*(T_0)$ But again $|\Delta\lambda_6| > 10^{-2}$. Since we require more precision, we must iterate again using $T_7 = 5^h\ 13^m\ 16^s$.

$GHA_D(T_7) = 67°.7933$ $GHA*(T_7) = 59°.7283$ $\delta_D(T_7) = -25°.3483$
$t_D(T_7) = 39°.58094136$ $t_*(T_7) = t_*(T_0)$ $\lambda_D(T_7) = -99°.30924136$
$\lambda_*(T_7) = -99°.31373343$ $\Delta\lambda_7 = 0°.004492076$ $\Delta T_7 = -26^s$. Note again that: $|\Delta\lambda_7| < |\Delta\lambda_6| < |\Delta\lambda_5| < |\Delta\lambda_4| < |\Delta\lambda_3| < |\Delta\lambda_2| < |\Delta\lambda_1| < |\Delta\lambda_0|$; and $\lambda_D(T_7) \lambda_D(T_6) > \lambda_D(T_5) > \lambda_D(T_4) > \lambda_D(T_3) > \lambda_D(T_2) > \lambda_D(T_1) > \lambda_D(T_0)$;
and $\lambda_*(T_7) > \lambda_*(T_6) > \lambda_*(T_5) > \lambda_*(T_4) > \lambda_*(T_3) > \lambda_*(T_2) > \lambda_*(T_1) > \lambda_*(T_0)$ But again $|\Delta\lambda_7| > 10^{-2}$. However, our desired accuracy still has not been reached. Therefore, we iterate again using: $T_8 = 5^h\ 12^m\ 50^s$.

$GHA_D(T_8) = 59°.6216$ $GHA*(T_8) = 67°.685$ $\delta_D(T_8) = -25°.35$
$t_D(T_8) = 39°.57879336$ $t_*(T_8) = t_*(T_0)$ $\lambda_D(T_8) = -99°.20039339$
$\lambda_*(T_8) = -99°.20543343$ $\Delta\lambda_8 = 0°.005040046$ $\Delta T_8 = -28^s.2$. $T_9 = 5^h\ 12^m\ 22^s$. Because $|\Delta\lambda_7| > |\Delta\lambda_7|$ we must abort the process of iteration and either accept T_9 as the final approximation or refine the procedure. (See next example.)

After having applied formula (20) as a final check, we may conclude that we can accept the above result and state the so obtained approximation as:

$T = 5^h 12^m 22^s$, and $\lambda \cong \lambda_*(T_9) \cong -99°.09 \cong 99°05'.4W$

COMPLETELY LOST—ROBINSON CRUSOE

Example #4   It started off as a friendly argument over a pint and soon degenerated into something more. Tiburon Tristan was on his third single-handed, circumnavigation of the globe. He put into Gibraltar to resupply and refit his sailboat. Tossing back a couple of ales at the Lord Nelson, he began talking sailing as sailors often do and eventually made the statement that, "If Robinson Crusoe were alive today, all he would need is this manual…" he tapped a book on the bar… "a programmable calculator, a nautical almanac and a sextant to know exactly where he was."

"… and a good chronometer," said a well-to-do looking man a few seats down from him.

"Wouldn't need one," Tiburon Tristan said.

"He could find his longitude without a watch or chronometer of some kind?" the well-to-do man asked.

"Exactly," Tiburon Tristan said.

"Even the British Admiralty can't do that," the well-to-do man said.

"Then they are behind the times," Tiburon Tristan said.

"You mean that I could drop you anywhere in the world with just that manual, your calculator, a nautical almanac, and a sextant and you could figure out where you were?"

2.21 On Finding Time and Longitude at Sea, the Equation ...

"Yep."

"I'll bet you a million dollars that it can't be done," the well-to-do man said.

"It would be like taking candy from a baby," Tiburon Tristan said. He thought the matter was settled. He finished his drink and left the pub. However, the next day, as he was inspecting the new bottom paint on his boat, two men walked up to him, threw a bag over his head and injected something in his arm. When he awoke on a strange bed in some kind of boat, he found the well-to-do man sitting next to him.

"Time to put up or shut up," the man said. "There are your books and sextant. I'll give you a million dollars if you can tell me where you are."

Tiburon Tristan stood up and walked to the cabin door. He opened the door and stepped out into the bright sunlight. He was on a boat, anchored off a distant shore. He turned back to the well-to-do man. "Piece of cake," he said.

THE PROBLEM: Tiburon Tristan is completely lost. He has no idea where in the world he is.

The Solution: The first thing he does is to use his sextant to determine true north and the Sun's altitude at approximately Local Apparent Noon in order to compute its declination. Later that night, after observing several stars and Moon, he computes his latitude by using this book and the Nautical Almanac and determines his latitude to be: $\varphi = -33°.68$, and the date to be 12/14/08.

Since the night is clear, Tristan decides to use **BETELGEUSE** for his calculations. Next he measures the azimuth and altitude of the star and Moon within seconds of each other using the counting method Mrs. Dranguinis taught him in the first grade: one-hippopotamus; two-hippopotamus; three-hippopotamus... for seconds. Using the methods outlined in Sects. 3.1–3.3, and Algorithm D, which he had preprogrammed into his calculator, he determines that the altitude of the Moon is $H = 18°.82$, and its azimuth $Z_n = 40°$. He then concludes that the observed altitude of **BETELGEUSE** should be $H_* = 41°.95$.

Smiling, Tiburon Tristan computes the approximate declination of the Moon by using formula (13) from the manual to obtain $\delta_D = 25°7'.2$. Then by consulting the daily pages of the NA, he finds that this declination corresponds to approximately $T_0 = 03^h\ 23^m\ 00^s$ UTC on 12/14/08.

Next he verifies that he had used a value for the horizontal parallax of the Moon that was consistent with the approximate value of T_0.

With an approximation T_0 for the time of observation so procured, he follows the exact steps of the iterative procedure as described in the manual's Examples #1–#3 to obtain:
SHA* = 271°.0766 δ_* = 7°.41 β = −0.11333 γ = 0.68666
$\delta_D(T_0)$ = 25°12 $GHA_D(T_0)$ = 29°.37166 GHA * (T_0) = 133°.87

By inspecting and interpolating the data listed in the 2008 NA and by using formula (9), he finds: $t_D(T_0)$ = 42°.21604252 and $t_*(T_0)$ = 26°.2650431. By employing expression (11), he finds $\alpha_0 = \alpha(T_0)$ = 0.4750025, and by means of expression (18), he computes
κ = 0.740498033. Finally, by using the formulae (13) he finds: $\lambda_D(T_0)$ = −71°.58770252 and $\lambda_*(T_0)$ = −71°.2117031. Hence $\Delta\lambda_0$ = −0.37599942, and again by using expression (18), he finds: ΔT_0 = 0.507765589 = $30^m\ 28^s$, resulting in $T_1 = 5^h\ 53^m\ 28^s$.
But his criteria for accuracy has not been reached; therefore, he iterates again finding:
$GHA_D(T_1)$ = 36°.63833 GHA * (T_1) = 52°.6416
$\delta_D(T_1)$ = 25°.06233833 $t_D(T_1)$ = 42°.30289709
$t_*(T_1)$ = 26°.2650431 $\alpha_1 = \alpha(T_1)$ = 0.476269214
κ = 0.740643018 $\lambda_D(T_1)$ = −78°.94122709
$\lambda_*(T_1)$ = −78°.9066431 $\Delta\lambda_1$ = −0.03458399
ΔT_1 = 0.046691546 = $2^m\ 48^s$, resulting in $T_2 = 5^h\ 56^m\ 16^s$.
But still $|\Delta\lambda_1| > 10^{-2}$, therefore Tristan iterates one more time to obtain:
$GHA_D(T_2)$ = 37°.30660 GHA * (T_2) = 53°.34332667
$\lambda_D(T_2)$ = −79°.60949709 $\lambda_*(T_2)$ = −79°.60836977
$\Delta\lambda_2$ = −0.00112732. Hence $\Delta T_2 = 5^s.5$, resulting in $T_3 = 5^h\ 56^m\ 22^s$.
Because now $|\Delta\lambda_1| < 10^{-2}$, Tiburon Tristan accepts his T_3 approximation and concludes that:

T = $5^h 56^m 22^s$ and $\lambda = 79°36'.5\,W$

Confident with his results, Tristan double checks himself. He takes the manual and using the latitude check, the one based on the principal that if the computed time and longitude are correct, then formula (20)[21] (Sect. 2.12) must yield the correct latitude.
Using $\delta_* = 7°.41, H = H_* = 41°.95$ and $t = t_* = 26°.2650431$ he computes:

---

[21] $\varphi = -\sin^{-1}\left(\dfrac{\sin\delta}{C}\right) - \alpha, \alpha = \tan^{-1}\left(\dfrac{\cos A_z}{\tan H}\right)$, with $A_z = \sin^{-1}\left(\dfrac{\sin t \cos\delta}{\cos H^*}\right)$.
$C = \left(1 - \cos^2 H \sin^2 A_z\right)^{\frac{1}{2}}$ and $t = -(\lambda + GHA)$ or $t = t_*(T_0)$.

$A_z = 36°.15974, \alpha = 41°.93195783,$ and $C = 0.98570847$
$C = 0.898570847.$ Hence:

$\varphi = \mathbf{-33°.68 = 33°41' \cdot S}$

Walking into the boat's chart room, Tiburon Tristan looks for a specific chart and checks his latitude and longitude.

"Well?" the well-to-do man asked.

Tristan smiled and pointed to the island off their port beam. "It appears that not much has changed on the Islands of Juan Fernandez since Lt. Alexander Selkirk was marooned here back in 1704."

The well-to-do man looked to his own captain who nodded in confirmation.

"I'll take my winnings in Pounds Sterling," Tiburon Tristan said. "I'm not fond of Euros."

Conclusions:

Headlines

1. NSA reveals North Korea hack
   January 19, 2015, 6:33 PM. Officials say that U.S. intelligence knew North Korea was trying to infiltrate America's computer systems years before the Sony hack. according to the New York Times, the NSA was tracking North Korean computer activity as far back as 2010...
2. U.S. Cracked North Korea's Computer Systems ... same date. NBC News.
3. North Korea's Internet Down Again...
   December 27, 2014... It could have been China, through which North Korea's Internet activities flows. It could have been US government hackers—part of what President Obama declared would be a "proportional" response to the Sony hack, which the FBI says it traced to North Korea. Or it could have been what one expert told CNN is an amateur teen hacker working out of his bedroom —"More like a 15-year-old in a Guy Fawkes mask." *Christian Science Monitor*
4. U.S. Air Force Confirms Boeing's Electromagnetic Pulse Weapon
   May 26, 2015... Known as the **"CHAMP,"** or Counter-electronics High-powered Microwave Advanced Missile Project, the American military project is an attempt to develop a device with all the power of a nuclear weapon but without the death and destruction to people and infrastructure that such a weapon causes. Theoretically, the new missile system would pinpoint buildings and knock out their electrical grids, plunging the target into darkness and general disconnectedness. *Digitaltrends.com*

It's not a question of IF our GPS systems will be hacked, but when. And when they are, ships at sea and aircraft in flight will be helpless to determine their positions unless they are carrying cumbersome nautical almanacs, trigonometric tables, and astronomical tables... helpless, that is, unless, they have a copy of this book tucked into a cubby or into the cargo pocket of their pants.

As a mathematician, I conceived the idea to write about the scientific part of navigation long before GPS became available to the general public. From the outset, it was meant to be invariant with regards to the fast changing technology. It was also meant to be a manual for the navigator that would make him less dependent on the availability of other ephemeredes.

Even the most casual observer must admit that we have achieved a level in our general education that renders anyone who does not know how to operate a personal computer illiterate. Today, the average student aged twelve or above should be able to handle the algebraic symbolic language as employed in all advances calculators without even understanding the underlying mathematics of the formulae used in navigation.

This book also does not rely on the use of any special calculator or PC. Nor does it depend on any other algorithms than the generally accepted mathematical algorithms. In theory, this amounts to saying that in the event where no calculator is available, a navigator can solve his positioning problem by using the provided formulae and the old, standby Logarithmic Tables.

However, the reader should be aware and alerted to the fact that the computational method for finding time crucially depends on the availability of precise altitude measurements. Errors should be less than one arc minute, which calls for considerable accuracy. Strictly from the mathematical point of view, the underlying problem of proving the existence and finding a approximate solution of the ECT is straight forward. But it is imperative that the user fully realize that only the exact Eq. (14) with the true parameters has a unique solution based on the formulae of spherical trigonometry. therefore, whenever those true parameters are substituted by approximate values, i.e., by measured altitudes and ephemeral data, the resulting equation may or may not have a solution and even if these equations possess a solution, it will not be the exact solution that we are seeking only a more accurate approximation.

Similarly, in all cases where all attempts to find a convergent sequence of iterations fail and no definitive conclusions can be reached, the user has to fall back on finding two values, T_0 and T_1, so that $F(T_0) \cdot F(T_1) < 0$ to establish the existence of a solution to the "perturbed" ECT, i.e., of $\widetilde{F}(\widetilde{T}) = 0$, and subsequently try to find this value by using another method.

Appendix

It is only for the sake of completeness that we need to review the simplest approximation of the Lunar Distance Method (LDM) in order to provide us with another alternative for correcting our watches. The first truly simplified LDM is probably due to the work of Nathaniel Bowditch and here we refer to the version provided (without proof) by John S. Letcher, Jn. [2]. In this particular approximation, the oblateness of the Earth, augmentation of the Moon's semi-diameter, parallax of the Sun, and contraction of the Sun and Moon's semi-diameter by refraction have been neglected, resulting in essentially one formula for the correction/CLEARENCE of the measured distance between the Moon and the Sun or another CO due to refraction. The other formula provides a correction/CLEARENCE for the parallax as it affect the distance between those two bodies.

The formula as stated in the aforementioned literature for the correction of refraction is:

(21) $R' = 1.9 \cdot \dfrac{(x - \cos D_S)}{\sin D_S}$, with $x = 0.5 \cdot \left(\dfrac{\sin h_1}{\sin h_2} - \dfrac{\sin h_2}{\sin h_1}\right)$, where "$D_s$" denotes the sextant distance between the two COs and "h_1" and "h_2" the sextant altitudes respectively.

The corresponding formula for the parallax correction is given by:

(22) $P' = HP_D(y + 0.000145 \cdot HP_D \cdot \cot D' \cdot (\cos^2 h_D - y^2))$, with:

$y = (\cos D' \cdot \sin h_D - \sin h_\odot)/\sin D'$, where "$HP_D$" denotes the horizontal parallax of the Moon in minutes of arc, "D'" the sextant Lunar distance corrected for semi-diameter of the Sun and Moon, i.e., $D' = D_s \pm SD_D \pm SD_\odot$, h_D the Moon's altitude corrected for semi-diameter, i.e., $h_D = h_{SD} \pm SD_D$, $h_\odot$ the other body altitude corrected for semi-diameter, i.e., $h_\odot = h_{S\odot} \pm SD_\odot$. Then the true Lunar Distance D is:

(23) $D = D' + R' + P'$.

Note that the Lunar Distance Method requires an approximate time in order to determine the HP_* of the Moon and also required the azimuth Z_D of the Moon in order to find the parallax $\Pi = HP$ of it when it is not on the meridian. Moreover, the monthly tables, if available, would have to be interpolated, thereby introducing additional errors. Therefore, the LDM can only be classified as an "approximate" method subject to systematic errors.

CAVEAT

As the inventor of the concept of Computed Approximate Dynamic Time (CADT), I would like to point out that in theory, any pair of COs could be used to find this time, i.e. for instance, a star and a planet or even two stars could be substituted for a star and the Moon, since the solution for the system of transcendental equations (Sect. 2.21) exists provided that all the parameters are exact and round-off errors are negligible.

However, since in any type of application to navigation some or all parameters are subject to finite errors and all calculations and computers have a finite arithmetic, it is possible that these equations do not have a solution for those given parameters or they are Ill-Conditioned and hence cannot be solved numerically. Therefore, I suggest that the reader employ the Moon-Sun combination whenever possible because the Moon changes its position relatively quickly with reference to the position of all stars.

To consider those pathological cases mentioned above, first consider cases where those two equations do not have a solution at all. These are the cases where the erroneous parameter no longer prescribe spherical triangles. The reader may recall the example given in a previous section where the three sides a, b, and c no longer prescribe a plane triangle since c > a + b. The case where the system of those equations is Ill-Conditioned corresponds to erroneous parameters that result in tangent lines at the point of the intersecting solution that are almost parallel. Therefore, the solution, although it exists, cannot be found numerically by employing approximate methods.

In conclusions, if the reader arbitrarily chooses two stars, he or she will find out that the system of those equations will in all likelihood be Ill-Conditioned since the declination of those stars remains almost constant over a period of one day as does their SHAs which change at nearly the same rate.

Chapter 3
Methods for Reducing Measured Altitude to Apparent Altitude

3.1 Navigational Refraction that Includes Astronomical Refraction for Low Altitude Observations

In the previous chapters, exact mathematical methods have been provided always assuming that the parameters entering the calculations were correct. However, as has been pointed out on various occasions, the latter is hardly the case. We now need to ask ourselves a very important question, namely, what are the parameters that can cause the greatest errors in determining our position at Sea or Air by way of astronavigation? The answer becomes obvious once we look at the magnitudes of all parameters which enter our calculations. Besides the exact time, exact ephemeral data, the altitude h_0 is one of the crucial parameters needed for finding the position of a vessel.

The so-called "observed altitude (h_0)" is composed of six elements, namely:

 (i) The measured altitude by sextant (h_s).
 (ii) The index error of the sextant (I).
 (iii) The dip of the horizon (D).
 (iv) The semi-diameter of the Sun, Moon, and Planets (SD).
 (v) The horizontal parallax (HP) of those COs … and most importantly,
 (vi) The atmospheric refraction (R).

Relative to the others, the dip of the horizon and the refraction are, in magnitude, together with the measured altitude (h_s) the crucial element of observed altitude. How the measurements of the sextant altitude can be made as precise as possible will be explained in Sect. 3.5 in great detail where it will also be shown that the dip of the horizon depends primarily on refraction so that the most important component in the calculation of h_0 is indeed the atmospheric refraction which may give rise to variations in altitude measurements of almost a full degree. Therefore, it is not only important to explain what refraction is in mere physical terms, but also to describe this phenomenon quantitatively in terms of mathematical formulae. [4, 5, 22, 23, 29, 30, 41]

For centuries, fishermen have known that they could not spear a fish if they aimed at what they actually saw. Instead, they learned to aim at a point somewhat closet to their position. They recognized that the rays of light were broken when they entered the water. This led to the discovery of the laws of refraction, culminating in "Snell's Law" that quantifies refraction.

Although the ancient Moors knew that light which comes from a CO gets bent once it enters the atmosphere and appears to follow a curve rather than a straight line, it took many centuries before mathematical formulae and tables became available to aid in navigation. However, despite the great effort on the part of astronomers, physicists, and mathematicians, many aspects of the real atmosphere are still not fully understood and all the available models for quantifying refraction still do not give satisfactory results in all practical situations. It should be clearly understood that without a better understanding of the physics of gases in boundary layers, we can not expect to be able to develop the theory of refraction to an extent that it becomes an infallible tool in the hands of the navigator. Until that time comes, we will have to rely on what is already at hand and on experimental results as they become available.

It is generally accepted that the basis for a model of the Earth's atmosphere should be a spherically symmetric atmosphere [1]. Depending on how we describe the atmosphere physically, we will obtain different models for refraction and the mathematics is of secondary concern only. However, if we opt for a mathematical method for computing refraction by employing experimental data, we will be concerned with numerical methods of approximation as for instance Chebyshev's or Padé's and Max-Min approximation by functions in order to derive suitable formulae for the practical navigator. Because of this, we need to classify the available results into three main categories:

I. Analytic Methods based on analytical models of the atmosphere.
II. Numerical Methods based on analytical descriptions of the atmosphere.
III. Numerical Approximations based on experimentally obtained data.

As has been stated before, the base for a quantitative description of refraction is Snell's Law that relates the indices of refraction of medium one and medium two to the angles Θ_1 and Θ_2 of the refracted light as follows:

(1) $\mu_1 \cdot \sin \Theta_1 = \mu_2 \cdot \sin \Theta_2$, where the index of refraction is defined by $\mu = \frac{c}{v}$, "c" denoting the speed of visible light in a vacuum and "v" denoting the speed of light in the respective other medium. For instance, $\mu = 1$ in the upper atmosphere and $\mu_0 = 1.00029$ in the region closest to the surface of the Earth. If we then divide the atmosphere surrounding the Earth into shells[1] of arbitrary thickness but distinct indices μ_k, k = 0, 1, ... n and apply Snell's Law, then we can readily see that the light coming from a CO follows a curved path as the thickness of the shells approach zero and arriving at a fixed angle to the zenith of the observer on the Earth (see Fig. 3.1.1).

---

[1]At a later time, it may become necessary to adopt a more realistic model of the atmosphere that departs from the assumption of shell symmetry.

3.1 Navigational Refraction that Includes Astronomical Refraction ...

Definitions

$\Theta = A\widehat{Q}Q$; $\Theta' = A\widehat{Q}P'$
$\gamma = A\widehat{P}'B$; $\Delta\Theta = \Theta' - \Theta$
$r = \overline{OQ}$; $r' = \overline{OP'}$
$\Delta r = r' - r$; $\Delta\phi = \phi' - \phi$
$\Delta R = \alpha - \alpha'$ Refraction.

We may then conclude that $\Delta R = -\gamma$; $\alpha = \phi + \Theta$; $\alpha' = \phi + \Theta$; $\Delta R = -(\Delta\phi + \Delta\Theta)$; and also $dR = -(d\phi + d\Theta)$. r_o: Radius of the Earth.

By applying Snell's Law to the two shells with index of refraction μ' and μ respectively, we find that $\mu' \cdot \sin \phi = \mu \cdot \sin \Psi$ and by applying the SIN-TH of plane trigonometry to the triangle QOP' we find that $\dfrac{\sin \phi}{r'} = \dfrac{\sin \Psi}{r}$. By eliminating $\sin \Psi$ on the right hand side and by employing the first Eq. (1), we deduce that:

$\mu \cdot r \cdot \sin \phi = r' \cdot \mu' \cdot \sin \phi'$.

This equation hold for any contiguous layers independent of their height above the Earth's surface. Hence we obtain the important expression:

(2) $\mu \cdot r \cdot \sin \phi = \mu_o \cdot r_o \cdot \sin \zeta$, the so called "Invariant Relation".

In this formula "μ_o" denotes the index of refraction at the surface of the Earth and "ζ" is the zenith distance of S measured by the observer at Z. Moreover for small values of $\Delta r \rho\, dr$ and $\Delta \Theta \rho\, d\Theta$, we deduce from the plane of triangle Q'QP' that:

$\tan \phi = \dfrac{r \cdot \Delta\Theta}{\Delta r} \simeq \dfrac{r d\Theta}{dr}$, or

(3) $\dfrac{dr}{r} = \cot \phi \cdot d\Theta$.

By taking the total differential of the invariant relation (2), we find that $\mu \cdot \sin \phi \cdot dr + r \cdot \sin \phi\, d\mu + r \cdot \mu \cdot \cos \phi\, d\phi = 0$, since the right hand side of (2) is

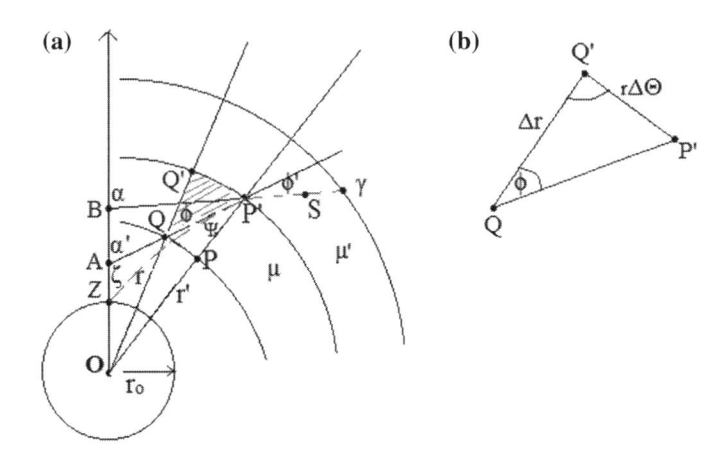

Fig. 3.1.1

a constant. Hence we can deduce that $\frac{dr}{r} + \frac{d\mu}{\mu} + \cot\phi \cdot d\phi = 0$, and by employing expression (3) and $dR = -(d\phi + d\Theta)$ from the expression above, we conclude that $\frac{d\mu}{\mu} = -\cot\phi \cdot (d\Theta + d\phi) = \cot\phi \cdot dR$, or

(4) $\frac{dR}{d\mu} = \frac{1}{\mu} \cdot \tan\phi$, the Differential Equation of Refraction that is based on our shell model.

By integrating (4), we find the Integral Representation of Refraction, namely:

(5) $R = \int_1^{\mu_o} \tan\phi \frac{d\mu}{\mu}$. Although the integrand is known only according to the invariant relation (2), approximations can be obtained by describing the physical state of the atmosphere quantitatively. This is the same as saying that once the dependence of μ on r is known analytically, we shall be able to integrate (5) and find an explicit expression for R as a function of ζ or h_o, and some physical parameters such as temperature, pressure of the air and water vapor, height of the observer above the water level, wavelength of visible light, and even the latitude of the observer. Therefore, the entire physical state of the atmosphere rests with the function $\mu = \mu(r)$, that also depends on the aforementioned parameters. Any attempt to approximate (5) without this vital information is bound to fail, as we shall see promptly.

If we express the factor $\tan\phi$ in the integrand in terms of $\sin\phi$ and use the invariant relation (3), we find that:

(6) $R = \mu_o \cdot r_o \sin\zeta \int_1^{\mu_o} \frac{d\mu}{\mu(\mu^2 \cdot r^2 - \mu_o^2 \cdot r_o^2 \sin^2\zeta)^{1/2}}.$

Although the integrand has a singularity at $r = r_o$; $\zeta = 90°$, the integral still exists.

Since all meteorology is built in the formula for $\mu(r)$, we must look for a relationship between the index of refraction and the density of the atmosphere. Such a relationship is provided by the Law of "Gladstone and Dales" that states explicitly:

(7) $\mu = 1 + c\rho$ with $c = 0.226$, ρ_o = density at the Earth's surface, $c\rho_o = 0.00029$. So far as the theory of refraction has been laid down in very general terms, however, it should be clear from those formulae that any realistic quantitative analysis must entail an accurate description of the physical state of the atmosphere and must therefore include the meteorology of the boundary layers on the surface of the Earth in order to yield results for low attitudes, i.e., $-2° \leq h_o \leq 5°$ that are compatible with the experimental results.

Now, let's consider two independent analytic methods for calculating atmospheric refraction.

I. Analytic Methods Based on Analytic Models of the Atmosphere.
Laplace Formula

For more than two centuries, mathematicians and astronomers have tried unsuccessfully to integrate the physics away from refraction, leaving the navigator with erroneous data to compute the observed altitude of COs near the horizon. As recorded in the relevant literature, authors like Lambert, Laplace, and later, Newcomb, Wooland and Clemence have attempted to integrate (6) by using Taylor series expansions for the integrand truncating the semi-convergent series after several terms and integrating them term by term and ignoring meteorology by merely using the unrealistic concept of mean atmosphere, or standard atmosphere at sea lever defined by T = 10° (C°) P = 1010 (mb). This mean refraction is commonly denoted by "R_0" and the temperature and pressure adjustment is simply made by introducing the meteorological factor M(T, P) that depends on the most important physical parameters.

Let us consider the concrete case that results in an inadequate approximation of refraction as it applies to CO observations that are close to the horizon, known in most of the text books and manuals on Celestial Navigation as the "Laplace Formula." Here, the underlying mathematical concept is that the integral in (6) can be evaluated approximately by introducing a small parameter "s" defined by: $\frac{r}{r_0} = 1 + s$, r and r_0 defined as before.

Since it is assumed that beyond 65 km above sea level the atmosphere no longer contributes to the integrand, for r_0 = 6,378,390 m and it follows that $0 \leq s \leq 0.0102$. Hence, by expanding the integrand in powers of s and then neglecting the terms containing s^2, s^3, ..., yields essentially only two terms that can be integrated by employing Eq. (7) in its differential form: $d\mu = c \, d\rho$, resulting in the expression known as "Laplace Formula", namely: $R_0 = A \cdot \tan \zeta + B \cdot \tan^3 \zeta + \cdots$.

The constants A and B corresponding to the mean refraction, i.e., to T = 10° (C°) and P = 760 (mm), are to be found by astronomical observations and are: $A = 58''.16$, $B = 0''.067$, so that the above formula becomes:

(8) $R_0 = 58''.16 \cdot \tan \zeta - 0''.067 \cdot \tan^3 \zeta, \zeta = 90° - h_s$.

This formula is said to constitute an adequate approximation for all zenith distances of up to $0 \leq \zeta \leq 75°$, $h_s \geq 15°$.

In case readers would like to refine the above formula by taking terms of higher than s^2 into consideration, they will obtain formulae of the type:

$R_0 = A \cdot \tan \zeta + B \cdot \tan^3 \zeta + C \cdot \tan^5 \zeta + D \cdot \tan^7 \zeta + \cdots$ [7], [30]

However for low altitudes, these formulae will also fail to provide reliable results for the navigator for the aforementioned reasons. The authors mentioned above and others then account for the variations in refraction due to changes in temperature and pressure by employing the law of Mariotte-Gay-Lussac that states:

$\frac{\rho_{T,P}}{\rho_0} = \frac{P}{760} \cdot \frac{1}{1+mT}$, where $m = \frac{1}{273}$ if T: (C°), P: (mb), at 10° (C°), 1010 (mb), thereby obtaining the refraction R at temperature T and pressure P by the expression:

(9) $R = M(T, P) \cdot R_0$ with $M(T, P) = \dfrac{0.28P}{273+T}$, P (mb), T (C°), and R_0 as above.

Cassini's Model

Here we consider a slightly more realistic approach to quantitative refraction that is based on a fairly simple model of the atmosphere and results in less complicated formulae. This model is due to Giovanni Domenico Cassini, an Italian astronomer [4]. Cassini simply treated the atmosphere as a uniform one, i.e., he assumed that the light gets refracted only once at the boundary layer between the so called "ether"-vacuum and the homogeneous atmosphere (see Fig. 3.1.2).

Obviously he simplifies things considerably but, at least, consistently so far as the application of Mariotte-Gay-Lussac's Law is concerned. On the other hand, what is it good for if one uses the shell model and then applies Mariotte-Gay-Lussac's Law only to the boundary layer of Earth's atmosphere? Nice mathematics, but poor physics.

By applying Snell's Law, Cassini deduced:
$\mu \cdot \sin z_1 = \sin z_2$ and using the SIN-TH. to OPZ he obtained:
$\dfrac{\sin z_1}{r_0} = \dfrac{\sin \zeta}{r_0 + H}$, i.e., $\sin z_1 = \dfrac{r_0}{r_0 + H} \cdot \sin \zeta$; and from the first equation, he then deduced:
$\sin z_2 = \dfrac{\mu \cdot r_0}{r_0 + H} \cdot \sin \zeta$. He also concluded that $R_0 = z_2 - z_1$, and hence:

(10) $R_0 = \sin^{-1}\left[\dfrac{\mu \cdot r_0}{(r_0+H)} \cdot \sin \zeta\right] - \sin^{-1}\left[\dfrac{r_0}{(r_0+H)} \cdot \sin \zeta\right]$.
$r_0 = 6{,}378{,}390$ m; $H - 65{,}000$ m.

Surprisingly, Cassini's formula (10) derived from a slightly oversimplified model of our atmosphere, produces results even more favorable than Lagrange's formula, since it may give a better approximation of to 80° zenith distance. Even more surprisingly, Cassini's formula does not appear in text books on Celestial Navigation with one exception where a formulae is being presented without proof and reference to the author. Explicitly, this formula is: [27]

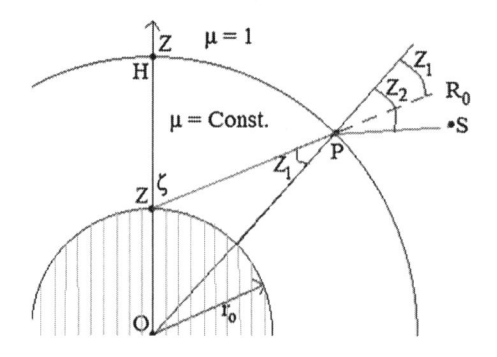

Fig. 3.1.2

3.1 Navigational Refraction that Includes Astronomical Refraction ...

(11) $R = \dfrac{P}{273+T}(3.430289\,(z - \sin^{-1}(0.9986047 \cdot \sin 0.9967614 \cdot z))$
$-0.01115929 \cdot z), z = 90° - h_s$.

Next, let us look at another class of approximations that produce acceptable results even for low altitude observations, i.e., $h_0 < 5°$, and which employs numerical methods for computing refraction.

II. Numerical Methods Based On Analytic Descriptions Of The Atmosphere.

Here we assume that a realistic physical description of the atmosphere–troposphere, stratosphere and boundary layers—is available, i.e., we assume that the function $\mu = \mu(r)$, depending also on the other parameters mentioned above, together with its derivative, has been found. Those two functions don't have to be continuous throughout, but should be, at least, piecewise continuous. Subject to these conditions, we will be able to compute the integral (5) by employing the invariant relation (2). [4]

The first step in this approach consists in transforming the integrand in (5), so that it can be computed later by using numerical quadrature formula (Sect. 4.1). Recalling that $\dfrac{d}{d\mu} \ln \mu = \dfrac{1}{\mu}$, we deduce that:

(12) $R_0 = \int_0^{\ln \mu_0} \tan \phi \, d(\ln \mu)$.

Next taking the logarithm of Eq. (2) and differentiating it with regard to μ, we find:

$\dfrac{d}{d\mu} \ln(\mu \cdot r) = -\dfrac{d}{d\mu} \ln \sin \phi = -\dfrac{d}{d\phi} \ln \sin \phi \dfrac{d\phi}{d\mu} = -\dfrac{1}{\tan \phi} \cdot \dfrac{d\phi}{d\mu}$, i.e.,

$\tan \phi \cdot d(\ln \mu) = -\dfrac{d(\ln \mu)}{d(\ln r \cdot \mu)} d\phi$. Hence :

(13) $R_0 = -\int_0^\zeta \dfrac{d(\ln \mu)}{d(\ln \mu \cdot r)} d\phi = -\int_0^\zeta \dfrac{d(\ln \mu)/d(\ln r)}{1 + d(\ln \mu)/d(\ln r)} d\phi = -\int_0^\zeta \dfrac{\frac{r}{\mu} \cdot \frac{d\mu}{dr}}{1 + \frac{r}{\mu} \cdot \frac{d\mu}{dr}} d\phi$.

It should be noted that the integrand in the last integral is a well-behaved function of δ and is known numerically because of the invariant relation (2), i.e., $\mu(r) \cdot r \cdot \sin \delta =$
$= \mu_0 \cdot r_0 \sin \zeta$, which for any given δ can be evaluated numerically for r by iterative methods, as for instance, by the well-known Newton-Raphson method as follows:

By defining the function:

$f(r) = \mu(r) \cdot r - \mu_0 \cdot r_0 \cdot \dfrac{\sin \zeta}{\sin \phi}$, we have: $f'(r) = \mu(r) + r \cdot \dfrac{d\mu}{dr}$. The Newton-Raphson method then consists in computing the sequence of values $\{r_k\}$, k = 1, 2, ... for r corresponding to δ, i.e., $r = r(\delta)$, by the formula:

(14) $r_{k+1} = r_k - \dfrac{f(r_k)}{f'(r_k)}$, k = 1, 2, ... With r_1 as the initial approximation.

Recalling that μ varies between 1 and 1.00029 we may use $r_1 \cong r_0 \cdot \dfrac{\sin \zeta}{\sin \phi}$ as an initial value in our iterative method defined by (14). Having calculated the value $r = \lim_{k \to \infty} r_k$, approximately, we know $\mu(r) = \mu(\phi)$ and also $\dfrac{d\mu}{dr}$ and therefore, the integrand in (13) is know at discrete points δ_k, satisfying $0 < \delta_k < \zeta$, k = 1, 2, ..., N. The latter enables us to use a suitable quadrature formula, as for instance, a Newton-Cotes formula [35] to compute the integral (13) numerically.

It should be noted that the lower limit of integration, i.e., $\phi = 0$ corresponding to $r = \infty$ can be replaced by a value of $\phi_0 \neq 0$ corresponding to a value of $\mu(\phi_0)$[2] for which the integrand differs distinctly from zero. However, by not using $\phi = 0$ in the numerical calculations, this is not necessary.

The numerical method described herein does not constitute something entirely new and only becomes a practical, viable method for computing refraction if a realistic model of the atmosphere is available. Depending on how one describes the atmosphere and, in particular the troposphere in and near the boundary layers, one obtains one or the other concrete method with the necessary formulae for computing refraction.

One method for treating the atmosphere more realistically can be found in a publication by the authors Lawrence H. Auer and E. Myles Standish [5] who divide the atmosphere into only two shells, namely the troposphere (0 ≤ 11,019 m) and the stratosphere (>11,019 m). For the two distinct regions, distinct analytic descriptions—formulae for the dependence of the density on the height above sea level—are being used and the physical parameters are adjusted according to the prevailing meteorological conditions.

One hopes that different and even more realistic models of the atmosphere will become available to the navigator in the near future. Until then, we will have to rely more on results obtained by astronomical observations if we want to include low altitude observation in the scope of practical navigation. [6]

My personal experience with low altitude observations goes back to the year 1973 when I held an appointment with the Data Institute of Fundamental Research in Bombay (now Mumbai), India. During my stay there, I frequently chartered local "Dhows" to take me out into the Arabian Sea, away from landmasses in order to practice low altitude navigation using the simple Arabic "Kamal" (described in Sect. 2.20). I had previously devised a formula for low altitude refraction based on tables of experimental data; however, my initial experiments with that formula frequently put me ten nautical miles off with my navigation. Repeated experiments showed that the given values had to be adjusted drastically for higher temperature, abnormal pressure, and, in particular, for the vapor pressure in the boundary layers. After making those adjustments, I was able to reduce my "fix-error" to about three NM. But I wasn't the only one who needed to make adjustments. I discovered that

---

[2]Such a value can be found by computing with (2), putting r = 12,000 m.

the local navigators also empirically adjusted the calibration of the Kamals to account for low altitude refraction.

Because of our rather limited knowledge of the physical state of the atmosphere, the experimental data obtained from astronomical observations is invaluable for the navigator and it makes perfect sense to condense that data into mathematical formulae that can be easily evaluated with a pocket calculator. Therefore, let's take a closer look at the numerical methods based on observational data.

III. Numerical Approximations Based On Experimentally Obtained Data.

As before, the starting point for the development of an analytic formula is the integral of refraction (5). Although the integrand has a singularity at $\phi = 90°$, we know that the integral itself exists. By replacing δ with α defined by $\phi = 90° - \alpha$, we find that: $R_0 = \int_i^{\mu_0} \cot \alpha \cdot \frac{d\mu}{\mu}$, where α depends on μ in a manner we don't really know. By taking the properties of the integrand into consideration, we may replace the integrand $\frac{\cot \alpha}{\mu}$ by a mean value $\cot \bar{\alpha} \cdot a$ thereby obtaining the relation:

(15) $R_0 = a \cdot \cot \bar{\alpha}$, where $\bar{\alpha}$ is still unknown by satisfies $0 < \bar{\alpha} \leq 90$.

The approach used here consists in approximating $\bar{\alpha}$ by a simple rational function:

(16) $\bar{\alpha}(h) = \frac{P_n(h)}{Q_m(h)} + E_{n,m}(h)$

"$E_{n,m}(h)$" denotes an error term to be determined later. Note that $P_n(h)$ and $Q_m(h)$—in many applications $n = m + 1$—are corresponding polynomials of degree n and m respectively. The ration of the two polynomials, i.e., the rational function, is expressed in terms of continued-fractions, as for instance:

$$\bar{\alpha}(h) = h + \cfrac{a_1}{h + \cfrac{a_2}{h + \cfrac{a_3}{h + \cfrac{a_4}{h + \cfrac{a_5}{h + a_6}}}}}$$

———— and so on.
………

If "R_0" denotes the mean refraction and "$M(T, P; p, H, \lambda)$" denotes the meteorological factor that depends on the temperature, the barometric pressure P, the pressure of water vapor p, the height of the observer at sea level H, the wavelength of visible light λ, and perhaps, the latitude of the observer, then the refraction is assumed to be related to R_0 by:

(17) $R(h) = M(T, P; p, H, \lambda) \cdot R_0(h)$.

The numerical methods described here differ radically from the previously described methods because in these methods, the coefficients $a_1, a_2 \ldots$ are determined by matching the resulting numerical values with the experimental data obtained by astronomical observations and denoted by:

$R_0^k = R_0(h_k)$, $k = 1, 2, \ldots, n + 1$. [9, 18, 35]

In the case of the approximation defined by (15), we are actually approximating $\cot^{-1}(a^{-1}R_0(h))$ by the rational function $\frac{P_{n+1}(h)}{Q_n(h)}$, for from (15) and (16) we deduce that:

(18) $\cot^{-1}(a^{-1}R_0(h)) = \frac{P_{n+1}(h)}{Q_n(h)} + E_{n+1}(h)$, where "h" is the independent variable.

Depending on how the requirements on the error function $E_{n+1}(h)$ are specified, you obtain one or the other type of approximation. For instance, by using a Padé approximation [18] for a given set of values R_0^k, you can obtain a specific set of coefficients $a_1, a_2, a_3, \ldots a_{n+1}$, and by employing a Chebychev approximation, for instance, you obtain another set of coefficients $\bar{a}_1, \bar{a}_2 \ldots \bar{a}_{n+1}$. However in all cases where $R_0(h)$ is not known analytically, but only for discrete values of R_0^k, we will have to solve at least a linear system of equations for the n + 1 coefficients $a_1, \ldots a_{n+1}$. The resulting formulae for the refraction then become:

(19) $R_0(h) = a \cdot \cot \frac{P_{n+1}(h)}{Q_n(h)}$, with "a" as a suitable scaling factor still to be chosen.

The meteorological factor $M(T, P; p, H, \lambda)$ is generally found by applying the laws of the theory of gases as they apply to well-defined layers in the troposphere and stratosphere and/or by utilizing the experimental data available. This permits adjusting those parameters so that the numerical results account for the actual physical state of the atmosphere.

Before we consider specific examples of this type of approximation, I would like to point out that from the mathematical point of view, the transition from the definite integral (5) to the approximation (15) could be completely ignored if we just take (15) as an a priori approach for the construction of an approximation of the desired type.

(The reader should understand that the scope of this book does not leave space for deriving explicit formulae for the above described approximations. Each specific approximation for a particular rational function in (16) and (19) and particular set of experimental data constitutes a major undertaking. Therefore in this context, I only refer to those concrete cases where the coefficients have already been computed and merely state the actual results as they can readily be used by the navigator.)

3.1 Navigational Refraction that Includes Astronomical Refraction ...

Explicit Formulae

Case (i) $a = 58''.2$, $n = 0$, $a_1 = 0$, corresponding to $\bar{\alpha} = h_s$,

$$R_0(h_s) = 58''.2 \cdot \cot h_s, \quad 45° \leq h_s \leq 90°, \quad M(T,P) = \frac{0.372P}{273+T}.$$

[7, 41]

Case (ii) $a = 0°.0167$, $n = 1$, $a_1 = 7.32$, $a_2 = 4.32$

$$R_0(h_s) = 0°.0167 \cdot \cot(h_s + \frac{7.32}{h_s + 4.32}), \quad 5° \leq h_s \leq 90°,$$

$$M(T,P) = \frac{0.28P}{273+T}, \quad R(h_s) = M(T,P) \cdot R_0(h_s). \quad [55]$$

Case (iii) $a = 1'$, $n = 1$, $a_1 = 7.31$ $a_2 = 4.4$, $M(T, P) = \frac{0.28P}{273.15+T}$

$$R_0(h_s) = 1' \cdot \cot(h_s + \frac{7.31}{h_s + 4.4}), \quad R(h_s) = M(T,P) \cdot R_0 \quad [44]$$

and the improved formula is

$$\overline{R}_0(h_s) = R_0(h_s) - 0.06 \sin(14.7 \cdot R_0 + 13), \quad M(T,P,R_0) =$$

$$\frac{(P-80)}{930} \cdot \frac{1}{1 + 8 \cdot 10^{-5} \cdot (R_0 + 39)(T-10)}, \quad -20° \leq T \leq 40$$

$970 \leq P \leq 1050$ mb, $0 \leq h_s \leq 90°$, $\overline{R}(h_s) = M(T,P,R_0) \cdot \overline{R}_0$

Case (iv) $a = \frac{1°}{62'.6}$, $n = 2$, $a_1 = 5.459$, $a_2 = 19.272$, $a_3 = 6.942$
$-0°.3258 \leq h_s \leq 90°$.

Case (v) $a = \frac{1°}{62.97411}$, $n = 4$, $a_1 = 3.86653$ $a_2 = 6.24727$ $a_3 = 8.56113$ $a_4 = 22.89592$ $a_5 = 7.15359$ with the formulae for M(T, P, p, H)—too complicated for the practical navigator. [5]

Note in all the formulae above "h_s" denotes the sextant altitude cleared for Index error, Dip, Semi-diameter and Parallax.

Aside from the formulae stated above, there still exists another class of formulae for practical applications that are based on direct approximations of the physical data obtained by astronomical observations. These types of approximations that are being use in direct approximations are either polynomial approximations, as for instance Chebychev approximation, or rational approximations of the Padé type or even Min-Max approximations [10]. Here I would like to draw attention to one particular formula for low altitude refraction as it appears in the "Refraction Calculator" by Google. This particular formula is claimed to be valid for altitudes below 15° and gives R(h_s) as follows:

Case (vi): $R(h_s) = \dfrac{P(0.1594 + 0.0196 h_s + 0.00002 \cdot h_s^2)}{(273+T)(1 + 0.505 h_s + 0.0845 h_s^2)}$ [14]

where, again "T" denotes the temperature in degrees C°; "P" the barometric pressure in mb; and "h_s" the sextant altitude in degrees; the refraction "$R(h_s)$" is then also given in degrees.

In all those direct methods, the physical data is obtained by inspecting accurate tables, as for instance, the Pulkovo Refraction Tables [34], or the Greenwich Obs. Tables—or perhaps, even your own tables. [15]

In conclusion, it can be said that with the present state of our knowledge of the physical structure of the atmosphere, and in particular, the theory of boundary layers, none of the approaches employed for calculating refraction accurately for all regions of the atmosphere is completely satisfactory. Therefore, the direct or indirect use of experimental data is imperative. However, it is hoped that progress is being made in the physics of our atmosphere that will result in even more accurate methods for calculating astronomical refraction.

For practical navigators, the nine explicit formulae provided in this section should give them the option of choosing the most suitable method for their particular applications.[3] Furthermore, the user now has the possibility of comparing several methods with regards to accuracy and efficiency.

Advisory for Aviators

The pilot and/or navigator of an airplane who wants to use a sextant in the dome or cockpit has to apply an additional refraction correction that is due to the refraction of the visible light on the plexi or safety glass of the dome or cockpit window. Anyone wishing to do so is advised to consult the manufacture of those windows to obtain the index of refraction and measure this refraction directly while on the ground.

Formulae for Conversions

(i) Celsius$\lfloor C^* \rfloor \Leftrightarrow$ Fahrenheit$\lfloor F^* \rfloor$

$$C^* = \frac{5}{9} \cdot (F^* - 32^*)$$

$$F° = \frac{9}{5} \cdot C° + 32°$$

$$50\lfloor F° \rfloor \cong 10\lfloor C° \rfloor$$

(ii) (mm Hg) $\Leftrightarrow$ (mb) $\Leftrightarrow$ (Pascal)

1 (mm Hg) = 1.33 (mb) = $1.333 \cdot 10^2$ (Pascal)

760 (mm Hg) = 1.01308 (bars) = $1.01308 \cdot 10^5$ (Pascal)

1 (Atmosphere) = 1.01308 (bars) = 760 (mm)

---

[3] Another formula for the astronomical refraction that uses Chebychev polynomial of up to the order nine is also available to the navigator [27]. However, I chose not to include it in this section because 1., I have not introduced those polynomials and 2., the author of this formula has not stated upon which analytical approximation this particular Chebychev polynomial expansion is based.

3.2 The Dip of the Horizon as a Function of Temperature and Pressure

It is generally recognized that the two most limiting elements to the accuracy of astronavigation are the determination of valid approximations for REFRACTION and DIP. In this section, I am chiefly concerned with finding a realistic approximation to the dip of the horizon, although dip cannot be analyzed without dealing with refraction as well which is the reason for not having been able to arrive at more satisfactory results before.

In most of the relevant literature, including the Nautical Almanac, the formula for calculating dip is:

(1) $D = 0.97 \cdot \sqrt{h}$, where "h" is the height of the observer above sea level measured in feet and "D" is given in arc minutes.

I want to clearly show that this expression for D is strictly an empirical formula based solely on geometry. Mathematically speaking, formula (1) implies that at any given height of the eye above sea level, dip is merely a constant which contradicts experimental results and also defies logic, since dip obviously also depends on refraction and therefore on temperature and pressure. It is all too often one hears a sailor ask, "Why should I be concerned about nonstandard low altitude corrections if I only select COs of higher altitude for my particular type of navigation?"

I want the reader to understand that the main reason for employing Eq. (1) is that no other simple expression or set of tables are available for computing dip. However, by using (1) because nothing else is available, navigators should be aware that the probable error can amount up to $2'.5$, and under extreme atmospheric conditions of up to $30'$. [8]

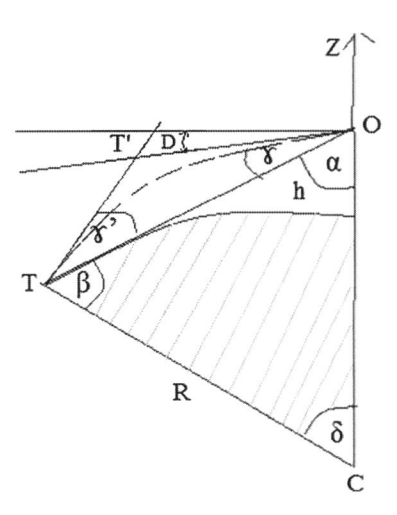

Fig. 3.2.1

At this point, our first task should be to analyze the hypotheses upon which formula (1) is based, and to derive some formulae that encompass the dependence of the dip on temperature and pressure.

In order to analyze formula (1), let's look first to Fig. 3.2.1 to understand the underlying geometry.

In the triangle OCT, we find that $\alpha + \beta + \delta = 180°$, with $\alpha = 90° - (D + \gamma)$, $\beta = 90° - \gamma'$. Then the two major assumptions made by the authors of formula (1) are:

(2) $\gamma' = \gamma$, and
(3) $\tau = \kappa\delta$. (Note that all of this is geometry and no physics.)

Next by applying the SIN-TH. of plane geometry to OTC we find that: $\dfrac{\sin \beta}{R+h} = \dfrac{\sin \alpha}{R}$. By substituting the expressions for α, β, and τ into the last equation and also in the previous one for the sum of the tree angles of OTC, we find:

(4) $D = \delta \cdot (1 - 2 \cdot \kappa)$, and
(5) $\dfrac{\cos(\kappa \cdot \delta)}{\cos(1 - \kappa)\delta} = 1 + \dfrac{h}{R}$. By taking into account that $\kappa\delta$ is of the order of arc minutes, we derive from (5), by using the Taylor-approximation $\cos x = 1 - \dfrac{x^2}{2} \sin^2 1'$, $x = \kappa\delta$, that:

(6) $\delta = \operatorname{cosec} 1' \cdot \dfrac{1}{\sqrt{R}} \cdot \sqrt{\dfrac{2}{1-2\cdot\kappa}} \cdot \sqrt{h}$. Next by suing the third assumption:

(7) $\kappa = \dfrac{1}{13}$ and the numerical values $\operatorname{cosec} 1' = 3437.749237$ and $R = 20968800$, we find:

(8) $\delta = 1.15\sqrt{h}$, and $D = \dfrac{11}{13}\delta = 0.97 \cdot \sqrt{h}$, i.e., formula (1).

Note that δ of expression (6) stands for the distance of the observed horizon in nautical miles.

Although it may be argued that the first assumption, namely $\gamma' = \gamma$, is a sufficiently close approximation for all values of, say h < 10 m, or so, the second hypothesis, namely (3), cannot be true for a wider range of temperature and pressure since the law of Maritotte-Gay-Lussac cannot be eliminated by using trigonometry. Therefore, I propose to modify the expression (3) treating it in the same manner as standard refraction. Standard refraction and non-standard refraction are related. Accordingly, the Eq. (3) now should read:

(9) $\gamma = M(T, P)\dfrac{d}{13}$, with $M(T, P) = \dfrac{0.28\,P}{273 + T}$ according to Boyle-Mariotte.

Again by using some trivial approximations, we arrive at:

(10) $D = 1.15\sqrt{h} \cdot \left(1 - \dfrac{2}{13} \cdot \dfrac{0.28\,P}{273 + T}\right)$, h (ft), P (mb), T (C°), D (arc min).

Of course, formula (10) is also bound to fail in extreme atmospheric conditions where the hypothesis of continuous dependence of density on temperature and pressure no longer holds. For such conditions where the theory of boundary layers and temperature inversion prevails, navigators still can only rely on experimental data compiled in tables, if available.

3.2 The Dip of the Horizon as a Function of Temperature and Pressure

The approach chosen above may be considered, by some, to be more empirical than analytical and therefore proves unsatisfactory in context of our investigations. Therefore, I will present a strictly theoretical approach that is based on the gas laws of Gladstone and Dales and Mariotte-Gay-Lussac and, of course, Snell's Law of refraction. [7]

The starting point for this approach is based on our model of refraction in Sect. 3.1. If we apply the invariant relation, i.e., Eq. (2.) Sect. 3.1, (see Fig. 3.2.1) to a shell of height h bounded by the surface of the earth from below, we have: $\mu \cdot (R + h) \cdot \sin \phi = R \cdot \mu_0 \cdot \sin \zeta$, and according to Fig. 3.2.1, we also have $\zeta = 90$, since $H_0 = 0$. Hence, the fundamental equation for our terrestrial refraction is:

(11) $\cos D = \dfrac{R}{R+H} \cdot \dfrac{\mu_0}{\mu}$. Note that according to Fig. 3.2.1 $\phi = 90° - D$.

Now what remains to be done is to express the indices of refraction in terms of densities and subsequently in terms of temperature and pressure. However, first let's consider the trivial case where refraction has been ignored, i.e., lets put $\mu = \mu_0$ in Eq. (11), and also let's denote the resulting Geometric Dip by "D_0". Then we conclude that $\cos D_0 = \dfrac{R}{R+h} = 1 - \dfrac{h}{R+h} \cong 1 - \dfrac{h}{R}$, and since $\sin D_0 = (1 - \cos^2 D_0)^{\frac{1}{2}}$, we deduce that:

(12) $D_0 = \dfrac{1}{\sin 1'} \cdot \sqrt{\dfrac{2h}{R}} = 1.06 \cdot \sqrt{h}$. Here we have used that $\cos^2 D_0 \cong 1 - \dfrac{2h}{R}$.

The reader should also notice that the value of D_0 differs only slightly from the value obtained by using the standard dip given by (1).

Next, let us take the important laws of physics into consideration. firstly, there is the law by Gladstone and Dales that states:

(13) $\mu = 1 + c\rho$, where $c = 0.226$ and $c\rho = 0.00029$ on the surface of the Earth. Hence with the help of (13), we deduce that:

(14) $\dfrac{\mu_0}{\mu} = (1 + c \cdot \rho_0) \dfrac{1}{1 + c \cdot \rho_0} = 1 + c \cdot \rho_0 \left(1 - \dfrac{\rho}{\rho_0}\right)$, since $c\rho_0 \ll 1$

Equation (14) expresses the ratio of indices of refraction in terms of ration of density—ρ_0 at the surface and ρ at height h above it. Substituting (14) into Eq. (11) yields: $\cos D \cong \left(1 - \dfrac{h}{R}\right) \cdot \left(1 + c \cdot \rho_0 (1 - \dfrac{\rho}{\rho_0})\right)$, and hence:

(15) $\sin D = \left(\dfrac{2h}{R} - 2 \cdot c \cdot \rho_0 (1 - \dfrac{\rho}{\rho_0})\right)^{\frac{1}{2}}$. This equation is know as the Laplace Formula.

Next we apply the well-known law by Boyle-Mariotte that states:

(16) $\dfrac{p}{p_0} = \dfrac{p}{p_0} \cdot \dfrac{1}{1+\epsilon(\tau - \tau_0)} \cong \dfrac{p}{p_0}(1 - \epsilon \cdot (\tau - \tau_0))$, with $\epsilon = 0.0036438$, and with β, β_0 measured in (C°), and p, p_0 in (mb).

Finally, substituting (16) in (15) yields:

$$D_{T,P} = \frac{1}{\sin 1'} \cdot \left(\frac{2h}{R} - 2 \cdot c \cdot \rho_0 \left(1 - \frac{p}{p_0}(1 - \varepsilon \cdot (\tau - \tau_0))\right)\right)^{\frac{1}{2}}$$

$$= D_0 \cdot \left(1 - \frac{2c\rho_0}{p_0 \sin^2 D_0}(p_0 - p(1 - \varepsilon(\tau - \tau_0)))\right)^{\frac{1}{2}}$$

where "D_0" is defined by (12).

Hence:

(17) $D_{T,P} = D_0 \cdot \left(1 - \frac{1}{D_0^2} \cdot \frac{2}{p_0 \cdot \varepsilon} \cdot C[p_0 - p(1 - \varepsilon[\tau - \tau_0])]\right)^{\frac{1}{2}}$, where $C = \frac{c \cdot \rho_0 \cdot \varepsilon}{\sin^2 1'} =,$

$= 12.48823085 = 12.49.$

The assumption that has been implicitly made about the density is: $c \cdot \rho = 0.00029$, $\varepsilon = 0.0036438$, $\tau, \tau_0 (C°)$, p, p_0 (mb), $D_{T,P}$ (arc min)

(18) $1 \cdot \frac{p}{p_0} > 1 - \frac{h}{c\rho_0 R}$, i.e., $\frac{p}{1+\varepsilon(\tau-\tau_0)} > p_0 - \frac{p_0 \cdot h}{c \cdot \rho_0 \cdot R}$

Any attempt to linearize the approximation (17) would require the additional assumption that the temperature differential $\tau - \tau_0$ would have to be of the order 10^{-4}, which is not a realistic hypothesis.

In actual applications either at sea or in the air, the skipper of the vessel hardly has the means to determine the pressure and temperature gradients sufficiently accurately and therefore the formula (17) is of very limited value to the navigator. However, it is of great value for the purpose of evaluating the error committed by using formulae (1), (10) and (12), above.

In conclusion, it can be said that similar to the results obtained in the previous section on refraction, the results for dip are also subject to further developments and can only be viewed as approximations. Therefore, the navigator will still have to depend on tables that are based on experimental data to determine dip under extreme, non-standard atmospheric conditions.

Because of some of the uncertainty in determining dip in general, other means and techniques have been conceived to eliminate dip all together as part of measuring the altitude of COs. Here are the five most popular ones:

Method One	Use a stable platform and a plumb away from big mountains.
Comment	This method works fairly well for the astronomer but is unsuitable for the navigator.
Method Two	This method employs an artificial horizon that consists of a tray that is filled with mercury, heavy oil, or water.
Comment	When used with shades and roof, altitudes of more than 15° can be measured fairly accurately, provided that the surface of the liquid remains stable. Again, this method is of very limited use since a relatively stable platform is hardly available at sea or in the air.

3.2 The Dip of the Horizon as a Function of Temperature and Pressure

Method Three Consists in employing an artificial horizon that incorporates a water-level spirit and can be mounted directly on a conventional marine sextant.

Comment Since the water-spirit level (bubble) is subject to rapid changes due to accelerated motions of the vessel, like rolling and pitching. Great care must be exercised when sights are being taken on a moving vessel and then only by taking a sequence of successive sights. The average of those sights will then represent a suitable approximation
(I have personally used an artificial horizon on two of my sextants with a degree of success and enjoy the freedom of being able to take sights at any time of night.)

Method Four This method consists in employing a special type of sextant that incorporates a bubble-spirit level device that is generally known as a Bubble Sextant. This sextant is the first choice for most air-navigators, but can also be adapted for use at sea. Some of the military-type bubble sextants have even incorporated an averaging device.

Comment Again, what has been said about the artificial horizon attachments to a standard marine sextant also applies to the Bubble Sextant
(I own a World War II British Mark IX and have been quite satisfied with the overall results subject to a prudent application of this instrument.)
One other type of Bubble Sextant that also utilizes the sea horizon is the U.S. Navy Mark V aircraft sextant that has all the features of a true marine sextant.

In addition to the errors introduced due to rolling, pitching and yawing, there occurs an additional error in Air-Navigation due to the relative high speed of the aircraft relative to a ship moving at sea. This error is due to the reflection of the vertical in a Bubble Sextant and is caused by the CORIOLIS FORCE that displaces all moving objects on the rotating Earth to the right of their paths if the moving object is in the northern hemisphere and to the left if it is in the southern hemisphere. The force expressed in vector form is:

(19) $\vec{c} = -m(\vec{\omega} \times \vec{v})$. Here "$\vec{\omega}$" denotes the vector that coincides with the direction of the Earth axis and which is in length equal to the angular velocity of the Earth. The vector "$\vec{v}$" denotes the velocity of the moving object. If one applies this force to the hydrostatics of the bubble inside the Bubble Sextant, the deflection Z_c of the vertical is found to be: [44]

(20) $Z_c = 2.62 \cdot v \cdot \sin \varphi + 0.146 \cdot v^2 \cdot \sin C \cdot \tan \varphi - 5.25 \frac{\Delta C}{\Delta t}$ and the observed Bubble Sextant altitude must be corrected by:

(21) $\Delta Z_C = Z_C \cdot \sin(A_z - C)$ [2]

The parameters in the last two expressions are:

Z_c Deflection of the vertical (arc min);
v $= |\vec{v}|$ of the aircraft (knots $\cdot\ 10^{-2}$);
φ Latitude of the observer (°);
C   Track or course angle (°);
$\frac{\Delta C}{\Delta t}$ Change of track/course angle (°/min);
ΔZ_c Altitude correction (arc min);
A_z Azimuth angle of observed CO (°).

For low speed aircraft only the first term of (20) is applied.

With a lot of practice and prudent application of averaging procedures, satisfactory results can be obtained in Air and Marine navigation. If used in small airplanes, those acceleration forces like rolling, pitching, and yawing can be as forceful as in the case of a sailboat.

Method Five   This method used by some sailors (and by me on occasion) utilizes reflectors—reflecting tape or similar material—that are strategically mounted on the stanchions of the vessel. Once the navigator determines the sea horizon under ideal meteorological conditions, he or she can determine the exact position of the sextant so that the line of sight of the reflected light from the reflectors coincides with the actual line of sight of the natural horizon. A more sophisticated technique consists in utilizing a laser beam as used in modern type of sprit water levels.

Comment   The major handicap of this method consists in aligning the vessel so that a particular star can be aligned so that the star, reflector, and observer lie on the same plane that also passes through the center of the Earth. If all the other measurements are correct, a fairly high degree of accuracy can be obtained.

Appendix

The reader will also find reference to the co-called dip of a shoreline or waterline of another vessel D_s in relevant literature. however, this type of dip has only limited usefulness since it requires accurate distances that, in the case of landmasses, can be ascertained by accurate charts or distance measuring devices. Nevertheless, for the sake of completeness, I have included the relevant formula below. The derivation follows directly from the application of plane trigonometry to the triangles in Fig. 3.2.1.

Denoting the distance of the observer from the shoreline or waterline of the other vessel by "d" in nautical miles, and the height of the observer's eyes above sea level by "h" in feet, the shore-ship-line dip in arc minutes is:

(22) $D_s = 0.565 \cdot \frac{h}{d} + 0.423 \cdot d.$ [1]

3.3 Planetary Parallax and Semi-diameter of the Sun and Moon

In what now follows, it is necessary to distinguish between what Navigators observe and what Astronomer, who provide the necessary data for determining the position of a CO, employ in their calculations. Both use their eyes and some kind of optical instrument. Navigators use sextants and astronomers use telescopes. Navigators determine the "Visual Position" (VP) of a CO; astronomers employ the "Apparent Position" (AP) in their calculations.

Next, we need to understand what is meant by "Apparent Position" of a CO. Since, for the purpose of positional astronomy, it is necessary to use only geocentric distances, it is necessary to reduce all distances to actual distances between the center of the Earth and the center of the CO, and also to reduce the measured altitude h_s to the actual altitude h with reference to an horizon parallel to the true horizon and passing through the center of the Earth. Furthermore, all these measurements have to be independent of the atmosphere and possible instrument errors. All this translates into requiring that the observer is to be positioned in the center of the earth and the surrounding space is to be void of an atmosphere. In addition, the observer's instruments are assumed to be free of errors.

Accordingly, the Apparent Position of a CO is defined as the position determined by an observer at the center of the Earth with reference to the center of the observed CO and void of atmospheric refraction, dip and instrument error.

It follows then that in order to reduce the visual position, determined by the observer to the apparent position, it is necessary to add some corrections to the visual, i.e., sextant altitude h_s. These corrections are:

(i) Instrument Error $\pm I$ (see Sect. 4.1)
(ii) Dip $-D$ (see Sect. 3.2)
(iii) Refraction $-R$ (see Sect. 3.1)
(iv) Parallax $+PA$ (see Sect. 3.3)
(v) Semi-diameter $\pm S$ (see Sect. 3.3)
(vi) Coriolis Correction $+Z_c$ (see Sect. 3.3)—For Air-Navigation only.

Hence:

(vii) $h = h_s \pm I - R - D + PA \pm S$.

In this section, I will derive exact and approximate formulae for calculating PA and S as well as for the augmentation of the semi-diameter S' of the Moon.

Until now, we have always assumed that the shape of the Earth is a sphere, but in order to develop accurate formulae for the planetary parallaxes, it is necessary to take into account that the approximate figure of the Earth is the GEOID and for our purpose it is sufficient to approximate its shape by a SPHEROID of revolution (see Fig. 3.3.1). [26, 29, 31, 41]

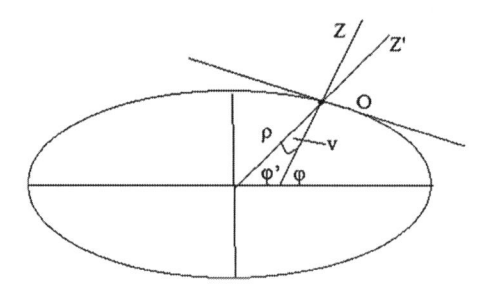

Fig. 3.3.1

It can be shown that:

(1) $v = \varphi - \varphi' \cong 695''.66 \cdot \sin 2\varphi - 1''.17 \cdot \sin 4\varphi + \cdots$ [1]

This expression for v is called the "Equation of the Vertical". Similarly, ρ—the distance of the observer O from the center of the spheroid can be found to be:

(2) $\rho \cong (0.99832005 + 0.00168349 \cdot \cos 2\varphi - 0.00000355 \cdot \cos 4\varphi) \cdot a$.

The quantities "v" and "ρ" are required in the formulae for the parallax p. In order to derive these formulae, we proceed as follows: (see Fig. 3.3.2)

First let us define the parallax PA with reference to the Earth conceived as a sphere. Then according to Fig. 3.3.2, we have:

(3) $\sin PA = \frac{a}{r_0} \cdot \sin \zeta = \frac{a}{r_0} \cdot \cos h_s = \sin HP \cdot \cos h_s$, i.e., $PA = HP \cdot \cos h_s$.

the equatorial horizontal parallax HP per definition corresponds to $h_0 = 0$, and is:

(4) $HP = \frac{a}{\sin 1''} \cdot \frac{1}{r_0}$, per expression (3) where "r_0" denotes the geocentric distance of P from the center of the Earth. If we define "$\bar{r}_0$" to be the "Mean Distance" of P from the center of the Earth, we find that the "Mean Equatorial Horizontal Parallax" ($\overline{HP}$) to be:

(4') $\overline{HP} = \frac{a}{\sin 1''} \cdot \frac{1}{\bar{r}_0}$. In the case of the Moon, for instance, $\bar{r}_0 = 60.2682 \cdot a$ km $= 60.2682 \cdot 6.37816$ km, i.e., 60.2682 times the radius of the Earth.

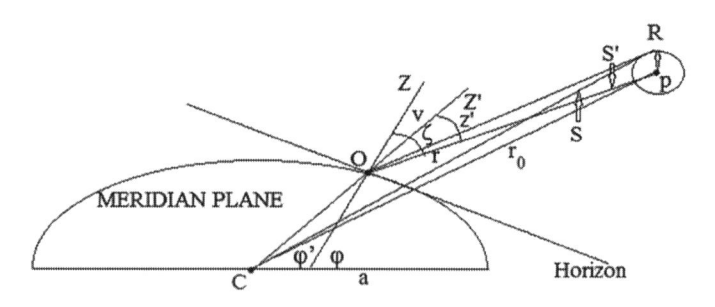

Fig. 3.3.2

3.3 Planetary Parallax and Semi-diameter of the Sun and Moon

$\overline{HP}_D = \frac{1}{\sin 1''} \cdot \frac{1}{60.2682} = 3''.42255 \; 10^3 = 57'.04.$

Next let us now take the true shape of the Earth into account. Then according to Fig. 3.3.2, for $A_z = 0$, i.e., in the meridian plane, the triangle POC permits the application of the SIN-TH. of plane trigonometry, resulting in:

$$\frac{\sin p}{\rho} = \frac{\sin z'}{r_0} = \frac{\sin(\zeta - v)}{r_0} = \frac{1}{a} \cdot \frac{a}{r_0} \sin(\zeta - v) = \frac{1}{a} \sin HP \cdot \sin(\zeta - v)$$

i.e., $\sin p = \frac{\rho}{a} \cdot \sin HP \cdot \sin(\zeta - v)$, or explicitly:

(5) $p = \frac{\rho}{a} \cdot HP \cdot \sin(\zeta - v) \cong HP \cdot \sin(\zeta - v)$, with v and ζ known because of (1) and (2).

For the purpose of navigation it is customary to express "p" in terms PA. Therefore, an additional approximation is to be applied in order to arrive at a simple formula. Taking into consideration that "v" is a relatively small quantity relative to ζ, we may employ the Taylor Expansion:

$\sin(\zeta - v) = \sin \zeta - \cos \zeta \cdot v \cdot \sin 1'' - \sin \zeta \cdot \frac{v^2}{2} \cdot \sin^2 1'' + \cdots$ to obtain:

$p = \frac{\rho}{a} \cdot HP \cdot \sin(\zeta - v) \cong \frac{\rho}{a} HP \cdot \sin(\zeta - \cos \zeta \cdot v \sin 1'' - \sin \zeta \frac{v^2}{2} \cdot \sin^2 1'')$

$\cong HP \cdot \cos h_s - (\sin h_s \cdot v \cdot \sin 1'' + \cos h_s \cdot \frac{v^2}{2} \cdot \sin^2 1'') \cdot HP$ since $\frac{\rho}{a} = 1$.

If we now define the correction OB to be:

(6) $OB = -(\sin h_s \cdot v \cdot \sin 1'' + \cos h''_s \frac{v^2}{2} \sin^2 1'') \cdot HP$, and approximate "v" according to expression (1) by $v = 695''.66 \sin 2\varphi$, we find that a suitable approximation for OB is:

(7) $\begin{aligned} OB &\cong -0.00337265485 \cdot \sin h_s \cdot \sin 2\varphi \cdot HP \\ p &\cong HP \cdot \cos h_s + OB \end{aligned}$

Next, let's consider a numerical example to demonstrate that the required correction (OB) is, indeed, extremely small in the case of the Meridian Parallax.[4]

Numerical Example

CO: Moon
Latitude of Observer $\varphi = 45°$
Sextant altitude $h_s = 33°$

Problem: Find the correction accounting for the oblateness of the Earth.
Solution:

First we approximate ρ by $\bar{\rho}$, the mean distance Earth-Moon and deduce that $HP_D \cong \overline{HP}_D = 57'.04$. substituting those values into the first part of expression (7) yields

---

[4]See appendix to this section.

$$\text{OB} = -0.00337265485 \cdot \sin 33 \cdot \sin 90 \cdot 57'.04 = 0'.104775605.$$ Of particular interest to the navigator is the horizontal parallax of the Sun which has been determined by astronomical observations to be: [30]

(8) $\text{HP}_\odot = \frac{a}{R_\odot} \frac{1}{\sin 1''} = 8''.794$, and $R_\odot = 1\,\text{a.u.} = 149{,}597{,}900\,\text{km}$.

If we express the geocentric distances of any of the other planets in astronomical units, i.e., if we put $r_0 = d \cdot R_\odot$ and also employ expression (8), we find that:

(9) $\text{HP} = \text{HP}_\odot \frac{1}{d} = \frac{8''.794}{d}$.

It follows then that $\text{HP} < 1'$ holds for all the navigational planets. For example, in the case "s" of **Venus** and **Mars**, HP varies between $0'.1$ and $0'.2$.

Next we still have to find formulae for computing the semi-diameters of the Sun, Moon and the Navigational Planets in order to find the apparent position of these COs. In other words, we have to find the remaining corrections to the sextant altitude h_s.

According to Fig. 3.3.2, we have:

(10) $\sin S = \frac{R}{r_0} = \frac{R}{a} \cdot \sin \text{HP}$, or $S = \frac{R}{a} \cdot \text{HP}$. Similarly, we deduce:

(11) $\sin S' = \frac{R}{r} = \frac{R}{a} \sin \text{HP} \frac{r_0}{r} = \frac{r_0}{r} \cdot \sin S$, or $S' = S \cdot \frac{r_0}{r}$.

In the above formulae, "R" stands for the semi-diameter (radius) of the Planet or Moon; and "S'" is commonly referred to as the augmented semi-diameter; and "$\frac{r_0}{r}$" as the augmented factor. This factor is known whenever "z" and "z'" are known. However only $z' = \zeta - v$ can be readily found by employing the Equation of the Vertical (1). The evaluation of "z" is a little more complicated and for our purpose, we will only approximate "z" in order to derive a fairly simple approximation for the augmented semi-diameter.

Again, let us look at the triangle POC (see Fig. 3.3.2) and apply the SIN-TH to it obtaining:

(12) $\frac{\sin z'}{r_0} = \frac{\sin z}{r}$ or $\frac{r_0}{r} = \frac{\sin z'}{\sin z}$, which when substituted into (11) yields:

(13) $S' - S \cdot \frac{\sin z'}{\sin z} - S \cdot \frac{1}{\sin z / \sin z'}$. In order to find a suitable approximation for "z", we assume that the augmentation take place in the Meridian Plane, i.e., for $A'_z = 0$ (see appendix to this section). We then readily deduce from the plane triangle POC that $z + 180° - z' + PA = 180°$ and hence:

(14) $z \cong z' - \text{HP} \cdot \sin z'$. By employing this approximation, we deduce that:

$$\frac{\sin z}{\sin z'} \cong \frac{(\sin z' \cdot \cos(\text{HP} \cdot \sin z') - \cos z' \cdot \sin(\text{HP} \cdot \sin z'))}{\sin z'}$$

$\rho\, 1 - \cos z' \cdot \text{HP} \cdot \sin 1'' \cong 1 - \frac{\sin h_s}{d}$, with $d = \frac{r_0}{a}$.

3.3 Planetary Parallax and Semi-diameter of the Sun and Moon

Then, it follows that:

$$\frac{1}{\sin z / \sin z'} \cong 1 + \frac{\sin h_s}{d},$$ and therefore, by substituting in Eq. (13), we obtain the desired result:

(15) $S' \cong S \cdot (1 + \frac{\sin h_s}{d})$. This formula can also be found in <u>The Almanac For Computers</u>, 1987, Nautical Almanac Office, United States Naval Observatory, Washington, D.C. [44]

Note that in the case of the Moon $S_D = 0.272476 \cdot HP$.

Appendix

It should be clearly understood that the formulae developed so far for the Earth as a figure of a spheroid are based on the assumption that the planet, Sun or Moon is observed on the Observer's meridian. [26]

However, in general cases where those COs are observed at any instant when their azimuths are different from zero and 180°, the underlying geometry is quite different since the position of the CO no longer lies in the plane of the meridian. Therefore, in this case, the geometry is now three-dimensional (see Fig. 3.3.3).

Although in principle not difficult to treat, from the mathematical point of view it is nevertheless very tedious to derive the corresponding equations and therefore, I prefer to omit that part of the mathematical derivations which have already been known since the times of Olbers and Chauvenet. [13]

Basically the three-dimensional case involves that we introduce two different coordinate systems, one at the point 0 of the observer, and the other at the center C of the Earth. Then by setting up the equation that transforms one coordinate system into the other, and by performing the tedious task of eliminating the unknowns by employing various approximations, we then arrive at the following two equations which enable us to find the correction OB to be applied to the parallax PA:

(16) $\sin(\zeta - \zeta') = \frac{\rho}{a} \cdot \sin HP \cdot \sin(\zeta - \gamma)$, and

(17) $\gamma = (\varphi - \varphi') \cdot \cos Z_n = v \cdot \cos Z_n$, where "$Z_n$" denotes the azimuth and "ζ" denotes the zenith distance as observed at 0. Since $p = \zeta - \zeta'$ is the

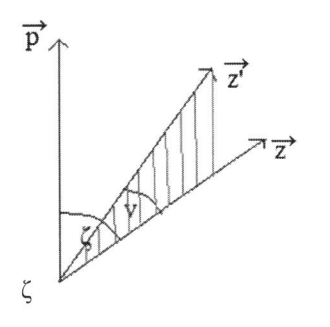

$\vec{z}$: Unit vector in direction of the astronomical zenith Z.

$\vec{z'}$: Unit vector in direction of the geocentric zenith Z'.

$\vec{p}$: Unit vector in direction of the CO.

Fig. 3.3.3

correction due to parallax that has to be added to the sextant altitude (h_s) in order to find the apparent altitude (h), we merely have to develop a formula for expressing this correction p in terms of PA and a term OB that is to be added to PA. But this follows exactly the same steps that have already been employed in deriving formulae (6) and (7) from expression (5).

In the case of expressions (16) and (17), the resulting approximations are:

(18) $P = \frac{\rho}{a} \cdot HP \cdot \cos h_s + OB$, where

(19) $OB = -\frac{\rho}{a} \cdot HP \cdot (695''.66 \sin 1'' \sin h_s \sin 2\varphi \cdot \cos Z_n$
$+ 695''.66^2 \cdot \sin^2 1'' \frac{1}{2} \cos h_s \sin^2 2\varphi \cdot \cos^2 Z_n)$

Since $\frac{\rho}{a} \cong 1$, we obtain the simplified approximation:

(18′) $P \cong HP \cdot \cos h_s + OB$, where

(19′) $OB = -0.00337265485 \sin h_s \cdot \sin 2\varphi \cdot \cos Z_n \cdot HP$
$- 0.568740038 \cdot 10^{-5} \cdot \cos h_s \cdot \sin^2 2\varphi \cdot \cos^2 Z_n \cdot HP$

In most practical applications, the second term can be neglected and thereby reducing (19′) to:

(20) $OB = -0.00337265485 \cdot \sin h_s \cdot \sin 2\varphi \cdot \cos Z_n \cdot HP$

Of course, formulae (18)–(20) have to coincide with the formulae (6) and (7) if we put $Z_n = 0$, which again confirms the correctness of our derivations. Furthermore, the results obtained by using the exact formula (16) differ only insignificantly from the results of the approximations (18′) and (20).

Advisory

The reader will notice that in all the previous sections I have used the symbols "a", "h_0", and "H_0" to denote the apparent altitude h, and I have also used the terms "true altitude" and "observed altitude" synonymously. Therefore, unless it is not the sextant altitude h_s, it is always meant to be the apparent altitude h (see also Algorithm D).

3.4 Time and Timekeeping

The concept of time is based on our brains ability to differentiate between the before and the after of any particular event. Prehistoric men had already used the diurnal movement of the Sun (Solar Time), and the stars (Sidereal Time), to develop a more rational concept of time based on the rotation of the Earth about its axis. The forerunners of our clocks were the Nocturnal of the sky and the Sundial. [37, 46]

With the advent of the invention of mechanical clocks and Chronometers, timepieces became available to the Astronomers and later to the ocean-going

3.4 Time and Timekeeping

Navigators. The basics of measuring time have changed very little since the 1600s. The rotation of the Earth is still used as the "base line" by the majority of current timepieces. The primary reason for this is that the angular velocity of the Earth's rotation about its axis was perceived to be of a constant value. It was not until the advent of quartz and Atomic clocks that it became obvious that the rotation velocity of the Earth was by no means invariant.

From a theoretical point of view, this was by no means a new discovery. During the last century, it was also established that the concept of time as conceived intuitively was no longer a physical reality. According to the Theory of Relativity, time is a relative concept depending on the frame of reference, i.e., NO MOTION = NO TIME.

However, even prior to the invention of the Atomic clock, a more accurate method for determining time had already been conceived. This method was based on the same concept of time as used in Celestial Mechanics and, in particular, on the dynamical theory of the motion of celestial objects such as the Moon. The motion of COs provide a more uniform concept of time and has been referred to as EPHEMERIS TIME (ET).

In this context, the reader is also reminded of the concept of COMPUTED TIME (as conceived by me) and applied to find the position of an observer without prior knowledge of an approximate position or an approximate time and without a proper timepiece. The reader may also recall that aside from having this manual, a calculator, sextant and ephemeris of the Moon or Nautical Almanac, nothing else was required to find the observer's position by computation (see Sect. 2.21).

Next, let us define the aforementioned concepts of time by beginning with the Sidereal Year.

Definitions

I. A SIDEREAL YEAR (SY) is equal to the time required by the Sun to complete a circuit of the ecliptic.
II. A TROPICAL YEAR (TY) is the average time required by the Sun to make two consecutive passages through the moving equinox.

Since the precessional motion of the equinox is $50''.3$ per tropical year, it follows that:
$$\frac{SY}{TY} = \frac{360°}{(360° - 53''.3)}, \text{ and}$$
(1) SY = 365.2564 MSD, TY = 365.2422 MSD.
 MSD: Mean Solar Day of Universal Time. Here I am using the approximation MSD ρ ED (Ephemeris Day).
III. SIDEREAL APPARENT TIME (SAT) is equal the local hour angle of the true equinox expressed in hours. Therefore, it is variable due to the nutation of the Earth and it is quantitatively expressed by the Equation of the Equinoxes (EE).
IV. SIDEREAL MEAN TIME (SMT) is equal the local hour angle of the mean equinox, i.e., of Aries (♈) expressed in time. Therefore, it is regulated by the

constant effect of precession on the Earth and it can be approximated by a regulated SIDEREAL CLOCK as used by Astronomers. Hence:

(2) SAT = SMT + EE.

V. APPARENT SOLAR TIME (AST) is equal to the local hour angle expressed in time plus 12^h of the true Sun.

Because the Earth orbits about the Sun, the orbital velocity of the gyro-Earth varies according to Kepler's Second Law. Therefore, the length of a solar day is variable, up to about ±15 min per month, making the solar day unsuitable for time measurements.

VI. MEAN SOLAR TIME (MST) is equal to the local hour angle expressed in hours plus 12^h of a fictitious mean sun that travels along the equator at a constant rate, terminating a complete revolution at the same time as the true Sun, implying that the RA of this fictitious sun is Equal to the Sun's mean longitude. The MST, therefore, can be determined by a suitable timepiece such as a Quartz watch, Atomic clock, or a Chronometer.

The AST differs from the MST due to the eccentricity of the Earth's orbit and can be expressed by the Equation of Time (ET) that is defined as the difference between the hour angle of the True Sun and the Mean Sun, i.e., ET = (HA ☉ − HA $\overline{\odot}$/15). Hence:

(3) AST = MST + ET.

VII. UNIVERSAL TIME (UT) is equal to the Greenwich hour angle (expressed in time) of the fictitious mean sun that moves around the mean equator but with a rate that is directly proportional to the angular velocity of the Earth and therefore is non-uniform. Hence, it can only be approximated with a timepiece.

Before it was discovered that the angular velocity of the Earth is non-uniform, Universal and Greenwich Mean Time (GMT) were the same thing.

VIII. LOCAL MEAN TIME (LMT) is equal to the local hour angle of the mean sun expressed in hours, plus 12^h. Hence:

(4) $\text{LMT} = \text{GMT} + \frac{\lambda}{15}$, $\lambda \geq 0$ if East and $\lambda \leq 0$ if West.

IX. COORDINATED UNIVERSAL TIME (UTC) is time based on the Atomic Time Scale and is broadcasted by time signals (TS). For navigational purposes, it is being used as an approximation to UT and GMT. It has been established that $|\text{UTC} - \text{UT}| \leq 0^s.9$

It merely remains to explain how the different concepts of time can be related quantitatively. From the very definitions of these and their magnitudes, it follows that: GMT $\cong$ UT; UT = UTC + ΔUT $\cong$ UTC; and ET = UT + ΔE. However, the extremely small quantities ΔUT and ΔE that only play an important role in astronomy cannot be predicted precisely. For the navigator who relies chiefly on a chronometer or watch that shows GMT, it suffices to know that UT $\cong$ UTC = GMT + ΔTS, where ΔTS is being broadcasted.

In order to convert GMST into GMT or conversely, the navigator must either rely on the NA or use the formula provided in the second part of this book to calculate GMST = GHAγ ♈ /15 = GMT + RA☉ − 12^h = GMT − 12 + RA ☉ + ET − 12^h.

3.4 Time and Timekeeping

Here I will derive a simple formula that enables the navigator to convert GMT to GSMT for the most relevant interval of 24^h, thereby avoiding the process of looking up the GHA of Aries ♈ more than once every 24-h.

Recall that once every year, the mean sun and the equinox coincide and after every tropical year, this happens again. during this period, the Earth has rotated about its axis 365.2422 times relative to the mean sun and once more with regards to the mean equinox. Therefore:

365.2422 MSD = 366.2422 SMD, i.e.,

(5) $\quad 24^h \text{GMT} = (1 + \frac{1}{365.2422}) \cdot 24^h \text{SMT} = 24^h + 3^m 56^s.556 \text{ SMT}$

(5') $\quad 24^h \text{SMT} = (1 - \frac{1}{366.2422}) \cdot 24^h \text{GMT} = 24^h - 3^m 55^s.910 \text{ GMT}$

From these relations, we can readily deduce that:

(6) $\quad 1^h \text{GMT} = 1^h + 9^s.8565 \text{ SMT}, \quad 1^h \text{SMT} = 1^h - 9^s.8296 \text{ GMT}$

$\quad\quad 1^m \text{GMT} = 1^m + 0^s.1643 \text{ SMT}, \quad 1^m \text{SMT} = 1^m - 0^s.1638 \text{ GMT}$

$\quad\quad 1^s \text{GMT} = 1^s + 0^s.0027 \text{ SMT}, \quad 1^s \text{SMT} = 1^s - 0^s.0027 \text{ GMT}$

With reference to the date or day of the year, we still depend on our CIVIL CALENDAR since the dates used in the NA correspond to this standard measure. However, as we will discover in the second part of this book, the civil calendar is unsuitable for astronomical purposes and it will be replaced by the day count know as JULIAN DATE.

Now, let's turn our attention to the physical task of time-keeping or monitoring time. To keep track of accurate time, highly accurate clocks or chronometers are employed on board ocean-going vessels. With the advent of Quartz watches the accuracy and reliability of timepieces has markedly increased and the prices have been reduced drastically. Some of those watches and calculators are SOLAR POWERED and the more sophisticated ones are automatically regulated by broadcast signals that are synchronized with the Atomic Clock. They are aptly known as "Atomic Watches".

Once a new watch has been acquired, its "rate", i.e., the accumulated error, has to be determined by the navigator. My personal suggestion is for navigators to use the signals broadcast worldwide as points of reference. For that, you would need a simple SINGLE SIDEBAND RECEIVER. As regards to watches, I would also suggest that you never leave the dock even on the smallest sea-going vessel with anything less than THREE WATCHES—(one wrist watch worn by the navigator, one stationary timepiece mounted inside the vessel, and one emergency watch tucked away in the emergency navigation kit.) This is independent of the stationary equipment on the vessel which may be highly sophisticated.

By now, it should be obvious that the most important tools for the practical navigator are the watch, the calculator, compass, and sextant. By carrying these basic navigational gear on board a vessel, the probability of losing time completely is less than the probability of losing the entire vessel. But if you do lose time completely, you can always refer to (Sect. 2.21) of this manual and re-set your clocks by using my concept of COMPUTED TIME.

3.5 On the Minimization Procedure for the Random Errors in Determining Altitude and Time

In Sect. 2.4, I developed the exact formulae for finding a position at sea or in the air and referred to it as the "exact" solution. However, this is only true provided that ALL parameters are accurate, which is seldom the case. The latter becomes obvious if the navigator takes more than one set of observations into consideration and consequently arrives at several distinct positions. It should be obvious that the most probable position so obtained is the CENTROIDAL POINT of the set of individual positions.

One of the crucial parameters in those exact equations is the Apparent Altitude (AA) that is found by computation only and is given by: $h = h_s \pm I - D - R \pm S + P$.

As previously noted, I have used the notations H_0, h_0, and H in previous sections and also terms like "observed altitude", but actually always meant the same thing, namely, APPARENT ALTITUDE.

In the above expression "h_s" denotes the measured sextant altitude that is subject to RANDOM ERRORS that may vary considerably depending on the experience of the navigator, sea and weather conditions, and the motion of the vessel. Therefore, it is imperative to minimize these types of random errors by devising methods for calculating and assessing these error. In order to do this, we must first define the underlying quantities that are being used.

Definitions

I. STANDARD ERROR (σ) of a set or measurements $\{x_i\}$ indicates the precision or the measurements relative to any other set of measurements.

(I') $\sigma = \sqrt{\frac{\sum_{i=1}^{n} v_i^2}{n-1}}$, where the deviations are defined as $v_i = x_i - \bar{x}$, with $\bar{x} = \frac{\sum_{i=1}^{n} x_i}{n}$.

II. PROBABLE ERROR (E) of any set of measurements $\{x_i\}$ and given by:

II' $E = 0.67456 \cdot \sigma$.

III. STANDARD ERROR OF THE MEAN of a set of like measurements σ_M:

III' $\sigma_M = \frac{\sigma}{\sqrt{n}} = \sqrt{\frac{\sum_{i=1}^{n} v_i^2}{n(n-1)}}$

IV. PROBABLEL ERROR OF THE MEAN defined as:

IV'. $E_0 = 0.67456 \cdot \sigma_M$.

V. ACCURACY is the closeness of the measurements of observations to the true value of the quantities being measured.

VI. PRECISION refers to the closeness with which the measurements agree with each other.

These definitions can be used to carry out an error analysis of the random errors that are always present in the measurements of sextant altitudes and universal time.

3.5 On the Minimization Procedure for the Random Errors ...

In order to understand the significance of standard and probable errors, it is necessary to employ some of the basic concepts of the Theory of Probability.

For the sake of brevity, let's start by referring to the curve that can be obtained by plotting the frequency of the occurrence p(v) of an error the size of v. This curve is also referred to as the "Normal Distribution Curve" or "Gauss Curve", and the function:

$p(v) = \frac{1}{\sigma\sqrt{\pi}} \cdot e^{-\frac{1}{2\sigma^2} \cdot v^2}$ (see Fig. 3.5.1) is called the "Probability Density Curve" satisfying: $\int_{+\infty}^{+\infty} p(v) dv = 1$.

Then the function $F(x) = \int_{-x}^{+x} p(v) dv$ (see Fig. 3.5.2) represents the probability that an error e is less than or equal to x in magnitude, i.e., $|e| \leq x$. In other words, the quantity $F(x) \cdot 10^2$ expressed in percentages are the errors which are less than or equal to x in magnitude. It is now easy to see what the significance of σ, σ_M, E and E_0 are. In particular one can easily verify the following results:

(i) The standard error corresponds to the probability of 0.6828, i.e., 68% of all errors |e| are less or equal to σ.
(ii) E, the probable error, corresponds to the probability of 0.5, i.e., 50 % of all the errors are less than E (see Fig. 3.5.1). [13, 19, 30]

With the help of the above definitions and explications, we can readily assess the precision of a one-dimensional set of observations. However, in our case where we are dealing with a two-dimensional set of observations—observed altitude and observed time—, the underlying mathematics is different. In our particular situation, we can either reduce the two-dimensional problem to an one-dimensional one, or we can introduce another measure for precision and accuracy. In either case, we will require the concept of REGRESSION and, in particular, the concept of LINEAR REGRESSION that can be applied to all situations where it is assumed that there exists a linear dependency of the two variables x and y of the given set $\{x_i, y_i\}$; in our particular case, time and altitude. At this point we need to make use of the approximation derived in Sect. 2.9, formula (4) that reads, if z is expressed in terms of h_s:

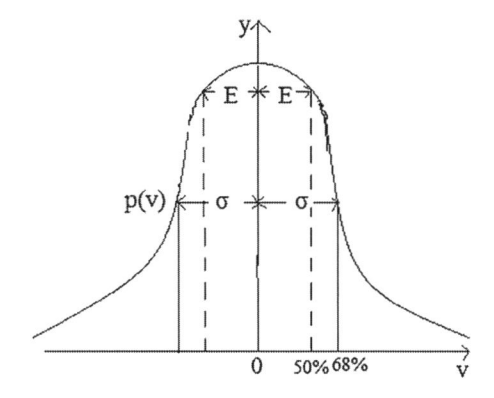

Fig. 3.5.1

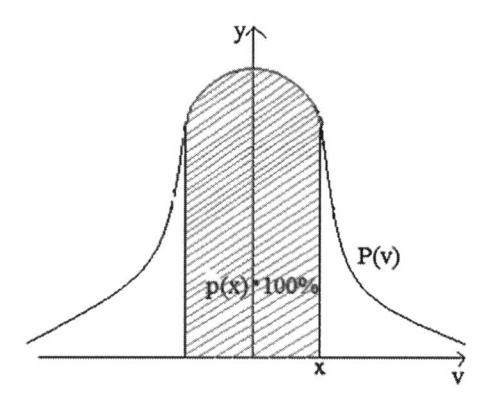

Fig. 3.5.2

(1) $h_s(t) = h_s(t_0) + 15 \cdot \cos \varphi \cdot \sin Z_n \cdot (t - t_0)$, for all $|t - t_0| \leq 5^m$.

This is the equation to use when assessing accuracy.

For now, let's assume that we are given a set of n pairs of measurements $\{x_i, y_i\}$, $i = 1, \ldots, n$ and let us assume that there exists a linear dependency between x and y. Then according to the Theory of Probability, the most probable straight line representing this dependence is the line given by the expression:

(2) $L^r : y^r = a + b \cdot x$ and subject to the condition that the sum of the squares of the errors between (2) and the points (x_i, y_i) assumes a minimum. Therefore, we must determine a and b in (2) from the given set $\{x_i, y_i\}$ so that:

$$P(a,b) = \sum_{i=1}^{n}(y_i - y_i^r)^2 = \sum_{i=1}^{n}(y_i - (a + b \cdot x_i))^2 =$$

$$= \sum_{i=1}^{n} y_i^2 - 2a \sum_{i=1}^{n} y_i - 2b \cdot \sum_{i=1}^{n} x_i \cdot y_i + 2ab \sum_{i=1}^{n} x_i + b^2 \sum_{i=1}^{n} x_i^2 + na^2$$

i.e., we have two equations to determine a and b, namely

(3) $\dfrac{\partial P}{\partial a} = -2n \cdot \overline{y} + 2na + 2nb \cdot \overline{x} = 0$, i.e.

(i) $a + b \cdot \overline{x} = \overline{y}$

$\dfrac{\partial P}{\partial b} = -2 \cdot \sum_{i=1}^{n} x_i \cdot y_i + 2an \cdot \overline{x} + 2b \sum_{i=1}^{n} x_i^2 = 0$, i.e.

(ii) $an \cdot \overline{x} + b \cdot \sum_{i=1}^{n} x_i^2 = \sum_{i=1}^{n} x_i \cdot y_i$, with $\overline{x} = \dfrac{\sum_{i=1}^{n} x_i}{n}$ and $\overline{y} = \dfrac{\sum_{i=1}^{n} y_i}{n}$

This linear system of equations for a and b has a unique solution:

(4) $a = \dfrac{\overline{y} \sum_{i=1}^{n} x_i^2 - \overline{x} \sum_{i=1}^{n} x_i y_i}{\sum_{i=1}^{n} x_i^2 - n \cdot \overline{x}^2}$ and $b = \dfrac{\sum_{i=1}^{n} x_i y_i - n \cdot \overline{x} \cdot \overline{y}}{\sum_{i=1}^{n} x_i^2 - n \cdot \overline{x}^2}$.

3.5 On the Minimization Procedure for the Random Errors ...

We also define the CENTROIDAL POINT C to be:

$C : (\bar{x}, \bar{y}) = \left(\frac{\sum_{i=1}^{n} x_i}{n}, \frac{\sum_{i=1}^{n} y_i}{n} \right)$, and conclude from Eq. (3)(i) that C lies on the line of regression (2).

Next we use the linear transformation, $X = x - \bar{x}$, $Y = y - \bar{y}$, to transform the centroidal point into the origin of our new system of coordinates (X, Y). In this new system, the equation of the line of regression (2) becomes:

(2′) $Y = b \cdot X$.

By straight forward calculations we find that: $\sum_{i=1}^{n} x_i y_i = \sum_{i=1}^{n} X_i \cdot Y_i + n \cdot \bar{x}\bar{y}$ and $\sum_{i=1}^{n} x_i^2 = \sum_{i=1}^{n} X_i^2 + n \cdot \bar{x}^2$. Substituting these expression into the second equation of (4), we find that:

(5) $b = \dfrac{\sum_{i=1}^{n} X_i \cdot Y_i}{\sum_{i=1}^{n} X_i^2} = \dfrac{\sum_{i=1}^{n} (x_i - \bar{x})(y_i - \bar{y})}{\sum_{i=1}^{n} (x_i - \bar{x})^2}$, the most important indicator for accuracy.

Suppose, now, that we know that the exact dependence of y on x can be expressed by the linear equation:

(6) $y^\alpha = \bar{y} + \alpha \cdot (x - \bar{x})$, α: parameter. Then, $y_i^\alpha = \bar{y} + \alpha(x_i - \bar{x})$ represents the exact value of y at x_i and the centroidal point C lies on (6) (see Fig. 3.5.3).

In Fig. 3.5.3, the set $\{x_i, y_i\}$ is scattered along L^r and L^α, just about evenly which indicates that precision and accuracy are about the same. But suppose we have a second set of values, $\{\tilde{x}_i, \tilde{y}_i\}$ very close to L^r but further away from L^α. In this case we have higher precision than accuracy.

Next let us consider the actual problem of measuring precision and accuracy as applied to navigation. Again, from the viewpoint of the Theory of Probability, the deviations of the points x_i, y_i from the line of regression L^r provides a suitable criteria for the precision if it can be quantified. Again, the measure to be used is the sum of the squares of all individual deviations from the corresponding points on the line of regression. Therefore, the actual measure to be employed is:

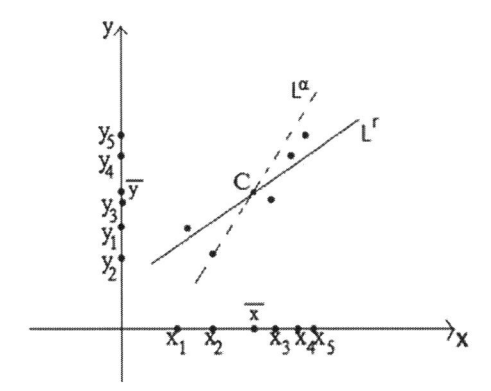

Fig. 3.5.3

$$\sigma_p^2 = \sum_{i=1}^{n} \frac{(y_i - y_i^r)^2}{(n-1)} = \sum_{i=1}^{n} \frac{[(y_i - \bar{y}) - (y_i^r - \bar{y})]^2}{(n-1)} =$$

$$\sum_{i=1}^{n} \frac{(y_i - \bar{y})^2}{(n-1)} + \sum_{i=1}^{n} \frac{(y_i^r - \bar{y})^2}{(n-1)} - 2 \cdot \sum_{i=1}^{n} \frac{(y_i - \bar{y})(y_i^r - \bar{y})}{(n-1)} =$$

$$\sigma_y^2 + b^2 \sigma_x^2 - 2 \cdot b \cdot \sum_{i=1}^{n} \frac{(y_i - \bar{y})(x_i - \bar{x})}{(n-1)}.$$

Because b is given by (5), we find that: $\sigma_p^2 = \sigma_y^2 - b^2 \sigma_x^2 \geq 0$.

The quantity:

(7) $\sigma_p = (\sigma_y^2 - b^2 \sigma_x^2)^{\frac{1}{2}}$, with b given by Eq. (5), and σ_x and σ_y defined as standard errors of the respective sets, is a suitable measure for the precision of a given set of observations.

In order to find a measure for the accuracy, we need to employ the exact or approximate linear dependence ... that is to say, we must compute: $\sigma_a^2 = \sum_{i=1}^{n} \frac{(y_i - y_i^\alpha)^2}{(n-1)} = \sigma_y^2 + \alpha \cdot \sigma_x^2(\alpha - 2 \cdot b)$. Minimizing this expression with respect to α yields again (7), i.e., $\sigma_\alpha(b) \geq \sigma_p$, and Min. $\sigma_a = \sigma_p$.

Hence the quantity:

(8) $\sigma_a = (\sigma_y^2 + \alpha \cdot \sigma_x^2(\alpha - 2 \cdot b))^{\frac{1}{2}}$ is the measure to be used or accuracy.

As I pointed out at the beginning of this section, one method for assessing the error consists in replacing the set of pairs of altitude and time by only a set of single values $\{z_i\}$. By doing this, you reduce the two dimensional problem to a one dimensional one. The latter can be accomplished by defining the reduced set $\{z_i\}$ as follows:

(9) $z_i = y_i - \alpha(x_i - \bar{x})$. Again (9) implies that $\bar{z} = \bar{y}$. Then it follows that
$z_i - \bar{y} = y_i - \bar{y} - (x_i - \bar{x})$ and hence
$(z_i - \bar{y})^2 = (y_i - \bar{y})^2 + (x_i - \bar{x})^2 - 2 \cdot (y_i - \bar{y})(x_i - \bar{x})$.

It follows that:

(10) $\sigma_0^2 = \sum_{i=1}^{n} \frac{(z_i - \sigma)^2}{(n-1)} = \sum_{i=1}^{n} \frac{(z_i - \bar{y})^2}{(n-1)} = \sigma_y^2 + \alpha \cdot \sigma_x^2(\alpha - 2 \cdot b) \geq 0$, and in particular that:

(11) $\sigma_0 = \sigma_a$ the simple relation between the accuracy of $\{x_i, y_i\}$ and the precision of $\{z_i\}$.

Again from Fig. 3.5.3, it is obvious that what really matters is how close the line of regression is to the lines of "exact" values L^α. This we can quantitatively assess by evaluating (5), i.e., by comparing b to α.

3.5 On the Minimization Procedure for the Random Errors ...

Finally, by applying the results of this section to navigation, we merely have to identify the following quantities:

(12)
$$\alpha = 15 \cdot \cos\varphi \cdot \sin Z_n, \quad \bar{t} = \frac{\sum_{i=1}^n t_i}{n}, \quad \bar{h}^s = \frac{\sum_{i=1}^n h_i^s}{n}, \quad C = (\bar{t}, \bar{h}^s)$$

$$\sigma_{h^s} = \left(\frac{\sum_{i=1}^n (h_i^s - \bar{h}^s)^2}{n-1}\right)^{\frac{1}{2}}, \quad \sigma_t = \left(\frac{\sum_{i=1}^n (t_i - \bar{t})^2}{n-1}\right)^{\frac{1}{2}},$$

$$\sigma_p = (\sigma_{h^s}^2 - b^2 \cdot \sigma_t^2)^{\frac{1}{2}}, \quad \sigma_a = (\sigma_{h^s}^2 + \alpha \cdot \sigma_t^2 (\alpha - 2 \cdot b))^{\frac{1}{2}}$$

$$\sigma_0 = \sigma_a, \quad b = \frac{\sum_{i=1}^n (t_i - \bar{t})(h_i^s - \bar{h}^s)}{\sum_{i=1}^n (t_i - \bar{t})^2}$$

A quick reminder—the formulae for σ_a and σ_0 are only valid for values of t that satisfy the condition $|t_i - \bar{t}| \leq 5^m$. Furthermore, the probable errors are obtained by multiplying the standard errors by the factor of 0.67456. Also note that "h^s" stands for the sextant altitude denoted in the previous sections by "h_s".

Contrary to the belief of some that σ_{h^s} serves as a measure for accuracy, the numerical example below will illustrate that by merely using standard errors of the given set of observations, not much can be said about the accuracy of the computed results.

Suppose now that there are $m \geq 2$ observers at the same place at the same time with the same equipment. Every observer takes exactly n observations of the same CO during an interval of 5 min and employs the same average azimuth Z_n. Then, we will have m different sets of observations $\{t_k, h_k^s\}_i$ $i = 1, \ldots, m$. How can we then determine which of the m sets of observations is the best?

It should be obvious that by merely evaluating the quantities σ_{h^s}, and σ_t we can say nothing about the precision nor the accuracy of each set since the σ_{h^s} are simply a measure for the mean deviations of the altitudes for the mean $\bar{h}_k^s$ that is different for each set in general. Therefore, in order to answer the above question, we must first evaluate the b_ks by using formula (12)(vii) and then employ formulae (12)(iv and v) to find the quantities σ_p and σ_a that represent precision and accuracy of the m distinct sets. Before we can compute (12)(v), we must calculate α by using formula (12)(i) and employ an approximate value for the azimuth of the observed CO. Therefore, we have found the measures for precision and accuracy for comparison of the m sets of observations.

Another and simpler method for comparing accuracy of distinct sets of observations consists in calculating the b_ks, and by using the absolute value of the deviation of the b_ks from α, we have another measure for accuracy. By employing the values $e_k = |b_k - \alpha|$, we can then establish a scale for the proficiency of the navigator.

As we have done before in most cases of theoretical analysis, let's also consider an actual numerical example:

Numerical Example:

Given the following sets of pairs of observation of altitude and time of the Sun, together with the azimuth and latitude, estimate the precision and accuracy of the navigator.

i	t_i	$t_i - \bar{t}$	$(t_i - \bar{t})^2$	h_i^s	$h_i^s - \bar{h}$	$(h_i^s - \bar{h})^2$	$(t_i - \bar{t})(h_i^s - \bar{h})$
1	$7^h\ 35^m\ 13^s$	−1.3669	1.8684	16°45′.5	10′.5	110.25	−14.3526
2	$7^h\ 35^m\ 54^s$	−0.6832	0.4667	16°40′.5	5′.5	30.25	−3.7576
3	$7^h\ 36^m\ 36^s$	0.0257	0.0006	16°34′.5	−0′.5	0.25	−0.0128
4	$7^h\ 37^m\ 16^s$	0.6834	0.4670	16°30′	−5′.0	25.00	−3.4171
5	$7^h\ 37^m\ 56^s$	1.3488	1.8193	16°24′.5	−10′.5	110.25	−14.1624
	$\bar{t} = 7^h 36^m 35^s$		$\Sigma = 4.6220$	$\bar{h}^s = 16°35′$		$\Sigma = 276$	$\Sigma = -35.7025$

$Z_n = 229°45′$, $\quad \varphi = 38°58′.53\,N$

By employing formula (12), we find that:

$\alpha = -8.89998 = -8.9$; $\quad \sigma_{h^s}^2 = 69$; $\quad \sigma_t^2 = 1.1555$; $\quad b = -7.7245$;
$\sigma_p = 0′.23$; $\quad E_p = 0′.16825$; $\quad \sigma_a = 1′.301 = \sigma_0$; $\quad$ and $\quad E_a = 0′.981125$.

We also find that $e = |b - \alpha| = 1.17548$.

These results show that the set of observed quantities is a very good one and that the sextant altitude and time should be approximated by:

$$\bar{h}_s = 16°35′ \quad \text{and} \quad \bar{t} = 7^h 36^m 35^s$$

The above error estimate describes the precision and accuracy of the final result. In cases where the navigators like to rate themselves, they should compute the values for e for each set of observations they take. This record will then reflect their proficiency.

Algorithm D

The Reduction of the Sextant Altitude h_s to the Apparent Altitude h.

1. Compute $h_s = \bar{h} = \sum_{i=1}^{n} \frac{h_i^s}{n}$, and $\bar{t} = \sum_{i=1}^{n} \frac{t_i}{n}$. If necessary or desired, access precision and accuracy of the set of measurements $\{t_i, h_i^s\}$ by using formulae I.3.5 #12. $\boxed{h_s =}$
2. Determine the index error by consulting the manual of your sextant. See I.4.2. $\boxed{I +}$
3. Calculate dip D by employing one of the three formulae depending on the atmospheric conditions near the surface of the ocean.

3.5 On the Minimization Procedure for the Random Errors ...

(i) $D = 0.97 \cdot \sqrt{h}$, h: height above sea level in feet.

(ii) $D = 1.15 \cdot \sqrt{h} \cdot \left(1 - \frac{2}{13} \cdot \frac{0.28 \cdot P}{273 + T}\right)$. Non-standard conditions.

(iii) D = I.3.2. (7). Extreme non-standard conditions. $\boxed{D=}$

4. Calculate atmospheric refraction by using one of following formulae:

(i) $R_0 = 0°\cdot 0167 \cdot \cot\left(h_s + \frac{7.32}{h_s + 4.32}\right)$, $M(T, P) = \frac{0.28P}{273 + T}$
   $R = M(T, P) \cdot R_0$ for all: $5 \leq h_s \leq 90°$.

(ii) $R = M(T, P, R_0) \cdot \bar{R}_0$, $M(T, P, R_0) = \frac{(P-80)}{930} \cdot \frac{1}{1+8\cdot 10^{-5}(R_0+39)(t-10)}$

$\bar{R}_0 = R_0 - 0.06 \cdot \sin(14.7 \cdot R_0 + 13)$, and

$R_0 = 1' \cdot \cot\left(h_s + \frac{7.31}{h_s + 4.4}\right)$ for all: $0 \leq h_s \leq 90°$, $970 \leq P \leq 1050$ mb, $-20 \leq T^0 \leq 40$

(iii) R = I.3.1 (iv) $\boxed{R=}$

5. Extract the equatorial horizontal parallax HP from the NA or other ephemeris or employ I.3.3 (4), i.e.

(i) $HP = \frac{1.315385814 \cdot 10^9}{r_0}$, where "$r_0$" denotes the geocentric distance of the CO from the center of the Earth.
Then compute the parallax PA = P by using I.1.3. (3)

(ii) $PA = HP \cdot \cos h_s$.
The distance r_0 can be computed by using the corresponding formulae in Part II. Alternatively, r_0 can be extracted from a suitable ephemeris.
In the case of the Moon, the oblateness of the Earth has to be taken into account resulting in a correction to the above expression and given by I.3.3. expressions (6) and (7).

(iii) $OB = -0.0033726548 \cdot \sin h_s \cdot \sin 2\varphi \cdot HP$, with φ: latitude.
$PA = HP \cdot \cos h_s + OB$.
Strictly speaking, the above formulae are valid only while the CO is being observed on the meridian. In all cases where a higher degree of accuracy is required, a slightly different correction is necessary expressing the dependence on the azimuth of the CO that is now different from 0° and 180°. Then you have to compute:

(iv) $\overline{OB} = -0.00337265485 \cdot \sin h_s \cdot \sin 2\varphi \cdot \cos Z_n \cdot HP$, Z_n: azimuth of CO.
$PA = HP \cdot \cos h_s + \overline{OB}$

The formulae (4) can be approximated by employing the mean distances $\bar{r}_0$ in lieu of r_0, resulting in the following simple expressions:

(v) Sun: $\overline{HP} = 8''.794$
Moon: $\overline{HP} = 57''.04$
Planets: $\overline{HP} = \frac{8''.794}{d}$, with $d = \bar{r}_O/R_O$. $R_O = 1.495979 \cdot 10^8$ km

$$\boxed{P = PA =}$$

6. Compute the semi-diameter and augmented semi-diameter of the Moon, Sun, and Planets as it may apply by employing I.3.3., expressions (10) and (15).

   (i) $SD = \frac{R}{a} \cdot HP$, R: radius of those COs; a: radius of Earth.

   (ii.) $SD' = SD \cdot \left(1 + \frac{\sin h_s}{d_0}\right)$, $d_0 = \frac{r_0}{a}$, with "r_0" and "a" defined as above.

$$\boxed{SD =}$$

7. Finally, compute the apparent altitude by using:
   $h = h_s \pm I - D - R + P \pm SD$ $\boxed{h =}$

Advisory

Make sure that all six quantities in 7. are expressed in either arc minutes or arc seconds.

For Air Navigation Only

8. Whenever a Bubble Sextant is used in an aircraft, the bubble sextant altitude has to be corrected to compensate for the effect of the Coriolis Force on the bubble. The necessary correction is given by formulae I.3.2 expressions (20) and (21) as:

   (i) $\Delta Z_C = Z_C \cdot \sin(A - C)$, with

   (ii) $Z_C = 2.62 \cdot v \cdot \sin \varphi + 0.146 \cdot v^2 \cdot \sin C \cdot \tan \varphi - 5.25 \cdot \frac{\Delta C}{\Delta t}$

Z_C Deflection of the vertical (arc min)
v $= |\vec{v}|$ of aircraft (Knots 10^{-2})
φ Latitude of aircraft (°)
C   Track or course angle (°)
$\frac{\Delta C}{\Delta t}$ Change of track/course angle (°/min)
A_z Azimuth angle of observed CO (°)
ΔZ_C Altitude correction (Arc min)

9. Furthermore, the calculated refraction has to be corrected for the refraction of the light as it passes through the window or dome of the aircraft. Consult the aircraft manual for correction.

$$\boxed{}$$

Chapter 4
Some of the Instruments and Mathematics Used by the Navigator

4.1 Some of the Formulae and Mathematics Used by the Navigator

This section is not meant to be a manual of elementary mathematics but is merely intended to show the reader how the formulae employed in the previous sections can be deduced by employing some of the concepts of the aforementioned discipline.

(a) **Plane Euclidian Geometry and Trigonometry**.

First let us consider the two dimensional rectangular triangle—Fig. 4.1.1.
By definition of the trigonometric functions we have:
$\sin \alpha = \frac{b}{c}$, $\cos \alpha = \frac{a}{c}$, $\tan \alpha = \frac{b}{a}$, and $\cot \alpha = \frac{a}{b}$. Then the Theorem of Pythagoras (Py.-TH.) yields:

(1a) $c^2 = a^2 + b^2$ and can be proven readily. From this theorem, it follows that: $\sin^2 \alpha + \cos^2 \alpha = 1$.

Next let us consider the oblique triangle with angles α, β, γ and sides a, b, c— Fig. 4.1.2.

By making use of the properties of the angles obtained by intersecting two parallels by an arbitrary straight line, we deduce that:

(2a) $\alpha + \beta + \gamma = 180°$. Further we may also readily deduce that $\sin \alpha = \frac{h}{b}$, $\sin \beta = \frac{h}{a}$, hence $\frac{\sin \alpha}{a} = \frac{\sin \beta}{b}$. Similarly, we may deduce that $\sin \beta = \frac{h'}{c}$, $\sin \gamma = \frac{h'}{b}$, hence $\frac{\sin \beta}{b} = \frac{\sin \gamma}{c}$, and the SIN-THeorem has been established.

(3a) SIN-TH. $\frac{\sin \alpha}{a} = \frac{\sin \beta}{b} = \frac{\sin \gamma}{c}$.

© Springer International Publishing AG 2018
K.A. Zischka, *Astronavigation*,
DOI 10.1007/978-3-319-47994-1_4

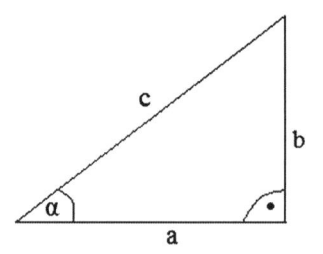

Fig. 4.1.1

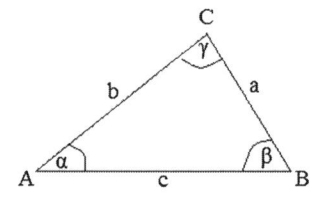

Fig. 4.1.2

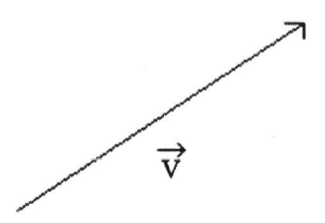

Fig. 4.1.3

By employing the Py.-TH. to triangle ABA', we deduce that $(a - b \cdot \cos \gamma)^2 + (b \cdot \sin \gamma)^2 = c^2$, and hence:

(4a) $c^2 = a^2 + b^2 - 2 \cdot a \cdot b \cdot \cos \gamma$, the COS-TH. and therefore the COS-TH. of plane trigonometry has been established.

(b) **Vectors and scalar products of vectors**.

Here I would like to reintroduce the very useful concept of vectors. The vector is defined in the most natural way as a quantity that comprises direction and length depicted by a segment of a straight line with an arrow attached to its pointed end and usually denoted by a letter with an arrow above it—Fig. 4.1.3.

It should be noted that the concept of vectors is more simple, i.e., less abstract, than the concept of real numbers. Besides many important physical quantities can be easily described by vectors such as velocities, forces, torques, and others. So far as the navigator is concerned, the position vector, the vector of velocities of vessels, current and wind are some of the most important vectors in navigational

4.1 Some of the Formulae and Mathematics Used by the Navigator

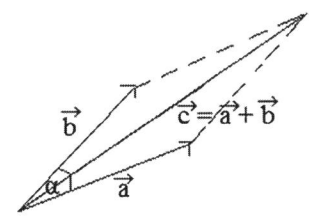

Fig. 4.1.4

applications. Anyone who wants to calculate the effect of current and wind on the course of a vessel or airplane has to know how to add vectors. The sum of two vectors, $\vec{a} + \vec{b}$, is defined as the vector that coincides with the diagonal of the parallelogram spanned by $\vec{a}$ and $\vec{b}$ and pointing away from the origin of these vectors—Fig. 4.1.4.

By defining the vector $-\vec{a}$ as the vector obtained from a by reversing its direction, it follows that $\vec{a} + (-\vec{a}) = 0$. We also define the difference of two vectors by $\vec{a} - \vec{b} = \vec{a} + \left(-\vec{b}\right)$. It follows then that the equation $\vec{a} + \vec{x} = \vec{b}$ has the unique solution: $\vec{x} = \vec{b} - \vec{a}$. Here we have made use of the associative and commutative law of addition since it can be easily shown that the laws of vector addition are the same as the laws of the addition of real numbers.

Since in many applications it is necessary to find the angle α subtended by $\vec{a}$ and $\vec{b}$ (see Fig. 4.1.4), it is advantageous to define the Scalar Product by:

(5b) $\vec{a} \cdot \vec{b} = |\vec{a}| \cdot |\vec{b}| \cdot \cos \alpha$, where "$|\vec{a}|$" denotes the length or magnitude of $\vec{a}$, and "$|\vec{b}|$" denotes the length or magnitude of $\vec{b}$.

Although for practical applications to sea and air navigation where most problems merely involve vector additions and calculating the angle between those vectors, it suffices to use graphical methods, i.e., plotting sheets, plotting boards, or flight computers-MB-4A. However, for more precise evaluations, it is necessary to resort to the uses of mathematical formulae. In order to describe a vector $\vec{x}$ in three dimensions, for instance, it is necessary to use three scalars, I.e., real numbers, namely the length $|\vec{x}|$ and two directional angles α and β. However, a much simpler

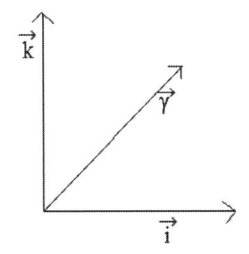

Fig. 4.1.5

scalar representation can be achieved by employing three base vectors $\vec{i}, \vec{j}$ and $\vec{k}$ of unit length that are mutually orthogonal. Hence $|\vec{i}| = |\vec{j}| = |\vec{k}| = 1, \vec{i} \cdot \vec{j} = \vec{i} \cdot \vec{k} = \vec{j} \cdot \vec{k} = 0$ (Fig. 4.1.5).

Next we define the product of any vector $\vec{x}$ and a scalar real number z as a vector $z\vec{x}$ of direction $\vec{x}$ and magnitude $|z| |\vec{x}|$, i.e., by $|z\vec{x}| = |z||\vec{x}|$. Then if "$\vec{x}$" denotes a vector with components x_1, x_2, x_3, meaning that $\vec{x} = x_1\vec{i} + x_2\vec{j} + x_3\vec{k}$, it follows then form the application of the Pythagorean Theorem that:

(6b) $|\vec{x}|^2 = x_1^2 + x_2^2 + x_3^2$.

It also follows from the definition of the sum of the vectors and the product of a scalar and vector that $\vec{x} \cdot (z + w) = z \cdot \vec{x} + w \cdot \vec{x}$, where "z" and "w" are scalars. By employing these results to the vector $\vec{c} = \vec{a} - \vec{b}$ (see Fig. 4.1.4), we deduce that with the help of the COS-TH.: $|\vec{c}|^2 = |\vec{a}|^2 + |\vec{b}|^2 - 2 \cdot |\vec{a}| \cdot |\vec{b}| \cdot \cos \alpha$. By expressing $\vec{a}$ and $\vec{b}$ in terms of their components, we find $\vec{a} = a_1\vec{i} + a_2\vec{j} + a_3\vec{k}$, and hence $\vec{c} = (a_1 - b_1) \cdot \vec{i} + (a_2 - b_2) \cdot \vec{j} + (a_3 - b_3) \cdot \vec{k}$. By applying the above formula for the square of the length of $\vec{c}$, we deduce that $|\vec{c}|^2 = (a_1 - b_1)^2 + (a_2 - b_2)^2 + (a_3 - b_3)^2$.

Substituting these component expressions into the above equation for $\cos \alpha$ we find:

(7b) $|\vec{a}| \cdot |\vec{b}| \cdot \cos \alpha = a_1b_1 + a_2b_2 + a_3b_3 = \vec{a} \cdot \vec{b}$, the equation for the scalar produce in terms of the components of $\vec{a}$ and $\vec{b}$. It follows then, that the scalar produce satisfies some of the laws of the product of real numbers, i.e., we have $\vec{a} \cdot \vec{b} = \vec{b} \cdot \vec{a}$, and $\vec{a} \cdot (\vec{b} + \vec{c}) = \vec{a} \cdot \vec{b} + \vec{a} \cdot \vec{c}$.

Next let us consider two important applications of the Scalar Product, namely, the derivation of the ADDITION THEOREM for trigonometric functions and the COS-TH.> of Spherical Trigonometry. Let "$\vec{a}$" and "$\vec{b}$" denote the two dimensional unit vectors expressed by $\vec{a} = a_x\vec{i} + a_y\vec{j}$; $\vec{b} = b_x\vec{i} + b_y\vec{j}$, with $a_x^2 + a_y^2 = 1$, and $b_x^2 + b_y^2 = 1$ (see Fig. 4.1.6).

Then by applying the COS-TH. of plane trigonometry to the triangle subtended by $\vec{a}$ and $\vec{b}$, we find that:

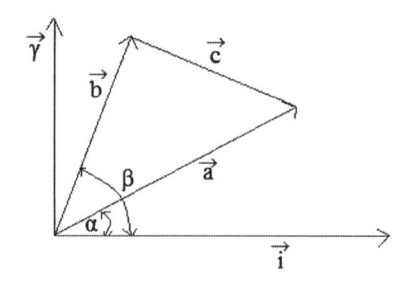

Fig. 4.1.6

4.1 Some of the Formulae and Mathematics Used by the Navigator

$\vec{c}^2 = 2 \cdot (1 - \cos(\beta - \alpha)) = (\vec{a} - \vec{b})^2 = |\vec{a}|^2 + |\vec{b}|^2 - 2\vec{a} \cdot \vec{b} = 2 \cdot (1 - \vec{a} \cdot \vec{b})$. Since the scalar product expressed by the components of $\vec{a}$ and $\vec{b}$ is $\vec{a} \cdot \vec{b} = a_x b_x + a_y b_y$, we obtain from the above expressions that $\cos(\beta - \alpha) = a_x b_x + a_y b_y$. Next we express the components of $\vec{a}$ and $\vec{b}$ by the cosines of the respective angles α and β, deducing that:

$$a_x = \vec{i} \cdot \vec{a} = \cos \alpha; \, a_y = \vec{j} \cdot \vec{a} = \cos(90° - \alpha) = \sin \alpha, \text{ and}$$
$$b_x = \vec{i} \cdot \vec{b} = \cos \beta; \, b_y = \vec{j} \cdot \vec{b} = \cos(90° - \beta) = \sin \beta.$$

By substituting these expressions into the above equations for $\cos(\beta - \alpha)$, we find that:

(8b) $\cos(\beta - \alpha) = \cos \beta \cdot \cos \alpha + \sin \beta \cdot \sin \alpha$.

Replacing α by $-\alpha$, we deduce $\cos(\beta + \alpha) = \cos \beta \cdot \cos \alpha - \sin \beta \cdot \sin \alpha$, since $\cos(-\alpha) = \cos \alpha$, and $\sin(-\alpha) = -\sin \alpha$.

Subsequently, we deduce that

$$\sin(\beta - \alpha) = \cos(90° - (\beta - \alpha)) = \cos(\alpha + (90° - \beta))$$
$$= \cos \alpha \cdot \cos(90° - \beta) - \sin \alpha \sin(90° - \beta) \text{ and hence:}$$

(8'b) $\sin(\beta - \alpha) = \sin \beta \cdot \cos \alpha - \cos \beta \cdot \sin \alpha$. Replacing α by $-\alpha$, we finally obtain:
$\sin(\beta + \alpha) = \sin \beta \cdot \cos \alpha + \cos \beta \cdot \sin \alpha$.

In the derivation of the second expression above, we have made use of the following obvious identities:

$$\cos(90° - \alpha) = \sin \alpha; \, \sin(90° - \alpha) = \cos \alpha; \, \cos(-\alpha) = \cos \alpha; \text{ and}$$
$$\sin(-\alpha) = \sin \alpha.$$

Next, let us derive the King-Pin of Spherical Trigonometry, namely the COS-TH. of said discipline. Once these formulae have been established, the rest of spherical trigonometry is just "algebra", i.e., all the other relevant formulae, as for instance the SIN-TH., the ANALOGUE-FORMULAE, and the FOUR-PARTS FORMULA, as well as numerous others, can be deduced from the COS-TH. by merely using algebraic manipulations.

By selecting three arbitrary points A, B, C on a unit sphere of radius one, and then connecting these point to the origin of the sphere by the use of three unit vectors $\vec{a}, \vec{b}, \vec{c}$, one has subtended a spherical triangle by these three vectors (see Fig. 4.1.7).

Noting that $|\vec{a}| = |\vec{b}| = |\vec{c}| = 1$, we deduce that $\vec{b} \cdot \vec{c} = \cos A$, $\vec{a} \cdot \vec{c} = \cos B$, and $\vec{a} \cdot \vec{b} = \cos C$. The unit tangent vectors at A are denoted by "$\vec{t}_B$" and "$\vec{t}_C$"

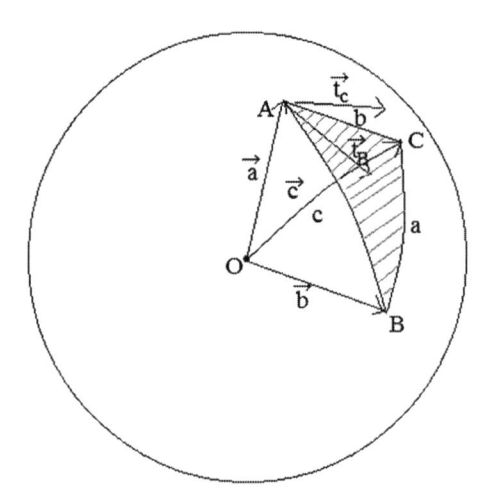

Fig. 4.1.7

respectively. Since these vectors are tangent vectors, we conclude that $\vec{a} \cdot \vec{t}_C = 0$, and $\vec{a} \cdot \vec{t}_B = 0$. We also note that $\vec{t}_B \cdot \vec{t}_C = \cos A$, where we have, as before, denoted the angle at A by the same symbol. Furthermore, since $\vec{t}_C$ lies in the plane spanned by $\vec{a}$ and $\vec{c}$, we may write $\vec{t}_C = \alpha \vec{a} + \beta \vec{b}$, with $\alpha, \beta \geq 0$, and $\vec{t}_B = \gamma \vec{a} + \delta \cdot \vec{b}$, with $\gamma, \delta \geq 0$. Note that α, β, γ, and δ are scalars.

Next we deduce that $|\vec{t}_C|^2 = \alpha^2 + \beta^2 + 2\alpha \cdot \beta \cdot \cos c = 1$, and $|\vec{t}_B|^2 = \gamma^2 + \delta^2 + 2 \cdot \gamma \cdot \delta \cdot \cos b = 1$. We also have: $\vec{t}_C \cdot \vec{a} = \alpha + \beta \cdot \cos b = 0$, and $\vec{t}_B \cdot \vec{a} = \gamma + \delta \cdot \cos c = 0$. It follows then that $|\vec{t}_C|^2 = \beta^2 \cdot \sin^2 b = 1$, and $|\vec{t}_B|^2 = \delta^2 \cdot \sin^2 c = 1$, i.e., we deduce that $\frac{1}{\beta \cdot \delta} = \sin b \cdot \sin c$. On the other hand, the scalar product of the two tangent vectors yields:

$$\vec{t}_B \cdot \vec{t}_C = \beta \cdot \delta \cdot \cos b \cdot \cos c - \beta\delta \cdot \cos b \cdot \cos c - \beta\delta \cdot \cos b \cdot \cos c + \beta\delta \cos a$$
$$= \cos A.$$

The last expression yields:

(9b) $\cos a = \cos b \cdot \cos c + \sin b \cdot \sin c \cdot \cos A$

And by interchanging the angles and sides, one finds the other two equations:

$\cos b = \cos a \cdot \cos c + \sin a \cdot \sin c \cdot \cos B$

$\cos c = \cos a \cdot \cos b + \sin a \cdot \sin b \cdot \cos C$

With the help of the COS-TH., we may now deduce another very important formulae, namely, the SIN-TH. Employing the first of the above expressions in (9b), we find:

$\sin b \cdot \sin c \cdot \cos A = \cos a - \cos b \cdot \cos c$, from which it follows that

4.1 Some of the Formulae and Mathematics Used by the Navigator

$\sin^2 b \sin^2 c \cos^2 A = \cos^2 a - 2 \cdot \cos a \cdot \cos b \cdot \cos c + \cos^2 b \cdot \cos^2 c$, hence, $\sin^2 b \sin^2 c \sin^2 A = 1 - \cos^2 a - \cos^2 b - \cos^2 c + 2 \cos a \cdot \cos b \cdot \cos c$.

Next we define a positive number $X > 0$ by $X^2 \cdot \sin^2 a \cdot \sin^2 b \cdot \sin^2 c = \sin^2 b \cdot \sin^2 c \cdot \sin^2 A$, resulting in $X = \frac{\sin A}{\sin a}$, since $X \geq 0$. Using the second equation of (9b), we deduce that $\sin a \cdot \sin c \cdot \cos B = \cos b - \cos a \cdot \cos c$. Taking again the square of the left-hand side of this equation and substituting "$\cos^2 B$" by "$1 - \sin^2 B$", we arrive at:

$\sin^2 a \cdot \sin^2 c \cdot \sin^2 B = 1 - \cos^2 a - \cos^2 b - \cos^2 c + 2 \cos a \cdot \cos b \cdot \cos c = \sin^2 b \cdot \sin^2 c \cdot \sin^2 A$. Hence $\frac{\sin^2 B}{\sin^2 b} = \frac{\sin^2 A}{\sin^2 a} = X^2$ and therefore, $X = \frac{\sin B}{\sin b}$, resulting in $\frac{\sin B}{\sin b} = \frac{\sin A}{\sin a}$. By employing equation three of (9b) and going through the same steps, we conclude that:

(10b) $\quad \dfrac{\sin A}{\sin a} = \dfrac{\sin B}{\sin b} = \dfrac{\sin C}{\sin c}$ SIN-TH.

Other very useful formulae are the ANALOGUE-FORMULAE which can be readily deduced from the COS-TH. by applying this theorem twice. For instance, if we replace in the first expression of (9b) cos c by the third expression, we obtain $\sin c \cdot \sin b \cdot \cos A = \cos a \sin^2 b - \sin b \cdot \sin a \cdot \cos b \cdot \cos C$ and therefore arrive at:

(11b) $\quad \begin{aligned} \sin c \cdot \cos A &= \cos a \cdot \sin b - \sin a \cdot \cos b \cdot \cos C^* \\ \sin a \cdot \cos B &= \cos b \cdot \sin c - \sin b \cdot \cos c \cdot \cos A^1 \\ \sin b \cdot \cos C &= \cos c \cdot \sin a - \cos a \cdot \sin c \cdot \cos B \end{aligned}$

From the first expression by cyclic permutations of the letters we obtain the other two expressions. These three equations constitute the ANALOGUE-FORMULAE of spherical trigonometry. [41]

Similarly we can deduce the FOUR-PARTS formulae of spherical trigonometry by employing the COS-TH. twice and the SIN-TH. once. For, from the first expression of (9b) and the third, we can deduce that:

$\cos c = \cos b(\cos b \cdot \cos c + \sin b \cdot \sin c \cdot \cos A) + \sin a \cdot \sin b \cdot \cos C$. Therefore, $0 = -\cos c \cdot \sin^2 b + \sin b \cdot \cos b \cdot \sin c \cdot \cos A + \sin a \cdot \sin b \cdot \cos C$. Next by dividing this expression by $\sin b \cdot \sin c$, we obtain: $0 = -\cot c \cdot \sin b + \cos b \cdot \cos A + \dfrac{\sin a}{\sin c} \cos C$. At this point, by employing the SIN-TH., we find that $\dfrac{\sin A}{\sin C} = \dfrac{\sin a}{\sin c}$, and hence:

---

[1] And also $\sin a \cdot \cos C = \cos c \cdot \sin b = \cos b \cdot \sin c \cdot \cos A$, and similarly for the other expressions.

$0 = -\cot c \cdot \sin b + \cos b \cdot \cos A + \sin A \cdot \cot C$, i.e.,

(12b) $\quad \begin{aligned} \cos b \cdot \cos A &= \sin b \cdot \cot c - \sin A \cdot \cot C \\ \cos c \cdot \cos B &= \sin c \cdot \cot a - \sin B \cdot \cot A \\ \cos a \cdot \cos C &= \sin a \cdot \cot b - \sin c \cdot \cot B \end{aligned}$

And again by cyclic permutation of the letters in the first equation, we obtain the other two. Other additional formulae of this type can be generated by cyclic permutations of the letters.

Although used less frequently in navigation, another set of formulae referred to as the COS-TH. of angles, can be obtained by employing the concept of the POLAR-TRIANGLE and then applying the COS-TH. to it. The results are:

(13b) $\quad \begin{aligned} \cos A &= -\cos B \cdot \cos C + \sin B \cdot \sin C \cos a \\ \cos B &= -\cos A \cdot \cos C + \sin A \cdot \sin C \cos b \quad \text{COS-TH. for ANGLES [36]} \\ \cos C &= -\cos A \cdot \cos B + \sin A \cdot \sin B \cos c \end{aligned}$

(c) **Trigonometry and Trigonometric Functions**.

In some applications to navigation, one frequently encounters the problem of solving an equation of the type:

(14c) $\quad a \cdot \sin \alpha + b \cdot \cos \alpha = c$, where a, b, and c are real numbers and "α" denotes an angle.

First let us consider the simple case where $c = 0$. In this case, we assume that $a \neq 0$ and $b \neq 0$, otherwise we have a trivial case. We conclude that: $\frac{a}{b} \cdot \tan \alpha = -1$, or $\tan \alpha = -\frac{b}{a}$ resulting in $\alpha = \tan^{-1}\left(-\frac{b}{a}\right) = -\tan^{-1}\frac{b}{a}$; $c = 0$.

In the case where $c \neq 0$, we obtain $\frac{a}{c} \sin \alpha + \frac{b}{c} \cos \alpha = 1$. If we now introduce the quantities "β" and "γ" defined by $\gamma \cdot \cos \beta = \frac{a}{c}$ and $\gamma \cdot \sin \beta = \frac{b}{c}$, the above equation reduces to $\gamma \cdot \cos \beta \cdot \sin \alpha + \gamma \cdot \sin \beta \cdot \cos \alpha = \gamma \cdot \sin(\alpha + \beta) = 1$, where γ is given by $\gamma^2 = \left(\frac{a}{c}\right)^2 + \left(\frac{b}{c}\right)^2$ and since by hypotheses $a^2 + b^2 > 0$, it follows that for all $|\gamma| \geq 1$, i.e., for all a, b, and c satisfying $a^2 + b^2 \geq c^2$, and $a \neq 0$, the given equation possesses infinity many solutions given by $\alpha = \sin^{-1}\frac{1}{\gamma} - \tan^{-1}\frac{b}{a}$, since $\tan \beta = \frac{b}{a}$. If $a = 0$ and $b^2 \geq c^2$, then $\alpha = \cos^{-1}\frac{c}{b}$; but if $b = 0$, then it follows that $\alpha = \sin^{-1}\frac{c}{a}$, $a^2 \geq c^2$.

Summarizing these results, we have:

(15c) $\quad \alpha = \begin{cases} \sin^{-1}\frac{1}{\gamma} - \tan^{-1}\frac{b}{a}, & a \neq 0, \quad a^2 + b^2 \geq c^2, \quad \gamma^2 = \left(\frac{a}{c}\right)^2 + \left(\frac{b}{c}\right)^2 \\ -\tan^{-1}\frac{b}{a} & \text{if } c = 0, \quad a \neq 0 \\ \sin^{-1}\frac{c}{a} & \text{if } b = 0, \quad a^2 \geq c^2, \quad c \neq 0 \\ \cos^{-1}\frac{c}{b} & \text{if } a = 0, \quad b^2 \geq c^2, \quad c \neq 0 \end{cases}$

4.1 Some of the Formulae and Mathematics Used by the Navigator

Note that these formulae cover those special cases where the solution of the above equation is given directly in terms of the inverse trigonometric functions.

Because it is imperative for all applications to computational navigation to know the inverse trigonometric functions with all its ramifications, it behooves us to derive all relevant formulae that can be readily evaluated on any inexpensive scientific calculator.

By applying the Addition-TH. of plane trigonometry, we deduce that $\sin(x + 2 \cdot n \cdot 180°) = \sin x$, and $\sin((2n+1)180° - x) = \sin x$, for all real value x and all positive and negative integers n. But by definition of the inverse to sin x, we conclude that $\sin^{-1}(\sin x) = n\,180° + (-1)^n x$, n = 0, ±1, ±2, ... This may also be expressed by stating that:

(16c) $\sin^{-1} x = n\,180° + (-1)^n \underline{\sin^{-1} x}$, $0 \leq x \leq 1$, n = 0, ±1, ...

The index "n" defines the n-th BRANCH of $\sin^{-1} x$, and this equation by be interpreted as saying that it defines all branches in terms of one particular one denoted by "$\underline{\sin^{-1} x}$".

Similarly, we deduce from $\cos(2 \cdot n \cdot 180° \pm x) = \cos x$ that $\cos^{-1}(\cos x) = 2 \cdot n \cdot 180° \pm x$, i.e.,

(17c) $\cos^{-1} x = 2 \cdot n \cdot 180° \pm \underline{\cos^{-1} x}$, $|x| \leq 1$, n = 0, ±1, ..., likewise, we find
(18c) $\tan^{-1} x = n \cdot 180° + \underline{\tan^{-1} x}$, $\forall x$, n = 0, ±1, ...

For the navigator, these expressions for the inverse trigonometric functions that refer to "BRANCHES" may be somewhat confusing. However, since the formulae of spherical trigonometry mainly encompasses values for arguments that lie between zero and plus-minus one hundred and eighty degrees, only three branches of the above functions are required for actual computations. Those branches correspond to n = 0, n = 1, and n = −1. However, in the light of the limitations imposed by the capacities of inexpensive calculators and tables, it is advantageous to define and subsequently use only the so-called "MAIN BRANCHES" of said functions that are denoted by $\sin^{-1}|x|$, $\cos^{-1}|x|$, and $\tan^{-1}|x|$. The exact definitions are:

$\sin^{-1}|x|$ Defined only for non-negative arguments x satisfying $|x| \leq 1$, and its range is restricted to $0 \leq \sin^{-1}|x| \leq 90°$.
$\cos^{-1}|x|$ Defined only for non-negative arguments x satisfying $|x| \leq 1$, and its range is restricted to $0 \leq \cos^{-1}|x| \leq 90°$.
$\tan^{-1}|x|$ Defined only for non-negative arguments x satisfying $|x| < \infty$, and its range is restricted to $0 \leq \tan^{-1}|x| \leq 90°$.

In order to derive the desired formulae, we must first consider the case n = 0, i.e., the zero branch, by employing the above formulae (16c), (17c) and (18c) with n = 0. When we do that, we find $\sin^{-1} x = \underline{\sin^{-1} x}$, and by using positive values for x, i.e., x = |x|, we find that $\underline{\sin^{-1} x} = \sin^{-1}|x|$, $0 \leq x \leq 1$. By using negative values for x, i.e., x = −|x|, we obtain $\underline{\sin^{-1} x} = -\sin^{-1}|x|$, since $\sin^{-1}(-|x|) = -\sin^{-1}|x|$. The latter can be readily deduced by noting that $\sin(-\sin^{-1}|x|) = -|x|$.

For the inverse of the cosine function, we deduce that $\cos^{-1} x = \pm \cos^{-1} x$, hence for $x = |x|$, we get $\cos^{-1} x = \pm \cos^{-1}|x|$, i.e., two values. If we put $x = -|x|$, we find that $\cos^{-1} x = \pm \cos^{-1}|x| = \pm(90° + \sin^{-1}|x|)$. The latter follows from the identity: $\cos(90° + \sin^{-1}|x|) = -|x|$.

Similarly, we find that $\tan^{-1}(-|x|) = -\tan^{-1}|x|$ since $\tan(-\tan^{-1}|x|) = -|x|$. Furthermore, for $n = 0$ we have $\tan^{-1} x = \tan^{-1} x$ and conclude that:

$\tan^{-1} x = \tan^{-1}|x|$ for all x satisfying $0 \leq x \leq \infty$, and
$\tan x^{-1} = -\tan^{-1}|x|$ for all x satisfying $-\infty < x \leq 0$.

Next let us consider the BRANCH $n = 1$.

By putting $n = 1$, we obtain $\sin^{-1} x = 180° - \sin^{-1} x$, hence for all $x = |x|$, we find that $\sin^{-1} x = 180° - \sin^{-1}|x|$. But if we substitute $x = -|x|$, we get $\sin^{-1} x = 180° + \sin^{-1}|x| \geq 180°$. Therefore, these values of this part of the branch do not fall in the prescribed range. By noting that $\sin(90° - \cos^{-1}|x|) = |x|$, it follows that for all $x \geq 0$, $\sin^{-1}|x| = 90° - \cos^{-1}|x|$, and hence, $\sin^{-1} x = 90° + \cos^{-1}|x|$, for all x: $0 \leq x \leq 1$.

In the case of $\cos^{-1} x$, we can readily see that for this particular branch, it does not fall in the prescribed range for all permissible x.

For $\tan^{-1} x$, we obtain the expression $\tan^{-1} x = 180° + \tan^{-1} x$, and hence only $\tan^{-1} x = 180° - \tan^{-1}|x|$ for all $x \leq 0$ satisfies the condition $|\tan^{-1} x| \leq 180°$.

Finally, let's consider the BRANCH corresponding to $n = -1$, for which $\sin^{-1} x = -(180° + \sin^{-1} x)$.

For all x satisfying $x = |x| \leq 1$, we obtain $\sin^{-1} x = -(180° + \sin^{-1}|x|)$, and for all values $x = -|x|$ and $|x| \leq 1$, we get $\sin^{-1} x = -(180° - \sin^{-1}|x|)$. Therefore, only the latter part of this branch satisfies the imposed condition $|\sin^{-1} x| \leq 180°$. Hence for $n = -1$, only $\sin^{-1} x = -(180° - \sin^{-1}|x|) = -(90° + \cos^{-1}|x|)$ counts.

For $n = -1$, we also have $\tan^{-1} x = -180° + \tan^{-1} x$ and therefore, only $\tan^{-1} x = -(180° - \tan^{-1}|x|)$ for all x satisfying $0 \leq x < \infty$ has the appropriate range. On the other hand, the corresponding branch of $\cos^{-1} x$ does not satisfy the condition $|\cos^{-1} x| \leq 180°$, and therefore, does not count.

Summarizing these results, we have found the desired formulae for the inverse trigonometric functions:

(19c)
$$\sin^{-1} x = \begin{cases} \pm \sin^{-1}|x|, & \text{plus }(+) \text{ if } 0 \leq x \leq 1 \\ \pm(90° + \cos^{-1}|x|), & \text{minus }(-) \text{ if } -1 \leq x \leq 0 \end{cases}$$

$$\cos^{-1} x = \begin{cases} \pm \cos^{-1}|x|, & \text{for all x: } 0 \leq x \leq 1 \\ \pm(90° + \sin^{-1}|x|), & \text{for all x: } -1 \leq x \leq 0 \end{cases}$$

$$\tan^{-1} x = \begin{cases} \pm \tan^{-1}|x|, & \text{for all x: plus }(+) \text{ if } 0 \leq x < \infty \\ \pm(\tan^{-1}|x| - 180°), & \text{for all x: minus }(-) \text{ if } -\infty < x \leq 0 \end{cases}$$

4.1 Some of the Formulae and Mathematics Used by the Navigator

Similar formulae can be derived for the other less frequently used trigonometric functions.

It should also be noted that for theoretical purposes, it is advantageous to employ the trigonometric functions with their arguments expressed in radians rather than in degrees. However, it is customary to use the same symbols for the two different functions. Most calculators now use distinct settings, degrees and radians, for these distinct functions. Mathematically, the relationship between those two classes of trigonometric functions is simply: $\sin^{rad} x = \sin x^0$, with $x^0 = \frac{180° x}{\pi}$, where "x^0" denotes the degrees and "x" the radians. here we have used the symbol "$\sin^{rad} x$" to make it abundantly clear that we are dealing with two different functions. What has been said about sin x also applies to all the other trigonometric functions.

In conclusion, it should be pointed out that the most frequently used trigonometric expressions in astronavigation are the COS-TH. in the form

$$a = \cos^{-1}(\cos b \cos c + \sin b \sin c \cos A) \text{ or in the form } A = \cos^{-1}\left(\frac{\cos a - \cos b \cos c}{\sin b \sin c}\right).$$

All the relevant formulae in Sect. 2.4 are of this type. It can be stated that the entire problem of finding your position at sea or in the air can be reduced to analytically evaluating the two KING-PIN formulae deduced above.

(d) **Interpolations and Approximations**.

Next, let's consider the problem of approximating a function either given analytically or merely in the form of discrete values numerically. This will include Lagrange approximation or interpolations, Padé approximations, as well as other approximations by rational functions, and also Taylor-type approximations.

In general, approximation are classified with reference to the type of elementary functions that are being used to approximate a certain class of functions, and also, most importantly, with regards to the requirements imposed upon the resulting error. For instance, the most widely used set of elementary functions are the polynomials, i.e., powers of x, and to a much lesser degree, the rational functions, i.e., the ratios of two polynomials (Sect. 3.1). In astronomy, where some of the functions are related to the periodic motions of a CO are periodic functions, one employs the set of trigonometric functions with periods of $2\pi/n$ (see also Fournier Analysis).

With regards to the error in terms of those approximations, the most often used requirements are that the error functions E(x) assume the value zero at a given set of points, $x = x_i$, $i = 0, 1, \ldots, n$, as, for instance, in interpolating with the set of elementary functions either the powers of x or rational functions. Another well-known class of approximations are the Taylor Expansions, in the case where the set of elementary functions is the set of all powers of x, or the Padé approximations with the set of elementary functions equal to the set of rational functions. In the case of Taylor

Expansions or Padé approximations, the requirements on the error function E(x) are that the error and all its derivatives of up to the order m vanish at merely one single point, $x = x_0$.

Let's consider the case of the interpolation of polynomials. Here we approximate f(x) by a polynomial, i.e., we have the relation $f(x) = \sum_{i=0}^{n} a_i x^i + E_{n+1}(x)$. Then for a given set of points, $P_k = (x_k, y_k)$, we require that $E_{n+1}(x_k) = 0$ for all $k = 0, 1, \ldots, n$, resulting in the system of n + 1 linear equations: $\sum_{i=0}^{n} a_i x_k^i = y_k$ with $y_k = f(x_k)$ for all $k = 0, 1, \ldots, n$.

Obviously, solving such a linear system on a small calculator can be a very tedious and time consuming task. Therefore, let us look at another approach based on the Lagrangen Polynomials $L_i(x)$ defined by:

(1d) $L_k(x) = \frac{p_{n+1}(x)}{(x-x_k)p'_{n+1}(x_k)}$, with $p_{n+1}(x) = \prod_{i=0}^{n}(x - x_i)$. From this, we then conclude that the $L_k(x)$ satisfy the identities:

(2d) $L_k(x_\mu) \begin{cases} 0 \text{ if } k \neq \mu \\ 1 \text{ if } k = \mu \end{cases}$, and therefore f(x) can be expressed by:

(3d) $f(x) = \sum_{i=0}^{n} L_i(x) y_i + E_{n+1}(x)$, with $y_k = f(x_k)$, $k = 0, 1, \ldots n$.

In general, f(x) will not be known analytically, but if it is, and furthermore, has derivatives of up to the order n + 1 with the n + 1st derivative still continuous, then it can be shown that:

(4d) $E_{n+1}(x) = \frac{p_{n+1}(x)}{(n+1)!} f(\zeta)^{(n+1)}$, with $\zeta \in [a, b]$, $a \leq x_k \leq b$, $k = 0, 1, \ldots n$.

It can be easily shown that the above representation of the interpolation polynomial in terms of the Lagrangen Polynomials is unique, i.e., if there exists another polynomial of the same degree n passing through the same set of pints, it must be identical to the polynomial designed by the first term of (3d), i.e., by the Lagrangen interpolation polynomial.

For any particular application to navigation and astronomy, it suffices to use an equal spacing of point of interpolation, i.e., we may assume that:

(5d) $x_{k+1} = x_k + h = x_0 + (k+1)h$, where $h = \frac{(x_n - x_0)}{n}$.

Furthermore, it is convenient to use the center of the interval spanned by x_0 and x_n as the point of reference, i.e., if we put

(6d) $x_c = \frac{x_0 + x_n}{2} = x_0 + \frac{n}{2} h$, and $x = x_c + m \cdot h$, with $m = \frac{x - x_0}{h} - \frac{n}{2}$, with the new variable m satisfying $-\frac{n}{2} \leq m \leq \frac{n}{2}$.

For our particular purpose, it suffices to consider only interpolation formulae of up to the fourth order, i.e., $1 \leq n \leq 4$.

Next, let us derive those relevant formulae explicitly.

4.1 Some of the Formulae and Mathematics Used by the Navigator   221

Examples

(i) Linear Interpolation—n = 1.

Here we have:

$$x_c = x_0 + \frac{h}{2}, h = x_1 - x_0, m = \frac{x - x_0}{h} - \frac{1}{2}, -\frac{1}{2} \leq m \leq \frac{1}{2},$$

$$L_0(m) = \frac{1}{2} - m, L_1(m) = \frac{1}{2} + m, \text{ and hence}$$

(7d) $$f(m) = \left(\frac{1}{2} - m\right) \cdot y_0 + \left(\frac{1}{2} + m\right) \cdot y_1, h = x_1 - x_0, m = \frac{x - x_0}{h} - \frac{1}{2};$$

$$-\frac{1}{2} \leq m \leq \frac{1}{2}.$$

(ii) Quadric Interpolation—n = 2.

Here we have:

$$x_c = \frac{x_0 + x_2}{2} = x_0 + h, h = \frac{x_2 - x_0}{2}, m = \frac{x - x_0}{h} - 1, -1 \leq m \leq 1$$

$$L_0(m) = \frac{m(m-1)}{2}, L_1(m) = -(m-1)(m+1), L_2(m) = \frac{m(m+1)}{2}, \text{ hence}$$

(8d) $$f(m) = \frac{m(m+1)}{2} \cdot y_0 - (m-1)(m+1) \cdot y_1 + \frac{m(m+1)}{2} \cdot y_2,$$

$$h = \frac{x_2 - x_0}{2}, m = \frac{x - x_0}{h} - 1, -1 \leq m \leq 1.$$

(iii) Cubic Interpolation—n = 3.
Here we have:

$$x_c = x_0 + \frac{3}{2}h, h = \frac{x_3 - x_0}{3}, m = \frac{x - x_0}{h} - \frac{3}{2},$$

$$L_0(m) = -\frac{(2m-1)(2m-3)(2m+1)}{48}, L_1(m) = \frac{(2m-1)(2m-3)(2m+3)}{16},$$

$$L_2(m) = -\frac{(2m-3)(2m+1)(2m+3)}{16}, L_3(m) = -\frac{(2m-1)(2m+1)(2m+3)}{48}, \text{ hence}$$

(9d) $$f(m) = -\frac{(2m-1)(2m-3)(2m+1)}{48} \cdot y_0 + \frac{(2m-1)(2m-3)(2m+3)}{16} \cdot y_1 - \frac{(2m-3)(2m+1)(2m+3)}{16} \cdot y_2 + \frac{(2m-1)(2m+1)(2m+3)}{48} \cdot y_3 \text{ with}$$

$$h = \frac{x_3 - x_0}{3}, \quad m = \frac{x - x_0}{h} - \frac{3}{2}, \quad -\frac{3}{2} \leq m \leq \frac{3}{2}.$$

(iv) Fourth-Degree Interpolation—n = 4.

Here we have:

$$x_c = x_0 + 2h, \quad h = \frac{x_4 - x_0}{4}, \quad m = \frac{x - x_0}{h} - 2, \quad x_{k+1} = x_k + h,$$

$$L_0(m) = \frac{(2m+2) \cdot m \cdot (2m-2)(2m-4)}{192}, \quad L_1(m) = \frac{-(2m+4)(2m-2)(2m-4) \cdot m}{48},$$

$$L_2(m) = \frac{(2m+4)(2m+2)(2m-2)(2m-4)}{64}, \quad L_3(m) = -\frac{(2m+4)(2m+2)(2m-4) \cdot m}{48},$$

$$L_4(m) = \frac{(2m+4)(2m+2)(2m-2) \cdot m}{192}, \text{ and therefore:}$$

(10d)
$$f(m) = \frac{(2m+2)(2m-2)(2m-4) \cdot m}{192} \cdot y_0 - \frac{(2m+4)(2m-2)(2m-4) \cdot m}{48} \cdot y_1 +$$
$$+ \frac{(2m+4)(2m+2)(2m-2)(2m-4)}{64} \cdot y_2 - \frac{(2m+4)(2m+2)(2m-4) \cdot m}{48} \cdot y_3 +$$
$$+ \frac{(2m+4)(2m+2)(2m-2) \cdot m}{192} \cdot y_4, \quad h = \frac{x_4 - x_0}{4}, \quad m = \frac{x - x_0}{h} - 2, \quad -2 \leq m \leq 2.$$

Of course, there are situations like interpolating the universe $x = g^{-1}(y)$ of a tabulated function $g(x)$ where it becomes necessary to resort to employing the general formula (3d) for the unequal spaced values of the y_ks. However, we will make an attempt to get by with formulae (7d)–(10d) in most, but not all situations. The construction of rational functions that approximate a certain class of functions is more difficult and involved and cannot be addressed in this treatise adequately. [35]

For example, in the most simple case where the desired approximation is an interpolation, one would expect it to be a straight forward computational problem, since the 2(n + 1) linear equations for the a_μ and b_μ are given by:

(11d) $\quad y_k = \frac{P_{n+1}(x_k)}{Q_n(x_k)} = \frac{\sum_{\mu=0}^{n+1} a_\mu \cdot x_k^\mu}{\sum_{\mu=0}^{n} b_\mu x_k^\mu}, \quad 0 \leq k \leq 2(n+1) - 1, \quad a_{n+1} = 1$

and explicitly by:

(12d) $\quad \sum_{\mu=0}^{n} (y_k b_\mu - a_\mu) x_k^\mu = x_k^{n+1}, \quad 0 \leq k \leq 2(n+1) - 1, \quad a_{n+1} = 1,$

appear to be solvable by well-known algorithms. However, this system of linear equations poses some difficulties since it may not have a solution for a given set of points (x_k, y_k), and a given n. [18]

Besides, even if this system posses a solution, it cannot be solved numerically easily on a small calculator. Nevertheless, the great advantage of rational approximation over polynomial approximations lies in the superior behavior of the error function $E(x)$. (Refer to Padé approximations in this section.)

4.1 Some of the Formulae and Mathematics Used by the Navigator

Next let us consider other types of approximations with the following requirements on the error function:

(13d) $E^{(\mu)}(x_0) = 0$, for all $0 \leq \mu \leq n$. Here "$f^{(\mu)}(x)$" demotes the μ-th derivative of f(x), and likewise for all E(x). It should be obvious that the underlying assumption for this type of approximation is that only functions that have n + 1 continuous derivatives can be approximated by this method.

In all cases where the elementary functions are again the powers of x, i.e., polynomials, we have the following representations:

$f(x) = \sum_{\mu=0}^{n} a_\mu (x - x_0)^\mu + E_{n+1}(x)$. Then because of $E_{n+1}^{(\mu)}(x_0) = 0$, we deduce that $a_\mu = \frac{f^{(\mu)}(x_0)}{\mu!}$, $\mu! = 1 \cdot 2 \cdot 3 \cdots \mu$, $0! = 1$. We also deduce that $E_{n+1}(x)$ must satisfy the ordinary differential equation: $E_{n+1}^{(n+1)}(x) = f^{(n+1)}(x)$ subject to the initial conditions: $E_{n+1}^{(\mu)}(x_0) = 0, 0 \leq \mu \leq n$. It can also be shown easily that the solution of this initial value problem is given by:

(14d) $E_{n+1}(x) = \int_{x_0}^{x} \frac{(x-t)^n}{n!} \cdot f^{(n+1)}(t) dt = \frac{(x-x_0)^{n+1}}{(n+1)!} \cdot f^{(n+1)}(\zeta), \zeta \leq (x, x_0)$. [18, 36]

The resulting approximation is then:

$f(x) \cong \sum_{\mu=0}^{n} \frac{f^{(\mu)}(x_0)}{\mu!} \cdot (x - x_0)^\mu$, usually referred to as Taylor Series Expansions and in all cases where $\lim_{n \to \infty} E_{n+1}(x) = 0$, is written as:

(15d) $f(x) = \sum_{\mu=0}^{\infty} \frac{f^{(\mu)}(x_0)}{\mu!} \cdot (x - x_0)^\mu$, TAYLOR SERIES.

Examples

$e^x = \sum_{\mu=0}^{\infty} \frac{x^\mu}{\mu!}, -\infty < x < \infty;$

$\ln(1+x) = x - \frac{x^2}{2} + \frac{x^3}{3} - \frac{x^4}{4} + \cdots, |x| < 1;$

$\ln(1-x) = -x - \frac{x^2}{2} - \frac{x^3}{3} - \cdots - \frac{x^n}{n} - ;$

$\tan x = x + \frac{x^3}{3} + 2\frac{x^5}{15} + \cdots, |x| \leq \frac{\pi}{2};$

$\frac{1}{1-x} = \sum_{\mu=0}^{\infty} x^\mu, |x| < 1;$

$\cos x = \sum_{\mu=0}^{\infty} \frac{(-1)^\mu x^{2\mu}}{(2\mu)!}, -\infty < x < \infty;$

$\sin x = \sum_{\mu=0}^{\infty} \frac{(-1)^\mu x^{2\mu+1}}{(2\mu+1)!}, -\infty < x < \infty$

$\sin^{-1} x = \sum_{\mu=0}^{\infty} \frac{(2\mu)! x^{2\mu+1}}{2^{2\mu} \cdot (2\mu+1) \cdot (\mu!)^2}, |x| < 1;$

$\cos^{-1} x = \frac{\pi}{2} - \sin^{-1} x;$

$\tan^{-1} x = \sum_{\mu=0}^{\infty} \frac{(-1)^\mu x^{2\mu+1}}{2\mu+1^2}, |x| < 1.$

PADÉ APPROXIMATIONS [18, 35]

In the case where the elementary functions are no long polynomials but rational functions $R_{m,n}(x) = \frac{P_m(x)}{Q_n(x)}$, the resulting approximations are the Padé Approximations for all functions f(x) that have a Taylor Series at $x_0 = 0$, i.e., for all f(x) that have the representation $f(x) = \sum_{\mu=0}^{\infty} c_\mu x^\mu$. The error function $E_{m,n}(x)$ is then given by

$$E_{m,n}(x) = f(x) - \frac{P_m(x)}{Q_n(x)} = \frac{\left(\sum_{\mu=0}^{\infty} c_\mu x^\mu\right) \cdot \left(\sum_{\mu=0}^{n} b_\mu x^\mu\right) - \sum_{\mu=0}^{m} a_\mu x^\mu}{\sum_{\mu=0}^{n} b_\mu x^\mu}, \quad b_0 = 1.$$

The requirements on the error function are that is must vanish together with all its derivatives of up to order $m + n$ at $x_0 = 0$. this requirement is obviously met if all the coefficients of the numerator (as specified below) are equal to zero, i.e., $\alpha_\mu = 0$ for all $0 \le \mu \le m + n$, with the α_μ defined by $\sum_{\mu=0}^{\infty} c_\mu x^\mu \cdot \sum_{\mu=0}^{n} b_\mu x^\mu - \sum_{\mu=0}^{m} a_\mu x^\mu = \sum_{\mu=0}^{\infty} \alpha_\mu x^\mu$. It follows then that the b_μs must satisfy the linear system of the n equations:

(16d) $\sum_{\mu=0}^{n} c_{m+n-1-\mu} \cdot b_\mu = 0$, for all $1 = 0, 1, \ldots, n - 1$, $c_i = 0$ if $i < 0$, $b_0 = 1$, and the a_μs are then given as linear combinations of the b_μs by:

(17d) $a_i = \sum_{\mu=0}^{i} c_{i-\mu} \cdot b_\mu$, for all $i = 0, 1, \ldots, m$, $b_k = 0$ if $k > n$.

The linear system of Eq. (16d) may or may not have a solution. Depending on the existence of a solution of said system, we obtain a Padé Approximation for a specific combination of m and n.

Examples

(i) $f(x) = \sin x$, $|x| \le \frac{\pi}{2}$, $m = n = 2$.

Since $\sin x = x - \frac{1}{6}x^3 + \frac{1}{120}x^5 - \cdots$ it follows that

$c_0 = 0, c_1 = 1, c_2 = 0, c_3 = \frac{1}{6}, c_4 = 0, \ldots$ and hence (16d) yields:

$\sum_{\mu=0}^{2} c_{4-\mu} \cdot b_\mu = c_4 + c_3 \cdot b_1 + c_2 \cdot b_2 = 0$, and it follows that $b_1 = 0$. We also have: $\sum_{\mu=0}^{2} c_{3-\mu} \cdot b_\mu = c_3 + c_2 \cdot b_1 + c_1 \cdot b_2 = 0$, and therefore, $-\frac{1}{6} + b_2 = 0$, and therefore $b_2 = \frac{1}{6}$.

Substituting in Eq. (17d) yields:

$a_0 = \sum_{\mu=0}^{0} c_{0-\mu} b_\mu = c_0 = 0$, hence $a_0 = 0$; $a_1 = \sum_{\mu=0}^{1} c_{1-\mu} b_\mu = c_1 + c_0 b_1 = 1 + 0 b_1 = 1$;

$a_2 = \sum_{\mu=0}^{2} c_{2-\mu} b_\mu = c_2 + c_1 b_1 + c_0 b_2 = c_2 = 0$.

Substituting these values into the Padé Approximation:

(18d) $f(x) \cong R_{m,n}(x) \cong \frac{P_m(x)}{Q_n(x)}$ yields $\sin x \cong \frac{x}{1 + \frac{1}{6}x^2}$, an approximation that constitutes a good one only for small values of x. For large values of x, as for instance $x = \frac{\pi}{2}$, the error is of the magnitude 10^{-1}.

(ii) $f(x) = e^x$, $m = n = 2$, we have $e^x = \sum_{\mu=0}^{\infty} \frac{x^\mu}{\mu!} = 1 + x + \frac{x^2}{2} + \frac{x^3}{6} + \frac{x^4}{24} + \cdots +$, hence:

4.1 Some of the Formulae and Mathematics Used by the Navigator

$c_0 = 1, c_1 = 1, c_2 = \frac{1}{2}, c_3 = \frac{1}{6}, c_4 = \frac{1}{24}$. Again, substituting these values in Eq. (16d) yields with $b_0 = 1$:

$$c_4 + c_3 b_1 + c_2 b_2 = \frac{1}{24} + \frac{1}{6}b_1 + \frac{1}{2}b_2 = 0, \text{ and}$$

$c_3 + c_2 b_1 + c_1 b_2 = \frac{1}{6} + \frac{1}{2}b_1 + b_2 = 0$, resulting in $b_1 = \frac{1}{2}$ and $b_2 = \frac{1}{2}$, and by substituting into (17d) we obtain $a_0 = c_0 = 1$; $a_1 = c_1 + c_0 b_1 = 1 - \frac{1}{2}$, i.e., $a_1 = \frac{1}{2}$; $a_2 = c_2 + c_1 b_1 + c_0 b_2 = \frac{1}{2} - \frac{1}{2} + \frac{1}{12} = \frac{1}{12}$. Therefore, the desired Padé Approximation is:

$$e^x = \frac{12 + 6 \cdot x + x^2}{12 - 6 \cdot x + x^2}.$$

Again, I would like to reiterate that the Taylor and Padé approximations for any given m and n are only accurate in the "SMALL", i.e., for values of x close to zero, but can be highly inaccurate in the "LARGE", i.e., for values of x away from zero. However, there are still other approximations by polynomials and rational functions that are "good" over the entire interval of approximation for relatively small values of m and n.

Next, let's deal with the problem of solving non-linear equations or system of non-linear equations as encountered in astronavigation.

(e) Solving Non-Linear Equations.

As has been mentioned in the preface of this book, any approach to solving the problems of astronavigation by computational methods involves a little more than just dealing with applied spherical trigonometry. In particular, we have employed various types of approximation and also have come across the problem of solving non-linear equations by iterative methods. Philosophically speaking, Navigation is a Science based on approximations and whenever there are approximations involved there exists the need for iterative processes for solving the resulting problem analytically.

Let us first consider the problem of solving the non-linear equation:

(19e) $f(x) = 0$, $x \in [a,b]$, where "f" is a continuous function with continuous first and second derivatives on $[a, b]$. This problem actually breaks down into solving two distinct problems, namely:

  (i) Determining whether or not (19e) possesses a solution—root on $[a, b]$ and
  (ii) If it does have a solution, how do you find it by calculation?

In what follows next, we must employ the concept of continuity that has been defined mathematically and that translates, in simplified terms, into saying that one must be able to plot the graph of the function f over the interval $[a, b]$ between points $(a, f(a))$ and $(b, f(b))$ without taking the pencil off the paper.

I shall now state a sufficient, but not a necessary condition for the existence of a root of (19e) on $[a, b]$:

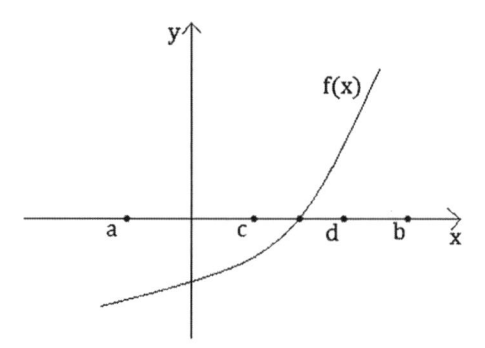

Fig. 4.1.8

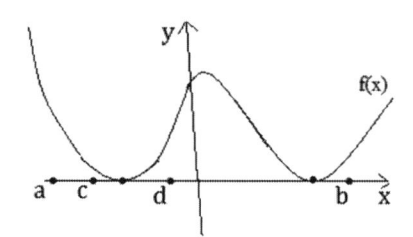

Fig. 4.1.9

"A sufficient condition for the existence of a root of (19e) on [a, b] is that f(x) is continuous on [a, b] and that there are two points c and d on [a, b] so that $f(c) \cdot f(d) < 0$."

Graphically speaking, this is very obvious (see Fig. 4.1.8) and according to Fig. 4.1.9, the fact that this condition is by no means necessary becomes obvious.

The mathematical proof of the above statement is a direct consequence of the INTERMEDIATE VALUE THEOREM of Calculus and states that if f is continuous on [c, d], then if there exists a number A such that $f(c) \leq A \leq f(d)$, then there exists a value of $\zeta \in [c, d]$ such that $f(\zeta) = A$.

Next we derive a procedure for actually calculating an approximation to the root r of (19e) subject to additional conditions on f, namely:

(i) There exists a root r on [a, b],
(ii) f is twice differentiable with continuous second order derivative on [a, b],
(iii) There exists a neighborhood of r defined by $|x - r| < \varepsilon$, and constants K and M such that $|f'(x)| \geq K$, and $|f''(x)| \leq M$, for all x: $|x - r| < \varepsilon$ for a positive number ε.

Then if x_0 lies in this neighborhood, i.e., $|x_0 - r| < \varepsilon$ and furthermore, if $N \cdot |r - x_0| \leq 1$, with $N = \frac{M}{2K}$, then the sequence $\{x_\mu\}$ defined by:

4.1 Some of the Formulae and Mathematics Used by the Navigator 227

(20e) $x_{\mu+1} = x_\mu - \dfrac{f(x_\mu)}{f'(x_\mu)}$, NEWTON-RAPHSON SEQUENCE, $\mu = 0, 1, \ldots$
converges to the root r, i.e., $\lim_{\mu \to \infty} x_\mu = r$.

To prove this statement of convergence to the root, we must employ Taylor's-TH. for n = 2 at $x = x_\mu$ to obtain:

$$f(x) = f(x_\mu) + f'(x_\mu)(x - x_\mu) + \dfrac{f''(\zeta)}{2}(x - x_\mu)^2.$$ Hence for $x = r$, we obtain

with $\zeta \in |r - x_\mu| : x_\mu - \dfrac{f(x_\mu)}{f'(x_\mu)} = r + \dfrac{f''(\zeta)}{2f'(x_\mu)}(r - x_\mu)^2$ for all $x_\mu \cdot |r - x_\mu| < \varepsilon$.

Substituting (20e) into this expression, we deduce that:

$$|r - x_{\mu+1}| = \left|\dfrac{f''(\zeta)}{2f'(x_\mu)}\right| \cdot |r - x_\mu|^2 \leq N \cdot |r - x_\mu|^2,$$ with "N" defined as above.

Therefore, we have:

(21e) $|r - x_\mu| = N^{-1}(N \cdot |r - x_0|^{2\mu}, \mu = 0, 1, \ldots$. Since $N \cdot |r - x_0| < 1$, we conclude that $\lim_{\mu \to \infty} x_\mu = r$, which was to be proved.

Note that (21e) enables us to determine how many iterations are required to make the error less than 10^{-K} for any positive K. Also note that the often used conditions $\lim_{\mu \to \infty} |x_{\mu+1} - x_\mu| = 0$, or $|x_{\mu+1}| < |x_\mu|$ for all $\mu \geq L$ are merely necessary but not sufficient conditions for the convergence of x_μ to r.

A typical application of this iterative method to Astor-Navigation consists in solving KEPLERS EQUATION, i.e., $E - e \cdot \sin E = M$.

In all cases where the derivatives of f are not given analytically, we must use an approximation instead, as, for instance, the backward difference quotient defined by

$f'(x_\mu) \simeq \dfrac{f(x_\mu) - f(x_{\mu-1})}{x_\mu - x_{\mu-1}}$. However, if we only replace the derivative in the Newton-Raphson formula (20e) by this approximation for the first derivative, we obtain the stationary iterative sequence $x_{\mu+1} = \dfrac{y_{\mu-1}}{y_{\mu-1} - y_\mu} \cdot x_\mu + \dfrac{y_\mu}{y_\mu - y_{\mu-1}} \cdot x_{\mu-1}, \mu \geq 1$,

x_0 and x_1 given.

This sequence so defined may not converge to the root r of (19e). Therefore, these formulae do not provide a suitable a!gorithm for computing the root of (19e).

As we have seen in Sect. 2.8, the Secant-Method is, in general, a non-stationary iterative method and only in cases where the function f is either convex or concave can it become a stationary iterative method. Hence, in order to assure convergence of our iterative process, we must use the formulae derived in Sect. 2.8 the latter entails subdividing the interval in which f is either convex or concave or choosing the interval about r sufficiently small to assure that f is either convex or concave. Then we may employ the formulae:

(22e) $x_{\mu+1} = \dfrac{y_\mu}{y_\mu - y_1} \cdot x_1 + \dfrac{y_1}{y_1 - y_\mu} \cdot x_\mu, \mu \geq 2, y_\mu = f(x_\mu)$, x_1 and x_2 given and

$f(x_1) \cdot f(x_2) < 0$ if f is CONVEX, or

(23e) $x_{\mu+1} = \dfrac{y_2}{y_2 - y_\mu} \cdot x_\mu + \dfrac{y_\mu}{y_\mu - y_2} \cdot x_2, \mu \geq 3$, otherwise as above but f is CONCAVE.

(f) **Differentials**.

As we have seen in several previous sections, the problems associated with finding error estimates in astronavigation are often related to approximations that are based on DIFFERENTIALS. In the cases where the error depend on more than one parameter, we are concerned with functions of several variables. In particular, as we have seen in Sect. 2.17, the dependent variable may be given implicitly. Therefore, let us consider the case where z is a function of x and y with the functional dependence given by $F(z) = G(x, y)$, where F is differentiable and G has partial derivatives of, at least, order one. Then, by defining a function H of the three variables x, y, and z by $H(x, y, z) = G(x, y) - F(z)$, we derive:

$dH = H_x \cdot dx + H_y \cdot dy + H_z \cdot dz = G_x \cdot dx + G_y dy - F'(z) dz = 0$, with $F'(z) \neq 0$. We find that:

(24f) $dz = \dfrac{1}{F'(z)} \cdot (G_x dx + G_y dy) = \dfrac{1}{F'} \cdot dG$, where $G_x = \dfrac{\partial G}{\partial x}, G_y = \dfrac{\partial G}{\partial y}$.

In cases where G depends on more than two variables, we find that:
$dz = \dfrac{1}{F'(z)} \cdot \sum_{\mu=1}^{n} \dfrac{\partial G}{\partial x_\mu} dx_\mu$, for all $F'(z) \neq 0$.

In order to enable Navigators to develop their own refraction formulae (see Sect. 3.1), it is also necessary to present some very important Quadrature Formulae, i.e., approximations to definite Integrals. Therefore, we need to address this particular aspect of Numerical Analysis.

(g) **Quadrature Formulae—Newton-Cotes Formulae**.

The majority of numerical quadrature formulae are based on the concept of approximating the integrand f(x) of the definitive integral $\int_a^b f(x) dx$ by functions for which the corresponding integrals are known, as for instance, in the case of polynomials. The difficult part in deriving such QUADRATURE FORMULAE consists in deriving explicit formulae for the error or error bounds. [18]

Here we shall only consider cases where the integrand f(x) is approximated by polynomials of degree n that pass through the n + 1 points (x_μ, y_μ), with $y_\mu = f(x_\mu)$. Furthermore, we will only consider such cases where the x's are spaced evenly on [a,b] and where $x_0 = a$ and $x_n = b$. Such types of formulae that are suitable for adoption on small calculators are referred to as Newton-Cotes-Closed Formulae. [18, 35]

4.1 Some of the Formulae and Mathematics Used by the Navigator

Newton-Cotes-Closed-Formulae—in the case of these types of formulae, the integrand f(x) is approximated by the Lagrangen Polynomials $L_\mu(x)$ of degree μ over the entire interval [a, b] and the abscissas x_μ are evenly spaced with the end point included. Hence $h = \frac{b-a}{n}$; $x_\mu = a + \mu \cdot h$; $\mu = 0, 1, \ldots, n$; $y_\mu = f(x_\mu)$. Integration of (3d) yields: $\int_a^b f(x)dx = \sum_{\mu=1}^n C_\mu y_\mu + \frac{1}{(n+1)!}\int_a^b P_{n+1}(x) \cdot f^{(n+1)}(\zeta)dx$, where the C_μs are:

(25g) $\quad C_\mu = \int_a^b L_\mu(x)dx.$

It can be shown [3] that the error E in the resulting approximation:

(26g) $\quad \int_a^b f(x)dx \cong \sum_{\mu=1}^n C_\mu y_\mu \text{ is given by:}$

(27g) $\quad E = \begin{cases} \dfrac{f^{(n+1)}(\eta)}{(n+1)!} \int_a^b P_{n+1}(x)dx \text{ if } n : \text{ odd } \eta \in (a, b). \\ \dfrac{f^{(n+2)}(\eta)}{(n+2)!} \int_a^b x \cdot P_{n+1}(x)dx \text{ if } n : \text{ even } \eta \in (a, b). \end{cases}$

Whenever the interval of integration [a, b] is large and our n is small, we may subdivide [a, b] into subintervals of length $n \cdot h$ and then apply the quadrature formula for a given n to each subinterval.

For the use on small calculators, the following explicit quadrature formulae are suitable:

TRAPAZOIDIAL RULE: By choosing n = 1 in formula (25g), (26g) and (27g) we find that

$C_0 = \dfrac{h}{2}, \quad C_1 = \dfrac{h}{2}, \quad y_\mu = f(x_0 + \mu \cdot h), \quad \mu = 0, 1, \ldots$ and

(28g) $\quad \int_{x_0}^{x_0 + h} f(x)dx \cong \dfrac{h}{2}(y_0 + y_1), \; E_1 = -\dfrac{h^3}{12}f''(\eta), \; \eta \in (x_0, x_0 + h).$

In the cases where [a, b] is large, we divide it into m > 1 subintervals, each of some length h and apply (28g) to each subinterval. The result of the summation of all subintervals is:

(29g) $\quad \int_a^b f(x)dx = h \cdot \left(\dfrac{y_0}{2} + \sum_{\mu=1}^{m-1} y_\mu + \dfrac{y_m}{2}\right), \; E_{1,m} = -\dfrac{h^2}{12}(b-a)f''(\eta); \; h = \dfrac{b-a}{m};$
$y_\mu = f(x_0 + \mu \cdot h), \; \mu = 0, 1, \ldots, m-1;$
$x_0 = a; \quad \eta \in (a, b).$

SIMPSON'S RULE: If we choose n = 2 in formulae (25g)–(27g), we obtain with

$x_\mu = x_0 + \mu \cdot h, \mu = 0, 1, 2$, that $C_0 = \dfrac{h}{3}$, $C_1 = \dfrac{4}{3} \cdot h$, $C_2 = \dfrac{h}{3}$ and with
$y_\mu = f(x_0 + \mu \cdot h), \mu = 0, 1, 2$.

(30g) $\displaystyle\int_{x_0}^{x_0 + 2h} f(x)dx \cong \dfrac{h}{3} \cdot (y_0 + 4 \cdot y_1 + y_2)$, $E_2 = -\dfrac{h^5}{90} f^{(4)}(\eta)$, $\eta \in (x_0, x_0 + 2h)$.

Again, in all cases where the interval of integration is large, we may apply the same procedures as in the case of the Trapezoidal Rule and obtain:

(31g) $\displaystyle\int_a^b f(x)dx = \dfrac{h}{3}(y_0 + 4 \cdot \sum_{\mu=0}^{m-1} y_{2\mu+1} + 2 \cdot \sum_{\mu=1}^{m-1} y_{2\mu} + y_{2m})$ $\quad E_{2m} = -\dfrac{h^4}{180}(b-a)f^{(4)}(\eta)$
$h = \dfrac{b-a}{2m}; \quad \eta \in (a,b);$
$m > 1$.

Note that the subdivision of [a, b] is only applied to the integration of the first term of (3d).

With the help of the above equations, together with some of the other formulae provided in this section, the reader can develop his or her own refraction formulae, provided that they use a realistic model of the atmosphere that can be expressed analytically by piecewise continuous functions.

(h) **Algorithm For Polynomials**.

Another useful algorithm that come into play for computing the value of a polynomial P(x) for any given value of x is this:

(32h) $P_n(x) = \sum_{\mu=0}^{n} a_\mu x^\mu$ with the n + 1 coefficients a_μ. Then let us define the sequence $b_\mu, \mu = 0, 1, \ldots n$, for a given x by:

(33h) $b_0 = a_n, b_\mu = b_{\mu-1} \cdot x + a_{n-\mu}, \mu = 1, \ldots n$, then $P_n(x) = b_n$.

Proof—by mathematical induction.

4.2 Some of the Instruments Used by the Navigator

By strictly adhering to the title and main objectives of this book, I should actually refrain from describing the various instruments and their uses in navigation. However, despite all rational reason for not doing it, I have decided to add a small section on the use of the major instruments for the sake of completeness and for the opportunity to share some of my experience in selecting those instruments that might interest the reader.

Beginning with, perhaps, the most important class of instruments, namely, the compasses. I have divided them into three categories:

4.2 Some of the Instruments Used by the Navigator

(i) MAGNETIC COMPASES—based on the electromagnetic field of the Earth and allowing a directional devise (needle) to point to the magnetic pole of the planet. I recommend that you buy only high quality units and also have, at least, two stationary compasses and one hand-bearing compass on board. I personally carry a miniature had-bearing compass on a lanyard around my neck.

Of course, it is imperative that the navigator checks the deviation and variation of the magnetic compass frequently. Nowadays, digital compasses with a higher degree of accuracy are also available to seafarers and aviators.

(ii) ASTRO AND SUN COMPASSES. Although used to a lesser degree by the mariner than by the aviator, the Astro Compass requires as input, the latitude of the observer, the local hour angle, and the declination of the celestial object. By pointing the sighting assembly towards the CO, the true bearing, i.e., heading is read at the lubber-line.

With the Sun Compass, this device shows the direction by means of a shadow cast by a pin when exposed to sunlight. First the course of the vessel is set opposite to the lubber-line which is aligned with the fore-aft axis of the vessel on an horizontal azimuth dial. Then, by means of another dial that is adjusted by a latitude scale, the pin is adjusted so as to be parallel to the polar axis and set to the local apparent time. When the vessel is on course, the shadow of the pin falls across the center of the local time dial and pointing at the preset local apparent time mark. [4, 46, 48]

As a former aviator, I have a personal preference for the versatile Astro-Compass that is used primarily for night observations.

(iii) GYRO COMPASS. Another class of compasses is based on the dynamics of a rotating gyro (discussed in the second part of this book). Because of the gyroscopic inertia, the gyro compass, once it has been set and activated, will continue to point in the same direction and the lubber-line will indicate the course.

(More about the big gyros of the universe, like the Sun, Earth, Moon, Planets, and Stars in the second part of this book.) [2]

Although standard equipment on all aircrafts, only bigger yachts afford the luxury of this device. Furthermore, this instrument requires a power source, i.e., batteries, alternator, etc.

Next in line of importance come the timepieces like clocks, watches, etc.

CLOCKS AND WATCHES constitute the most indispensable instruments for the navigator. Since we have already examined this item in great detail in Sect. 2.4, I would merely like to add another small and inexpensive timepiece to the previous list, namely, an independent small stopwatch that can be attacked directly to the sextant.

SEXTANTS, THEODOLITES, AND KAMELS. Here let me distinguish between the so-called BUBBLE SEXTANTS as used in Air Navigation, as for instance, the US Navy Mark V and the British RAF Mark IX, and the optical sextants that are based on two mirrors, a telescope and a graduated arc with a movable part that contains one of the two mirrors (see Fig. 4.2.1). The most expensive one also utilize a tangent screw with micrometer drum attached enabling the navigator to measure angles as small as $0'.1$ or less. [24]

SEXTANT AND ARTIFICAL HORIZONS

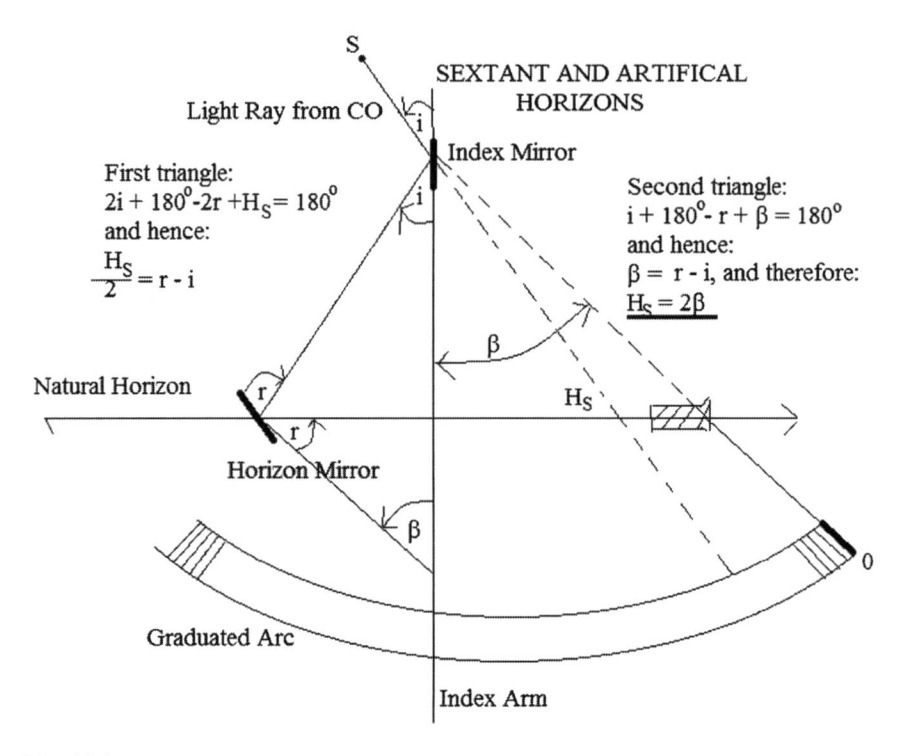

First triangle:
$2i + 180° - 2r + H_S = 180°$
and hence:
$\frac{H_S}{2} = r - i$

Second triangle:
$i + 180° - r + \beta = 180°$
and hence:
$\beta = r - i$, and therefore:
$\underline{H_S = 2\beta}$

Fig. 4.2.1

A very important accessory to the optical sextant is the Artificial Bubble Horizon that can be attached to almost any conventional sextant, enabling the navigator to make night observations. I have personally equipped all my optical sextants with bubble horizons.

For observation on land or coastlines, the most accurate instrument is the THEODOLITE as used by land surveyors and explorers. In case that such an instrument, which measures angles as small as 1″ or less, is unavailable, an artificial horizon that may consist of a tray filled with a high viscosity liquid can be used in conjunction with a marine sextant for measuring vertical angles. In an emergency, or for "non-instrument navigation" the KAMEL and ASTROLABE can be substituted for a sextant buy only if it is used as a calibrating device. [10, 49]

Next come the instruments necessary for determining Dip and Refraction for Non-Standard conditions, i.e., for almost any condition that the navigator will encounter at sea or in the air.

BAROMETER AND THERMOMETER. It is important that each vessel has at least one pair of these instruments on board for the purpose of calculating Refraction and dip, and also for meteorological purposes such as Weather Forecasts, etc. All leading marine and aviation supply stores carry a large assortment of these instruments and it us up to the navigator to decide what suits him or her the best. However, I strongly recommend that the navigator also acquires a

4.2 Some of the Instruments Used by the Navigator

pocket type Barometer-Thermometer as used, for instance, by mountain climbers. Nowadays, one can even buy a wristwatch that incorporates a digital barometer and thermometer along with the timepiece.

LOGS. Next comes the instruments for measuring the speed of the vessel relative to the water, or in the case of an airplane, the air speed, and also the distance traveled relative to those elements or relative to the ground. There are basically four different types of devices for measuring the magnitude of velocity.

The first consists in measuring the distance traveled together with the elapsed time which can be done, for instance, by throwing a line that is attached to a float over board and counting the time and distance traveled.

The second type of log utilizes an impeller and thereby measures the rotation of this device that corresponds to the distance and speed of the vessel.

The third type is based on Bernoulli's Law that relates pressure and speed in a suitable tube. Whenever the speed of the vessel changes, the pressure in this device also changes and because of said law, speed is determined by measuring the pressure in the tube. In the case of airplanes, air speed is measured in a similar way by means of the so-called "pitot" tube.

The fourth type of instruments are the electronic devices like transmitters and receivers used in RADARS. Ground speed is determined by utilizing the DOPPLER EFFECT.

DEPTH INDICATORS AND SONARS. Next to the logs come the instruments that measure the distance above the ground level, in the case of a vessel, the bottom of the sea. Besides speed and distance, the navigator also has to determine the distance the vessel or aircraft is above the ground level. In bygone times, the mariner used a "lead-line" with the distances marked on it. Modern depth sounders use ultra sound or SONAR to determine depth and, perhaps, drift. In the case of an airplane, barometric pressure, DOPPLER, and/or RADAR are all employed to determine height above ground.

CALCULATORS AND COMPUTERS. Finally, there is the brain of all these instruments, namely the frequently used and indispensable Calculator and/or Computer, and a set of good old Logarithm Tables to cover an emergency situation when all the electric and electronic devices have been deactivated, i.e., blown out. As we have seen on several occasions in this book, the calculator can be anything that has an algebraic logic and also has the elementary trigonometric functions as well as the exponential functions built in. This amounts to saying that any Scientific Calculator that can be had for $10.00 or less will do. On the upper scale of those devices, the limits are only set by the manufacturers and the purse of the buyer.

However, in order to perform all the necessary calculations in a speedy way for the EPHEMERIS as developed in the second half of this book, a more powerful and programmable calculator is highly recommended. This is particularly true when calculating orbits. [50, 54]

STAR FINDERS AND ASTROLABES. In order to facilitate the identification of stars and planets (see Sect. 2.19), and at the same time provide a picture of the sky overhead at any given instant, a STAR FINDER and/or an ASTROLOBE are very handy. [11, 37, 40, 46]

GPS. Last, but not least, is a GPS receiver which is obviously the first choice of the Coastal-Navigator. Its accuracy and simplicity with regards to its use are unsurpassed by all the other electronic navigational systems. Since virtually every serious boater has a GPS and also know hot to use it, I will refrain from elaborating on it. however, I am compelled to point out the shortcoming and limitations of this System.

The major shortcomings of this system are its dependence on artificial satellites that can be sabotaged and/or eliminated, and its dependence on electronic transmitters and receivers, both of which can be knocked out by an electro magnetic pulse. It may never happen. But then, it could happen tomorrow.

An Astro-Compass

A Mark II Astro Compass (Stock photo) (*Image credit* www.ebay.com)

4.2 Some of the Instruments Used by the Navigator

The British RAF Mark IX Bubble Sextant

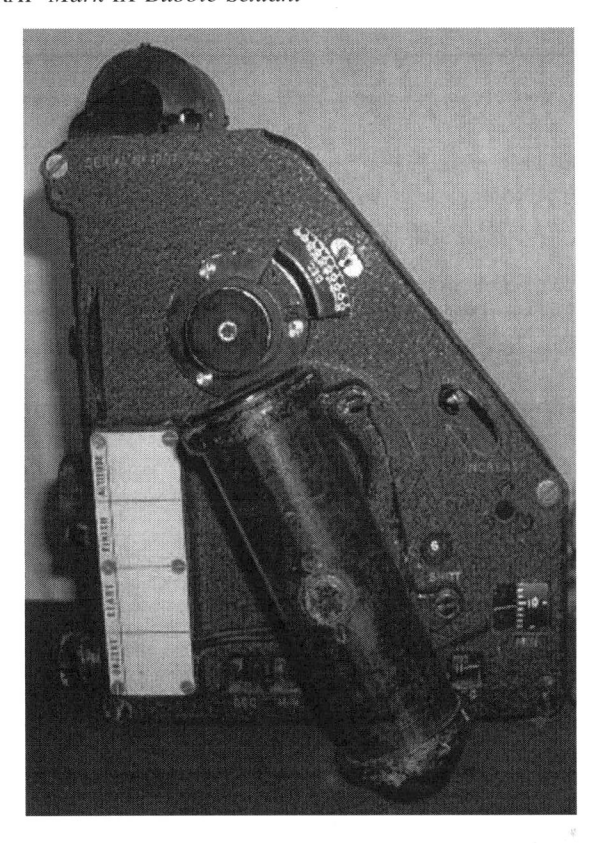

A RAF Mark IX Bubble Sextant courtesy of Military Autographs UK (*Image credit* militaryautographs.com)

Various Sextants

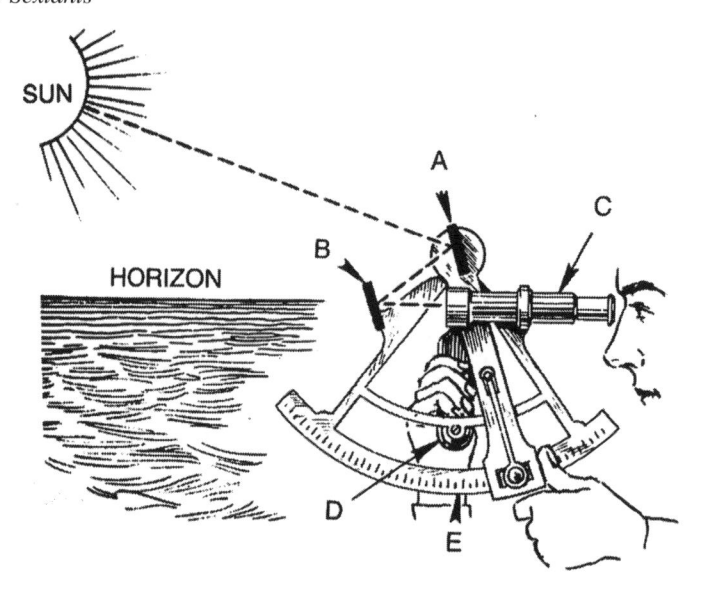

Marine Sextant in use (*Image credit* Pearson Scott Foresman)

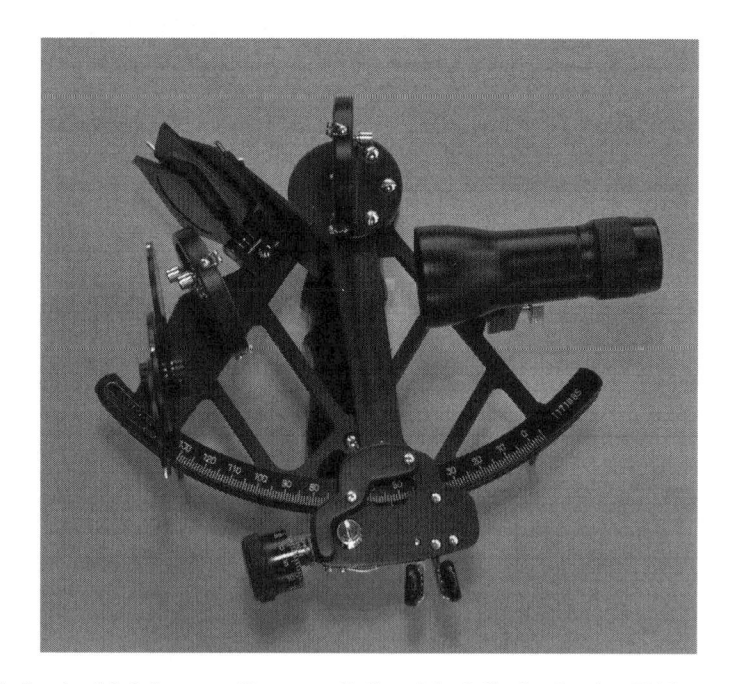

A Stanly London Mark 3 sextant (*Image credit* Copyright © Stanley London 2016)

A Sun Compass

Solar Compass (*Image credit* Dustin Plunkett, originally posted to Flickr as William Burt Solar Compass, CC by 2.0)

Star Finders and Astrolabes

Commercial Star Finder (*Image credit* physics.csbsju.edu)

Part II
Formulae and Algorithms of Positional Astronomy

In which the navigator will find the necessary formulae and algorithms employing the basics of Positional Astronomy to enable him or her to find their position without having a Nautical Almanac at their disposal. Included is an ephemeris for the navigational stars and an abridged, low or medium low ephemeris for the sun.

Chapter 5
Elements of Astronomy as Used in Navigation

5.1 Some Basic Concepts Describing the Motion of the Earth Around the Sun

In this section we shall develop some approximations for the most basic aspect of an ephemeris for the Sun and the stars which will enable the navigator to calculate approximations to the required Greenwich Hour Angle (GHA) and Declination δ of the Sun and stars. Those results can be classified as low precision ephemerides for the Sun and stars. In the following sections those formulae will be replaced by more accurate ones that result in improved low precision and intermediate precision ephemerides.

Contrary to what a navigator may find in text books on Astronomy, all the formulae provided in this manual come with all the necessary data for the actual numerical evaluation. What good is it for a navigator to be provided by an approximate formulae for the equation of time or the equation of the center if no explicit formulae or values for the mean anomaly and mean longitude are given? Furthermore, all formulae provided herein can be evaluated by using an inexpensive scientific calculator.

Since this book is not meant to be a comprehensive text on Positional Astronomy, only the basics of the underlying theory will be explained allowing the reader who has only a very limited knowledge of Astronomy to apply the provided formulae intelligently. Detailed derivation of the formulae will sometimes be omitted.

The starting point for deriving a practical ephemeris must, of course, be to define a reverence point on the celestial equator. This reference point is called the first point of Aries or the point of the vernal equinox, similar to the referenced system on the globe referred to as the Greenwich Meridian.

As has already been explained in the first part of this book, the first point of Aries (♈) is defined as the first point of intersection of the celestial equator and the ecliptic. (See Fig. 5.1.1.)

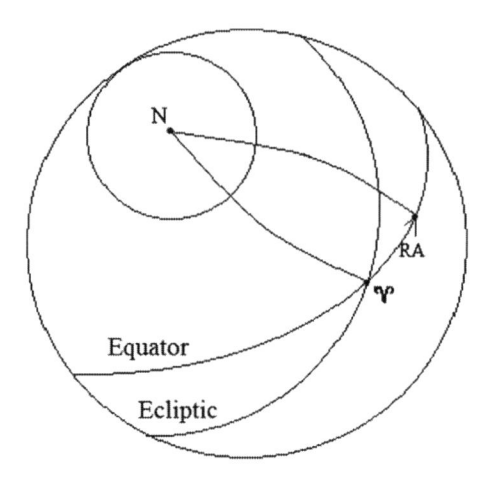

Fig. 5.1.1

It is obvious that ♈ can also be defined as one of the points where the declination of the Sun is equal to zero. Of course, ♈ is not necessarily occupied by a real star but can be identified with the help of a real one. Take **ALPHERATZ** with SHA 357°.47. Once ♈ has been determined, all Celestial Objects (CO) are assigned coordinates in terms of the hour angle of the CO relative to ♈, referred to as Right Ascension (RA) and the angle measured from the equator to the CO and referred as Declination δ. Therefore, if we can find the hour angle subtended by the Greenwich Meridian, i.e. GHA ♈, we can then find the GHA or the star by adding the Sidereal Hour Angle (SHA) to the GHA ♈. In other words, GHA (CO) = GHA ♈ + SHA (CO). Also recall at this point how the RA is related to the SHA, namely RA = 360° − SHA.

It follows then that the RA is the angle measured from ♈ to the intersection of the celestial equator with the meridian of the CO by going to the right (see Fig. 5.1.1).

Since our objective is to find GHA ♈ at any instant of Universal Time (UT), we must introduce the concept of a fictitious mean Sun, denoted by ⊙, that travels along the equator from left to right at a constant angular velocity of n°, and another fictitious Sun, referred to as the dynamic mean Sun and denoted by $\overline{\odot}$, that travels with the same constant velocity n° along the ecliptic. The two fictitious Suns are synchronized by meeting at the Vernal Equinox T at the same instant. Therefore, the RA of the fictitious mean Sun ⊙ is equal to the longitude ℓ of the dynamic mean sun $\overline{\odot}$; i.e. RA ⊙ = ℓ. Because of the latter, we shall refer to both fictitious Suns as the mean Sun ⊙. In order to visualize the three Suns, Fig. 5.1.2 shows the position of these three Suns after they have passed the first point of Aries at the same instant.

Here "D" denotes the position of the dynamic mean Sun on the ecliptic; "S" denotes the position of the true Sun on the ecliptic; "M" denotes the position of the fictitious mean sun on the equator; and "T" indicates the point of intersection of the hour circle of the true Sun with the equator. Therefore we can conclude as per the definition of ⊙ that:

5.1 Some Basic Concepts Describing the Motion ...

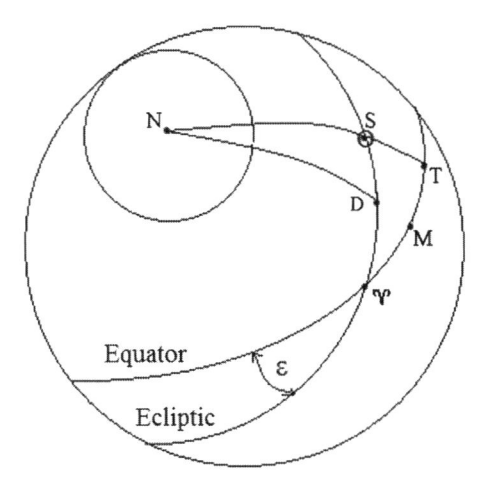

Fig. 5.1.2

$\sphericalangle \Upsilon D = \sphericalangle \Upsilon M = RA_\odot = \ell;\quad \sphericalangle \Upsilon T = RA_\odot = \ell;\quad \sphericalangle \Upsilon S = \lambda;$ where "ℓ" denotes the longitude of the mean Sun and λ the longitude of the true Sun.

The dynamic mean sun has been introduced with reference to the true Sun that moves according to Kepler's Second Law with a variable angular velocity around the Earth's geocentric system. However, the two Suns are no more than about 16 min apart at the extreme. Or course, the spacing of D, S, M and T in Fig. 5.1.2 are highly exaggerated for the sake of clarity.

An immediate consequence of the introduction of the mean Sun is the equation:

(1) $\quad$ GHA Υ = $RA_\odot + GHA_\odot = \ell + GHA_\odot$

Since the GHA $\odot$ is directly related to the universal time UT and therefore to the UTC (also known as Coordinated Mean Time, or clock time) finding the GHA Υ amounts to finding the RA $\odot$ or ℓ.

Although we can't observe the transit of Υ directly, a nearby star whose SHA is known can be used to time the transit at any specific meridian, say the Greenwich meridian. Another way of obtaining the time of transit of Aries it to use compiled data as found in Almanacs. Once the time of transit of the first point of Aries has been determined, the GHA Υ can be calculated easily since the rate of rotation of the earth is 24 h Sidereal Time. Of course, we have assumed so far that the first point of Aries does not move, which is by no means correct. Only if we can describe the movement of Υ on the celestial sphere quantitatively we will be able to obtain suitable results for our ephemeris. However, the discussion of the Equation of the Equinox will be deferred until later when we deal with more exact equations for the GHAs and the declinations δ of the COs.

5.2 An Approximation to the Time of Transit of Aries at Greenwich and the Greenwich Hour Angle GHA of ♈

For the purpose of this section, it will suffice to assume that ♈ remains stationary. As with regard to finding an approximate value for the Greenwich Hour Angle of ♈—GHA ♈—at any instant during a specific year, we can employ the method of simple interpolation, provided that we know the GHA ♈s at two distinct days at 0^h:00 UTC during this specified year. This method of interpolation is valid because we can deduce from the general theory [see Sect. 6.2 and (3)] that the dependence of GHA ♈ on the time can be approximated sufficiently accurately by a linear function of the time. For now, assume that we have extracted those two values for the GHA ♈s from the NA of the year in question. (In Chap. 8, we will examine another simple formula for the GHA ♈ that does not required those two values and is therefore independent of the NA.)

In order to derive a suitable approximation for the GHA ♈, we must assume that we know the value of GHA ♈ on two distinct days at 0^h:00 UTC. Let us denote those values by GHA ♈$(\overline{N}_0)$ and GHA ♈$(\overline{N}_1)$, respectively. Furthermore, we have imposed the condition that:

GHA ♈$(\overline{N}_0)$ < GHA ♈$(\overline{N}_1)$ < 360°, where "$\overline{N}_0$" and "$\overline{N}_1$" denote the distinct days by their year—number—(see example below). Then if $\overline{N}$ is any day of the specified year in question at 0^h:00 UTC, we deduce that:

GHA ♈$(\overline{N})$ = GHA ♈$(\overline{N}_0)$ + m · $(\overline{N} - \overline{N}_0)$ with m = (GHA ♈$(\overline{N}_1)$ − GHA ♈$(\overline{N}_0))/(\overline{N}_1 - \overline{N}_0)$.

This is the simple interpolation formula referred to above (see also Sect. 4.2). It follows then that on day N at UTC denoted by "$\overline{N}$" we have:

(1) GHA ♈$(\overline{N})$ = GHA ♈$(\overline{N}_0)$ + m · $(\overline{N} - \overline{N}_0)$ + 15.04106865 UTC

m = (GHA ♈$(\overline{N}_1)$ − GHA ♈$(\overline{N}_0))/(\overline{N}_1 - \overline{N}_0)$, GHA ♈$(\overline{N}_0)$ < GHA ♈$(\overline{N}_1)$ < 360°.

Note that $\overline{N}_0$ and $\overline{N}_1$ and the values of GHA ♈ required, together with m, have to be specified only once per year (see example below).

Next, let's consider two examples:

Example #(1) Find the GHA ♈ on 07/14/13 at 17^h:00 UTC.
Solution:

i. Choose $\overline{N}_0$ and $\overline{N}_1$ within 2013—to be done only once in 2013.

$$03/21/13; 0^h : 00, N_0 = 80$$
$$06/21/13; 0^h : 00, N_1 = 172$$

ii. From NA 2013 take GHA ♈$(\overline{N}_0)$ = 178°.676666, and GHA ♈$(\overline{N}_1)$ = 269°.356666

5.2 An Approximation to the Time of Transit of Aries …

iii. Calculate m.

$$m = 0.985652166$$

iv. Count $\overline{N} = 195 (07/14/13 \text{ at } 0^h : 00 \text{ UTC})$
v. Substitute those values into Eq. (1) to obtain:

$$\text{GHA} \Upsilon (\overline{N}) = 547°.7248328 - 360° = 187°.7248328$$
$$= 187°43'.49 (\text{NA value is } 187°43'.2)$$

Example #(2) Find the GHA Υ on 07/15/13 at 11^h:00 UTC.

Solution:

i. Count $\overline{N}$, i.e., $\overline{N} = 196$ for (07/15/13 at 0^h:00 UTC).
ii. Substitute these values together with those values calculated in Example 1. into Eq. (1) to obtain:

$$\text{GHA } \Upsilon (\overline{N}) = 485°.4640728 - 360° = 98°.46407285$$
$$= 98°27'.84 (\text{NA value is } 98°27'.68.)$$

In conclusion, it should be noted that Eq. (1) also enables the navigator to obtain a simple approximation for the RA of the mean Sun—for by definition of the RA we have RA $\odot$ = GHA Υ (N) − (UTC − 12) · 15. Hence by substituting the expression given by (1) in this equation above we obtain:

(2) $\text{RA} \odot = \text{GHA} \Upsilon (\overline{N}_0) + m \cdot (\overline{N} - \overline{N}_0) + 0.04106865 \cdot \text{UTC} + 180°.$

Note that in RA $\odot$ is given in degrees instead of hours, immediately providing that the navigator with the mean longitude ℓ, i.e., $\ell(N) = $ RA $\odot$. Also note that the time of transit of Υ at Greenwich is given implicitly by (1)—put GHA $\Upsilon = 0$ and solve for UTC.

5.3 The Right Ascension of RA of the Mean Sun, Mean Longitude, Mean Anomaly, Longitude of Perigee, Longitude of Epoch and Kepler's Equation

Next in importance to having an explicit formula for the GHA Υ is to have a formula for the right ascension of RA $\odot$ of the true Sun, or which amounts to the same, a formula for the difference in the right ascension of the true and the mean Sun, i.e., the so-called Equation of Time (ET). By definition, the equation of time is given by:

(1) $ET = RA\underline{\odot} - RA \odot$.

Although the RA $\underline{\odot}$, which can readily be calculated by employing Eq. (1) in Sect. 5.2, furnishes one part of this equation, the other component, namely the RA $\odot$, cannot be calculated without knowledge of the parameters of the orbit of the Sun about the center of the Earth. Furthermore, Kepler's equation, or the Equation of the Center, has to be used in order to obtain a formula for the RA $\odot$ of the true Sun. In order to enable navigators to understand the underlying theory to the extent that they can assess the degree of accuracy of the resulting approximations, it is necessary to develop a clear picture of the underlying dynamics of the orbital motion of the Earth around the Sun and Sun around the Earth, respectively.

Starting with Kepler's first law, we know that the Earth moves along an ellipse of major semi-axis a, and minor semi-axis b. The sun, itself, occupies one of the foci of the ellipse. This true model is referred to as the heliocentric model. However, we will use the geocentric model instead, where the Sun is assumed to move around the Earth on the elliptic and the Earth occupies one of the foci. These two models correspond to the same physical situation and generate the same results with one exception, namely, the longitude of the Earth differs 180° from the longitude of the Sun, i.e. $\lambda_\odot = 180° + \lambda$. The geocentric model offers the advantage in that it relates directly to the celestial sphere where the Sun moves along the ecliptic. Here we are using the same analogy for the Earth as when we pretend that the celestial sphere rotates about the axis of the stationary Earth.

The geocentric model of the orbit is depicted in Fig. 5.3.1. It also lets us interpret the first point of Aries geometrically, for the line connecting the center of the Earth with Aries turns out to be the projection of the axis of the gyro-Earth onto the ecliptic.

In order to describe the motion of the Sun analytically, it will be necessary to define and compute the relevant parameters of the orbit as indicated in Fig. 5.3.1.

The parameters which define the Sun's orbit are:

a: semi-major axis of the ellipse.
$a = 1AU = 149.6 \times 10^6$ km
e: eccentricity of orbit. e = f/a, where "f" denotes the distance of the center of the Earth from the center of the ellipse. e = 0.016718, at epoch 1980.
ε_0: the obliquity of the ecliptic. $\varepsilon_0 = 23°.4418$ at epoch 1980.
T_p: period in tropical years. $T_p = 1.00004$ at epoch 1980.
$\tilde{\varepsilon}$: longitude at epoch. $\tilde{\varepsilon} = 278°.83354$ at epoch 1980.
ω: longitude at perigee. $\omega = 282°.596403$ at epoch 1980. [63]

In addition, we require:

v: true anomaly of the Sun measured from perigee.
M: mean anomaly of the mean Sun also measured from perigee and synchronized with v by $v(\tilde{\tau}) = M(\tilde{\tau}) = 0$, where "$\tilde{\tau}$" denotes the instant when v passes perigee.
Finally,
E: eccentric anomaly defined as shown in Fig. 5.3.1.

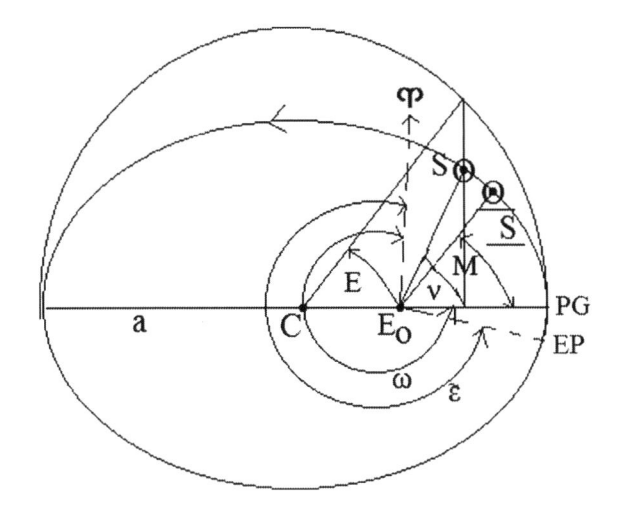

5.3 The Right Ascension of RA of the Mean Sun ...

Fig. 5.3.1

E: Eccentric Anomaly

M: Mean Anomaly

C: Center of Ellipse

S: Sun

$\underline{S}$: Mean Sun

E_0: Earth

v: True Anomaly

a: Semi-Major Axis

ω: Longitude at Perigee

PG: Perigee

EP: Epoch

$\tilde{\varepsilon}$: Longitude at EP

As can readily be deduced from Fig. 5.3.1, the mean longitude ℓ, which is the angle subtended by $\overline{E_0 \text{Aries}}$ and $\overline{E_0 \odot}$ is given by:

(2) $\ell = M + \omega = RA\underline{\odot}°$.

Similarly, the true longitude λ of the Sun is given by:

(3) $\lambda = v + \omega = l + C$

If "t" denoted the time, measured in days, which has passed since the Sun has passed the epoch, it follows that:

(4) $M = n° \cdot D + \tilde{\varepsilon} - \omega = n° \cdot (t - \tilde{\tau})$; where $n° = \frac{360°}{365.25} = 0.985626283$ and "D" denotes the time in days and fractions of days from the given epoch. It follows then from the definition of $\tilde{\tau}$ that:

(5) $\tilde{\tau} = \frac{\omega - \tilde{\varepsilon}}{n°}$.

It can also be easily shown that the eccentric anomaly E satisfies Kepler's equation:

(6) $E = M + e \cdot \sin E$, where E and M are expressed in radians. [58], [formula 1 and 3, 63]

Once a solution to Kepler's equation has been found by iterative methods, for instance, v, the true anomaly, can be calculated by using the formula:

(7) $\tan \frac{v}{2} = \sqrt{\frac{1+e}{1-e}} \cdot \tan \frac{E}{2}$ [59, 60]

5.4 The Equation of the Center, Equation of Time and True Longitude of the Sun

Instead of solving Kepler's equation by an iterative procedure, let's expand v in terms of e as a power series, or since it is a periodic function of M, develop it as a Fourier Series in M resulting in the equation of the center, namely:

(1) $C := v - M = \frac{180}{\pi}[(2e - \frac{e^3}{4} + \frac{5}{96} \cdot e^5) \sin M + (\frac{5}{4}e^2 - \frac{11}{24}e^4) \cdot \sin 2M + (\frac{13}{12}e^3$
$- \frac{43}{64} \cdot e^5) \cdot \sin 3M + \frac{103}{96} e^4 \cdot \sin 4M + \frac{1097}{960} e^5 \cdot \sin 5M + \cdots]$, [58, 63]

Similarly, we can calculate the distance from the center of the Earth to the center of the Sun (required for finding the semi-diameter of the disc of the Sun) by:

(2) $\frac{r}{a} = 1 + \frac{e^2}{2} - (e - \frac{3}{8}e^3 + \frac{5}{192}e^5) \cdot \cos M - (\frac{e^2}{2} - \frac{e^4}{3}) \cdot \cos 2M - (\frac{3}{8}e^3 - \frac{45}{128}e^5)$
$\cdot \cos 3M - \frac{e^4}{3} \cdot \cos 4M - \frac{125}{384} \cdot e^5 \cdot \cos 5M + \cdots$ [68, 72]

As will be explained later, for the purpose of navigation, it will suffice to truncate these formulae, i.e., consider only terms containing of up to the power of three in e. The above formula for the distance r between the Sun and Earth is based on the relation:

(3) $r = a \cdot \frac{(1-e^2)}{(1+e \cdot \cos v)} = a \cdot \frac{(1-e^2)}{(1+e \cdot \cos(M+C))}$. [59, 60]

Similarly, the Sun's angular size, i.e., its angular diameter Θ can be fund to be:

(4) $\Theta = \Theta_0 \cdot \frac{(1+e \cdot \cos v)}{(1-e^2)} = \Theta \cdot \frac{(1+e \cdot \cos(M+C))}{(1-e^2)} = \frac{\Theta_0}{r}$ in AU where Θ_0 is the angular diameter when r = a. For instance at epoch 1980 $\Theta_0 = 0°.533128$.

Before we address the problem of how to calculate or extract the elements of the orbit and its parameters, let us first consider the very relevant problem of finding an

5.4 The Equation of the Center, Equation of Time ...

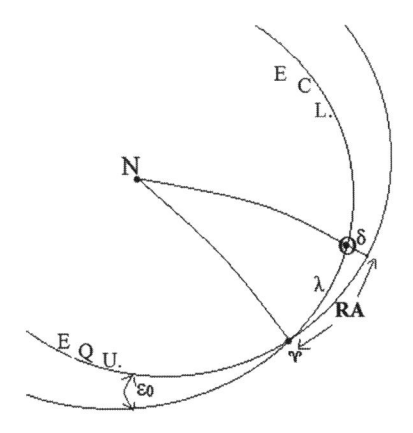

Fig. 5.4.1

approximation to the Equation of Time. According to (1) in Sect. 5.3, we must first calculate the right ascension RA ⊙ of the true Sun. The latter is given by:

(5) $\tan RA_\odot = \cos \varepsilon_0 \cdot \tan \lambda$ or $\cos \lambda = \cos \delta \cdot \cos RA_\odot$. [66]

and the declination of the true Sun δ is given by:

(6) $\sin \delta = \sin \varepsilon_0 \cdot \sin \lambda$. (See Fig. 5.4.1.)

Since the longitude λ of the true Sun ⊙ is given by formula (5), Eqs. (1) and (2) in Sect. 5.3 yield the exact equation of same as:

(7) $ET = \ell - \tan^{-1}(\cos \varepsilon_0 \cdot \tan \lambda)$.

In order to find a crude approximation to ET, we use a crude approximation to C by neglecting in (1) all terms that contain powers of e higher than one, i.e., we approximate C and therefore ET by their linear components. Specifically, we employ the approximation $C \simeq \frac{360}{\pi} \cdot e \cdot \sin M$. Using this approximation together with the fact that $\frac{\tan^2 \varepsilon}{2} \ll 1$ and also $RA_\odot = \lambda - \alpha$. Where α is a very small quantity, we may derive the following expression for the ET:

(8) $ET \simeq \frac{720}{\pi} \cdot (\tan^2 \frac{\varepsilon_0}{2} \cdot \sin 2\ell - 2 \cdot e \cdot \sin M)$—given in minutes of time. [63]

With the help of this low precision expression, formulae (1) and (2) in Sect. 5.2, and (5) and (6) in Sect. 5.4, and a simple formula for calculating the SHA of the navigational stars (as provided in a subsequent section), the navigator can still function without a NA. However, one obstacle in applying the above formula consists in finding the mean anomaly M that appears in this formula. In case that we can find a good approximation to $\tilde{\tau}$, we may use formula (4) in Sect. 5.3, where t is counted in days and fractions of a day from the epoch in question. If not, we may have to extract again values from the NA—once a year only, and calculate approximations to ω and e, as shown in Sect. 5.6, or use the polynomial expressions provided in Sect. 6.3.

In Sect. 5.6, I shall address this particular problem and eventually provide an exact formula for M(t). But first, let us consider some numerical examples and definitions.

5.5 Numerical Examples and Other Concepts of Time

First, let us consider an example for finding an approximation to the Equation of Time (ET) that is based on a crude estimate for the time of perigee.

Example #1 Find the ET on 07/15/13 at 12^h:00 UTC given the following estimate for $\tilde{\tau} = 01/03/13$ at $0^h : 00$ UTC.

Solution:

1. Calculate t and $\tilde{\tau}$, i.e., $t = -(365 - 196) = -169$. $\tilde{\tau} = 3$.
2. Calculate M(t):

$$M(t) = n°(t - \tilde{\tau}) = - n° \cdot 172 = -0°.985626283 \cdot 172 = -169°.5277207 + \\ + 360° \\ = 190°.4722793$$

3. From Example 1 in Sect. 5.2, take the values: $N_0 = 80$ and $\overline{N} = 196$ and

   calculate $RA\odot = 178°.67666 + 0.985652166 \cdot 116 + 0.04106865 \cdot 12 + \\ + 180° = 473°.505411 - 360° = 113°.5057411$, hence—
   $\ell = 113°.5057411$

4. By employing the approximate values: $\frac{\epsilon_0}{2} = 11.72$ and $e = 0.01666$ and the equation for ET, we find:

$$ET = \frac{720}{\pi} \cdot (\tan^2 11.72 \cdot \sin 227.0102821 - 2 \cdot 0.01666 \\ \cdot \sin 190.4722793) = -5^m.854398, \text{ i.e.,}$$

$$ET = -5^m 49^s.61 = -5^m 50^s. \text{ The NA value is } - 5^m 59^s.$$

However, the reader should realize that the relatively large error in the result of this example depends strongly on the crude estimate of $\tilde{\tau}$, i.e., the time of transit at perigee.

In the next sections we will see how not only the formula for the ET can be considerably improved upon, but also how the necessary parameters can be determined more accurately. Before we improve upon the basic Eqs. (1) and (2) in Sect. 5.2 and (8) in Sect. 5.4, we shall also examine how the equation for the ET that is valid not only for the apparent noon instant (AN), but also for any instant defined by UTC, can be readily employed to find the GHA $\odot$ and the declination δ of the true Sun. Hence, with the help of those basic equations and expressions (1) and (2) below, navigators have a "low precision" ephemeris for the Sun at their disposal.

5.5 Numerical Examples and Other Concepts of Time

Specifically, the GHA and the declination δ are then computed as follows:
From (1) in Sect. 5.3, we deduce that:
$RA\odot^0 = \ell - \frac{ET}{4}$ and hence:
$SHA \odot = 360° - RA \odot^0 \frac{ET}{4} - \ell$.
Since GHA $\odot$ = GHA Υ + SHA $\odot$ it follows that:

(1) $GHA \odot = GHA \Upsilon + \frac{ET}{4} - \ell$.

By applying the Four-Parts formula of spherical trigonometry (see Sect. 4.2) to the triangle of Fig. 5.4.1, we deduce that the declination δ is given by:

(2) $\tan \delta = \tan \varepsilon_0 \cdot \sin RA\odot^0 = \sin(\ell - \frac{ET}{4}) \cdot \tan \varepsilon_0$.

Again, let's consider a numerical example in order to illuminate the application of the ET to navigation.

Example #2 Find the GHA $\odot$ and the declination δ for the Sun on 07/15/13 at $12^h : 00$ UTC.

Solution:

1. First compute ET by employing formula (8)—see previous example—to obtain:

   $ET = -5^m.826854694$, and $\ell = RA\odot^° = 113°5051411$.

2. Next, compute GHA Υ by employing formula (1), Sect. 5.2 together with the values $N_0 = 80$, $N_1 = 172$ and $m = 0.985652166$, as used in Example #1, Sect. 5.2 and computed only once a year.

   $GHA \Upsilon = 178°.676666 + 0°.985652166(196 - 80) +$
   $+ 15°.04106865 \cdot 12 = 473°.5051411 - 360° = 113°.5051411$.

3. Compute GHA $\odot$ by employing formula (1) to find:

   $GHA\odot = 113°.5051411 - 113°.5051411 - \frac{5°.826854694}{4} =$
   $= -1°.456713674 + 360° = 358°.5432863 =$
   $= 358°32'.6$.

4. Using the value $\varepsilon_0 \cong 23°.44$ and formula (2), we find:

   $\tan \delta = \tan 23.44 \cdot \sin 114.9618548 = 0.393067738$, hence
   $\delta = 21°.45818852 = 21°27'.5$
   (The values given in the NA are :
   $GHA\odot = 358°30'.1$ and $\delta = 21°26'.9$)

The reader may recall that one of the basic assumptions in the derivation of Eq. (8) was that C, the Equation of the Center, contains merely the first term of order one in powers of e. Accordingly, one would expect that much better approximations can be obtained by taking terms of up to power three in e into account. This is in deed possible and intermediate precision formulae can be obtained, as will be shown in the next few sections. However, it should also be obvious at this point of these presentations, that any formula provided herein will depend on the parameters that enter it and in particular on the elements of the orbit that are time dependent. As we have seen in the examples above, parameters such as the eccentricity e, the obliquity of the ecliptic ε_0, and, in particular, the mean anomaly M enter all relevant calculations and must be found if not already given.

Therefore, in the next section of this book, we will address the actual problem of procuring all the necessary parameters that enter the equations for finding the GHA ⊙ and the declination δ of the Sun. At that time, I shall also present a more accurate expression for the Equation of the Time based on a more accurate equation for the center C.

So far, I have only used the concept of UTC (Coordinated Universal Time) and not considered other time frames as used in astronomy. Although, for the navigator, this UTC is the most important and only "relevant" time since it is the time provided by time signals and chronometers. However, we must not overlook the importance of other time frames used in positional astronomy. Therefore, let's briefly review the most important concepts of time in use in accordance with their proper definitions.

DEFINITIONS OF TIME

(1) SMT: Sidereal Mean Time—defined by the LHA ♈/15 with reference to the mean equinox.
(2) GSMT: Greenwich Sidereal Mean Time = GHA ♈/15, referred to the mean equinox.
(3) MST: Mean Solar Time—defined by the local hour angle of the mean Sun, measured from Mean-Noon = LHA⊙/15.
(4) GMT: Greenwich Mean Time = MST + 12^h = GHA ⊙/15 + 12^h.
(5) UT: Universal Time—defined by the equation for the GHA ♈.
(6) TAI: Atomic Time—basis of all modern time measurements.
(7) UTC: Coordinated Universal Time—time signals as used by all navigators and referred to the GMT adjusted for the variable angular velocity of the gyro-Earth. UT + ΔUT. |ΔUT| ≤ $0^s.9$.
(8) EphT: Ephemeris Time—based on the dynamics of the orbital motion as defined by the equation for the mean longitude $\ell(T)$. TAI + $32^s.1841$.
(9) TDT: Terrestrial Dynamic Time = EphT.

5.5 Numerical Examples and Other Concepts of Time

Quite often it is necessary to convert MST into SMT and conversely. The values provided below are sufficiently accurate for the required conversions.

CONVERSION OF MEAN SOLAR TIME (MST) TO SIDEREAL MEAN TIME (SMT)

$$24^h - \text{MST} = 24^h 3^m .94248 = 24^h .065708 \text{ SMT}$$
$$1^h - \text{MST} = 1^h .00273781 \text{ SMT}$$
$$1^h - \text{SMT} = 0^h .99726967 \text{ MST}$$
$$1^h - \text{MST} = 15°.041068 \text{ Rev.}$$

5.6 An Approximate Method for Finding the Eccentricity, the Longitude of the Perigee and the Epoch

As we have seen in the previous section, all the formulae provided there required the numerical value of the element of the Earth-Sun orbit for any quantitative evaluation. This implies that no ephemeris for the Sun can be obtained without knowing directly or indirectly the following elements of the orbit:

(i) T_1: Period of the orbit.
(ii) e: Eccentricity of the ellipse
(iii) a: Semi-major axis of the ellipse = 1 AU = $1.495985 \cdot 10^8$ km.
(iv) ω: Ecliptic longitude of perigee.
(v) $\tilde{\varepsilon}$: Ecliptic longitude of the epoch.
(vi) ε_0: Obliquity of the ecliptic.

Except for the semi-major axis of the orbit which remains constant for any practical application, all the other orbital elements change slightly from year to year and are therefore time dependent.

The elements of the orbit as listed above can all be found in the corresponding issues of the Astronomical Ephemeris (AE) which is accessible either in print or on line. At one time, prior to the age of the internet, it was sometimes difficult to find the required issue of the AE when you actually needed it. Frustrated by this difficulty, I devised a method for calculating those parameters by using the Nautical Almanac (NA), current or outdated, it did not matter.

The idea to calculate approximate values for some of the orbital elements from the NA data was triggered by an attempt to find a simple formula for calculating the eccentricity of the Earth orbit by using the approximate time of the two equinoxes. I was, of course, familiar with the method devised by F. Gauss, but I was also intrigued by the failure of Copernicus to come even close to the true value of the required eccentricity. It was obvious that the only part of astronomy that I knew and that Copernicus could not have know was the Kepler's Second Law. It became the key to developing my simple formula.

For the moment, let's assume that we have somehow found the exact or approximate times of the equinoxes T_s, T_F and the times of the solstices T_{ss} and T_{sw}. Then, following the tradition of our ancestral astronomers who depicted the orbit of the Earth about the Sun by a circle of radius a about the center C, we draw a circle of radius one about the center C. (See Fig. 5.6.1.) Next we fix a point F very close to C on the axis through C, indicating the position of the Sun. (Of course, this will result in an highly distorted picture of the real situation, but it will serve our purpose well.) Now, we draw a line through F, inclined by an angle φ towards the semi-major axis $\overline{CF}$, showing the ecliptic longitude of the perihelion. (Since we have chosen the heliocentric model, perigee now becomes perihelion.) This line through F constitutes the direction of the first point of Aries from the Sun and therefore, is perpendicular to the projection of the axis of the gryro-Earth on to the plane of the ecliptic. This line intersects the circle at points E_s and E_F of the equinoxes. Next we draw a line perpendicular to it through C that intersects $\overline{FC}$ at an angle φ. Now we can plainly see that the ecliptic longitude of perihelion ω is directly related to φ by:

$$\omega = 360° - (90° - \phi) = 270° + \phi.$$

Let us also draw a line parallel to $\overline{FE_s}$ passing through C. These two parallel lines cut out a section of area A of the circle that is given by:

$$A = 2 \cdot \Delta r \cdot \cos \phi.$$

Furthermore, we also draw a line parallel to the line which is perpendicular to $\overline{FE_s}$ through C. These two parallel lines cut out an area A' of the circle that is given by:

$$A' = 2 \cdot \Delta r \cdot \sin \phi.$$

If we then identify the intersections of the line through F that intersects $\overline{FC}$ at an angle φ with the circle by S_{ss} and S_{sw}, respectively, we may apply Kepler's Second Law to derive expressions for e and ω. Note that by identifying those two points of intersection with the summer and winter solstice respectively, we shall obtain only approximations for the eccentricity e and longitude of perihelion ω (Fig. 5.6.1).

By applying Kepler's Second Law first to the area A, we find that:

$$\frac{\frac{\pi}{2} + A}{T_F - T_s} = \frac{\frac{\pi}{2} + 2\Delta r \cdot \cos \varphi}{T_F - T_s} = \frac{\pi}{365.25} \text{ or } \cos \varphi = \frac{\pi}{4\Delta r}\left[\frac{2(T_F - T_s)}{365.25} - 1\right].$$

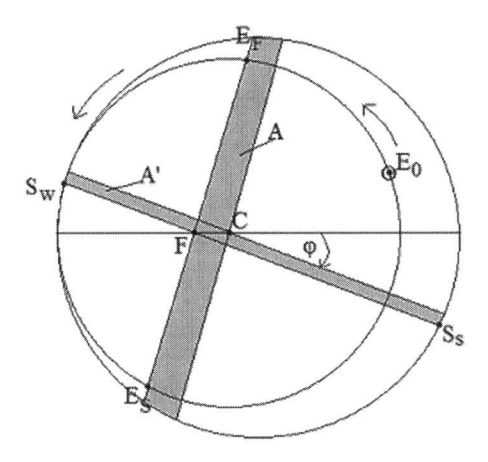

5.6 An Approximate Method for Finding the Eccentricity ...

Fig. 5.6.1

By applying the same law to area A' we find that:

$$\frac{\frac{\pi}{2} + A'}{T_{sw} - T_{ss}} = \frac{\frac{\pi}{2} + 2\Delta r \cdot \sin \varphi}{T_{sw} - T_{ss}} = \frac{\pi}{365.25} \text{ or } \sin \varphi = \frac{\pi}{4\Delta r}\left[\frac{2(T_w - T_s)}{365.25} - 1\right]$$

(1) $\phi = \tan^{-1}\left\{\left[\frac{2(T_W - T_S)}{365.25} - 1\right] / \left[\frac{2(T_F - T_S)}{365.25} - 1\right]\right\}; e = \frac{\Delta r}{1}$, hence...

(2) $e = \frac{\pi}{4 \cdot \cos \varphi} \cdot \left[\frac{2(T_F - T_S)}{365.25} - 1\right]$ and

(3) $\omega = 270° + \phi$.

Therefore, we have found two important elements of the Earth-Sun orbit. In order to calculate the third important element, namely $\tilde{\varepsilon}$, the ecliptic longitude of the epoch, we require one additional piece of information: namely the mean longitude ℓ at a specific instant T_0, which we may assume to be equal an epoch. According to Eqs. (2) and (4), Sect. 5.3, we have:

$\ell(T) = n^0 \cdot (T - T_0) + \tilde{\varepsilon}(T)$. It follows that:

(4) $\tilde{\varepsilon} = \ell(T_0), \quad T_0 = T_{EP}$.

and since we have already approximated the RA $\odot$ by means of formula (2), we have found an approximation to $\tilde{\varepsilon}$ by virtue of Eq. (4), Sect. 5.3.

Once e, ω and $\tilde{\varepsilon}$ are known, the mean anomaly $M(T)$ can be calculated by employing Eq. (4), Sect. 5.3, and the ecliptic mean longitude $\ell(T)$ can be found by using Eq. (2), Sect. 5.3. Conversely, if the equations for $M(T)$ and $\ell(T)$ are given as functions of time T, then no other values other than e and ε_0 are required for the calculation of an approximation to the ephemeris for the Sun.

It should be clearly understood that so far, we are only concerned with approximations even when we can take the specific elements of the orbit from an AE. Take, for instance, the following explicit data from the issue of ELEMENTS OF ORBIT EARTH—1980: [74]

$$T_p = 365.25 \text{ days.}$$
$$e = 0.016718$$
$$\omega = 282°.596403$$
$$\tilde{\varepsilon} = 278°.833540$$
$$\varepsilon_0 = 23°.441884$$
$$a = 1 \cdot AU = 1.495985 \cdot 10^8 \text{ km}$$

However, we are still not able to find ℓ, M, and λ at any instant apart from the epoch precisely since those parameters are dependent on the time due to the retrograde movement of the first point of Aries along the ecliptic. (The causes and extent of this movement of ♈ will be explained in detail in the next section.)

Before we derive another improved formula for the equation of time, we shall apply the approximate formulae (1)–(4) for finding e, ω and $\tilde{\varepsilon}$ for the epoch 2008. In order to apply our formulae (1)–(4), we extract from the NA (2008) the following data:

$$T_F : 09/22/08@15^h.77 \, UTC = 266^d.6570$$
$$T_S : 03/20/08@06^h.00 \, UTC = 80^d.25$$
$$T_W : 12/21/08@12^h.00 \, UTC = 356^d.4383$$
$$T_S : 06/21/08@00^h.00 \, UTC = 173^d$$

Substituting these values in formula (1) we find that:

$$\phi = \tan^{-1} 0.21504257 = 12°.13621147 \text{ and hence by formula (3)} :$$
$$\omega \cong 282°.1362115. \text{ Then with the help of formula (5), we find:}$$
$$e \cong 0.016636712$$

Finally by recalling that T corresponds to the epoch 2008, i.e., $T_0 = 366^\alpha$, and therefore according to (2) in Sect. 5.2, and (4) above, we find:

$$\tilde{\varepsilon}(T_0) = \ell(T_0) \cong 279°.79697$$

Of course these results clearly show that the approximations that we have employed merely constitute a crude approximation to the true values. Nevertheless, it is significant that by simply invoking Kepler's Second Law and using some approximate dates for our four major astronomical events, we were able to obtain tangible results... results that Copernicus could not obtain.[1]

---

[1] Obviously he did not have an AE, but had the data available through observation.

5.7 Some Improved Formulae for the Equation of Time and Center

The reader may recall that the derivation of the approximation (8) in Sect. 5.4 for the Equation of Time was based on the rather crude approximation $C = \frac{360}{\pi} \cdot e \cdot \sin M$ of the Equation of the Center. Therefore, in order to derive higher order approximations for the Equation of Time, it is necessary to employ higher order approximations for the Equation of the Center. However, for the purpose of navigation, because of the smallness of the eccentricity $e \cong 10^{-2}$, it is not necessary to take any terms in formula (1), Sect. 5.4 into account that contain powers of e higher than three. Therefore, the basis for our intermediate precision ephemeris will be the following Equation of the Center:

(1) $C = \frac{180}{\pi} \cdot (2 \cdot e \cdot \sin M + \frac{5}{4} \cdot e^2 \cdot \sin 2M - \frac{1}{4} \cdot e^3 \sin M + \frac{13}{12} \cdot e^3 \cdot \sin 3M)$

Also note that a similar approximation in powers of e up to powers of three can be obtained for the magnitude of the radius vector $r = |\vec{r}|$, namely:

(2) $r = 1 - e \cdot \cos M + \frac{1}{2} \cdot e^2 \cdot (1 - \cos 2M) + \frac{3}{8} \cdot e^3 \cdot (\cos M - \cos 3M)$, in units of AU. (Also see Eq. (2), Sect. 1.4.)

By employing the same techniques as used in the derivation of formula (8), Sect. 5.4 but by carrying higher terms of e along in the formula for the center, we readily obtain an improved formula for the Equation of Time, namely:

(3) $ET = \frac{720}{\pi} \cdot (\tan^2 \frac{\varepsilon_0}{2} \cdot \sin 2\ell - 2 \cdot e \cdot \sin M + 4 \cdot e \cdot \tan^2 \frac{\varepsilon_0}{2} \cdot \sin M \cdot \cos 2\ell$
$- \frac{1}{2} \cdot \tan^4 \frac{\varepsilon_0}{2} \cdot \sin 4\ell - \frac{5}{4} \cdot e^2 \sin 2M)$, in minutes of time.

This improved equation will only yield improved results if the parameters e, M, ℓ, and ε_0 are sufficiently accurate. (See Sect. 2.19.)

So far, we have made no attempt to overcome the two major obstacles for calculating the orbit of Sun-Earth, namely not having the exact formulae for the time dependent parameters of M(T), ℓ(T), e(T), and ε_0(T) available and not having introduced a suitable time frame, namely the Julian calendar as opposed to the Gregorian calendar. Aside from the requirement of having exact formulae for the aforementioned parameters, it is also necessary to have an exact formulae for the GHA ♈(T). Once these formulae are available it will not be necessary to employ the Equation of Time again since GHA ☉ and the declination δ of the Sun can be calculated directly by employing the Equation of the Center.

An ephemeris for the Sun can be calculated by following the steps listed below:

A. Compute GHA ♈ by using Eq. (1) in Sect. 5.2.
B. Compute RA ☉ by using Eq. (2) in Sect. 5.2.
C. Find the elements of the orbit. (See Sect. 5.6.)
D. Compute M by using Eq. (4), Sect. 5.3.
F. Compute ET by using Eq. (8), Sect. 5.4.
G. Compute GHA ☉ by using Eq. (1), Sect. 5.5.
H. Compute δ by using Eq. (2), Sect. 5.5.

In the next chapter, I will introduce a more suitable time frame for measuring time in JD dates and centuries of JDs denoted by T. Then we will discover the formulae for GHA ♈(T), M(T), ℓ(T), e(T), and ε_0(T)—all of them in the form of polynomials of powers of T up to the grade of three; i.e. $P(T) = a_0 + a_1 \cdot T + a_2 \cdot T^2 + a_3 \cdot T^3$. Readers will also find a brief treatise on the underlying theory of astronomy to the extent that they will be able to evaluate the effects of precession, nutation, proper motion, aberration, and annual parallax on the right ascension and declination of stars.

Chapter 6
Qualitative Description: The Relevant Astronomical Phenomena

6.1 On the Change of the Elements of the Orbit with Time

As a consequence of being a gyroscope, the axis of rotation of the Earth slowly changes its position relative to the normal of the plane of the ecliptic. Furthermore, the normal of the ecliptic and therefore the plane of the ecliptic itself also changes its position. (Readers who might be interested in the underlying dynamics of the gyro-Earth are advised to jump ahead and read the corresponding section of Sect. 8.6. They will then understand where the limitations of astro-dynamics as applied to positional astronomy enter the picture.) Since our objective is to find the coordinates of celestial objects that are based on the well defined coordinate systems of the celestial sphere such as the celestial equator and the equinox or the ecliptic and the equinox as opposed to merely finding trajectories of mass-points defined either by the two, three or multi-body problem of classical mechanics, we are compelled to consider the Earth to be a spheroid.

Let us, for a brief moment, visualize the replica of the celestial sphere our grandfather bought us a long time ago. By trying to deduce from that sphere the current coordinates of a star, we would certainly not get the correct coordinates since the old equator of that replica is no longer the current equator on the celestial sphere. (Of course, our geographic equator has hardly changed.) The explanation for this phenomenon is simple. Because of the movement of the axis of rotation and therefore of the figure axis of the Earth about the pole of the ecliptic, the celestial equator changes continuously as does the plane of the ecliptic. Since the horizontal component of the figure axis of the Earth is perpendicular to the intersection of the equator with the plane of the ecliptic, i.e., the line passing through the center of the Earth and the first point of Aries, the first point of Aries also moves accordingly. The first point of Aries moves retrograde along the ecliptic. Because of the change in the position of the normal to the ecliptic, an additional movement of the first point of Aries ensues, but in the opposite direction. The first displacement of the first point of Aries is caused by the gravitational forces of the Sun and Moon and is referred to as the luni-solar precession. The second displacement is due to the gravitation forces for the planets and is referred to as planetary precession. The combined effect of these changes on the position of the first point of Aries is referred to as the general precession and amounts to approximately $50''.22$/year. This means that within about 25,800 years, the figure axis and therefore the axis of rotation will complete one revolution relative to the normal of the plane of the ecliptic and hence to the pole of the elliptic.

In addition to the aforementioned secular displacements of the first point of Aries, there are also other periodic displacements of relative short duration due to gravitational forces of the Moon and Sun which can readily be attributed to the periodic changes in the relative position of the Earth with regard to the positions of the Moon and Sun. Those changes that result in changes of RA and declination of any celestial objects are referred to as Nutation.

Since the effects or precession on the coordinates of a CO can be determined fairly accurately by observation, and the mathematics required to develop explicit formulae merely involves spherical trigonometry and elementary calculus, the derivation of such formulae is straight foreword. However, the situation with regard to finding explicit formulae, together with the necessary data, for the effects of nutation on the coordinates is a great deal more difficult since it entails dealing with more than two celestial bodies that mutually attract each other. They are referred to as multi-body problems and result in unsolved theoretical problems for which only approximate solutions can be found. (See Theory of Perturbations of Celestial Dynamics.) [74], [75], [78].

Specifically, if one were to try to develop said formulae strictly by using the methods of Celestial Mechanics, one would have to deal with the theory of gyroscopes, (see Sect. 8.6.) and would ultimately be faced with the formidable task of determining the distribution of masses in the interior of the Earth referred to as moments of inertia. As we all know, the latter is an undertaking physicists and geologists still have not solved satisfactorily.

As a result of this theoretical assessment, the approach being used in this book is strictly based on classical position astronomy and rests quantitatively on results obtained over a long period of time by many distinguished astronomers. [57], [59], [60], 65].

6.2 The Concept of the Julian Date (JD) and Time Expressed by Julian Centuries (T)

As has been pointed out already that in order to develop suitable formulae for the effects of precession and nutation on the coordinates of celestial objects, it is imperative to first introduce a suitable frame of time, referred to as the Julian Calendar with the Julian date (JD) and subsequently with the time (T) expressed in Julian Centuries. The reader should be reminded that our calendar is known as the Gregorian Calendar and therefore it is necessary to know the algorithm for converting our calendar days into Julian dates. Hence, let us first define JD by:

6.2 The Concept of the Julian Date (JD) …

Definition—JD The Julian date (JD) is a continuous count of days from the 1st of January 4,713 BC (−4,712, Jan. 1. at Greenwich Noon (12^h:00 UTC).

Examples January 1, 1978 AD, 0^h:00 UTC = 2443509.5 JD
July 21, 1978 AD, 15^h:00 UTC = 2443711.125 JD

We also need the definition of the truncation operator (int <x>) to extract from a decimal, or rational number x, its integer part.

Definition—int <x> Given any real number x in decimal representation, int < x > is equal that part of x that precedes the decimal point.

Examples int <13.75> = 13, int <−1.8> = −1, int <6/5> = 1.

Next, let me state an algorithm for actually converting any date of the Gregorian calendar into its corresponding JD. For the sake of brevity, I will abbreviate the year by Y, the month by M and the day of the month, with its fractions, by D. Then we may state the algorithm as follows:

ALGORITHM - JD: i. Calculate: $y = \begin{cases} Y-1, \text{ and } m = M+12 \text{ if } M \leq 2 \\ Y, \text{ and } m = M \text{ if } M > 2 \end{cases}$

1.)

ii. Calculate: A = int<y/100>, and B = 2 - A + int<A/4>,

if $Y \geq 1582$, and B = 0 if Y < 1582.

iii. Calculate:

$$JD = \text{int}<365.25 \cdot (y + 4716)> + \text{int}<30.6001 (m+1)> + D + B - 1524.5$$

$$JD^0 = JD \text{ with } D = d, \quad D = d + \frac{UTC}{24}, \quad d: \text{ full days.}$$

Once more, the reader is reminded that the Julian date commences at mid-day, i.e., at
12^h:00 UTC.

Examples: 1 Given Gregorian date: Y = 1979; M = 12; d = 31; UTC = 12^h:00—Find JD
  Solution:

(i) y = 1979, m = 12, and since Y > 1582 we have:
(ii) A = int < $\frac{1979}{100}$ >= 19; B = 2 − 19 + int < $\frac{19}{4}$ > = −13
(iii) JD = int < 365.25 (1979 + 4716) > + int < 30.6001 · 13 > + 31.5 − 13 −1524.5 = 2445348 + 397 + 31.5 − 13 − 1524.5 = 2444239.

2. Given Gregorian date: Y = 1980; M = 1, d = 1, UTC = 12^h:00—Find JD

Solution:
  (i) y = 1979, m = 13, since M < 2.
  (ii) A = int < $\frac{1979}{100}$ >= 19, B = 2 − 19 + int < $\frac{19}{4}$ > = −13
  (iii) JD = int < 365.25 (1979 + 4716) > + int < 30.6001 · 14 > + 1.5 − 13 − 1524.5 = 2445348 + 428 + 1.5 − 13 − 1524.5 = 2444240.

3. Given Gregorian date: Y = 1987, M = 4, d = 10, UTC = 12^h:00—Find JD
Solution:
  (i) y = 1987, m = 4, d = 10.5, since M > 2.
  (ii) A = int < $\frac{1987}{100}$ >= 19, B = 2 − 19 + int < $\frac{19}{4}$ > = −13
  (iii) JD = int < 365.25 (1987 + 4716) > + int < 30.6001 · 5 > + 10.5 − 13 − 1524.5 = 2446896

Because of the considerable magnitude of JD ($\approx 10^6$), in our century, it is obvious that JD itself is unsuitable as a measure for time in astronomical calculations. Therefore, it is logical instead to use Julian Centuries (T), from a specific epoch on, and denoted by JD_0 as the actual measure for elapsed time. Accordingly, the definition of time T is given by:

2. (i) T = $\frac{(JD - JD_0)}{36525}$, and
   (ii) $T_0 = \frac{(JD^0 - JD_0)}{36525}$ if referred to UTC = 0^h:00.

For example, if we choose the year 2000AC as epoch, the JD_0 corresponding to this date is: JD_0 = 2451545, and T becomes:

3. T = $\frac{(JD - 2451545)}{36525}$.

By adopting the concept of Julian Centuries, we are bound to deal with astronomically small numbers representing the hours, minutes and second as fractions of Julian Centuries. This becomes obvious if one notes that: 1 day = 2.7378 10^{-5} JC; 1 h = 1.14075 10^{-6} JC; 1 min = 1.90125 10^{-8} JC; 1 s = 3.16875 10^{-10} JC.

Since $0 \leq T < 1$, we can readily see that any calculator with a ten digit arithmetic will not be good enough to account for the seconds in our calculations unless we resort to using a different representation of our decimal numbers. Actually, we run into a similar situation if we choose to express hours, minutes, seconds in terms of regular days, and therefore, of JD, for note that: 1 JD = 1 day; 1 h = 0.4166 JD; 1 min = 0.0006944 JD; 1 s = 0.000011574 JD.

In cases where the user has a calculator with fifteen or higher digit arithmetic, or with a sufficiently high floating point arithmetic available, there is no reason not to carry the seconds along in the calculation of T. The same holds if the user has a calculator with a scientific number representation at his disposal.

If, however, no such sophisticated calculator is available, then we must resort to using the 10^x function of our ordinary scientific calculator to represent all the

6.2 The Concept of the Julian Date (JD) ...

decimal number as products of powers of ten and decimal number that have only one digit before the decimal point. Ultimately, one should avoid using seconds in the calculations of T all together. The latter is quite possible in the context of our treatise with the exception, perhaps, in calculating the equation of the equinox.

Again, let's consider some examples to numerically demonstrate what is actually involved in finding T by using a simple scientific calculator with a ten digit arithmetic.

Examples: 1 Given: Y = 2001, M = 1, d = 1, UTC = $14^h:27^m$, Epoch: 2000AC.
Calculate T.
Solution:

(i) Calculate JD: Since M < 2, we have: y = 2000, m = 13
A = int $<\frac{2000}{100}>$, B = 2 − 20 + int $<\frac{20}{4}>$ = −13 and
D = 1.6020833, hence:
JD = int <365.25 · (2000 + 4716) + int <30.6001 · 14> + 1.6020833 − 13 − 1524.5 = 2451911.102

(ii) Calculate T using (3)
$T = \frac{(2451911.102 - 2451545)}{36525} = 0.010023326$, i.e.,
T = 1.0023325 10^{-2}

2. Given: Y = 2001, M = 1, d = 1, UTC = $14^h:27^m:50^s$, Epoch: 2000AC.
Calculate T.
Solution:

(i) Calculate JD: Again, since M < 2, we have: y = 2000, m = 13 and D = 1.602662003.
However, since the data, except for the UTC, is identical to the data in example 1, the result will differ only by the difference of the fifty seconds in UTC expressed in JC. First we calculate the difference in JC and find that:
JD = 2451911.102 + 0.000578703 = 2451911.102578703.
This result cannot be obtained by using a calculator that has only a ten digit arithmetic straight forward. Nevertheless, by splitting this number into its integer part, 2451911, and its fractional part, 0.102578703, we find that:
JD − JD_0 = 366 + 1.02578703 · 10^{-1}. Hence:
$T = \frac{366}{36525} + \frac{1.02578703}{3.6525} 10^{-5} = $
1.0020533 · 10^{-2} + 2.80845182 · 10^{-6} =
1.002334145 · 10^{-2}.

The last example clearly shows the arithmetic problems arising from the seconds in JD in cases where only a simple calculator is available. Later, I will show you how to get around this problem. But for now, let's consider some more illustrative examples.

Examples: 3 Given: Y = 2013, M = 11, d = 19, UTC = $18^h\ 47^m$, Epoch 2000AC
Compute T.
Solution:

(i) Calculate JD using (1)
Since M > 2, we find: y = 2013, m = 11, and D = 19.782638888
hence: A = int $\langle\frac{2013}{100}\rangle$ = 20, B = 2 − 20 + int $\langle\frac{20}{4}\rangle$ = −13,
JD = int <365.25 (2013 + 4716) + int <30.6001 · 12>
+ 19.782638888 − 13 − 1524.5
JD = 2456616.283

(ii) Calculate T using (3)
$$T = \frac{(JD - JD_0)}{36525} = \frac{5071.283}{36525}$$
T = 1.38844161 10^{-1}

4. Given: Y = 2013, M = 11, d = 19, UTC = $18^h\ 47^m\ 36^s$, Epoch 2000AC
Compute T
Solution:

(i) Compute JD using (1)
Since M > 2 we find that: y = 2013, m = 11, D = 19.78305555 and
A = int $\langle\frac{2013}{100}\rangle$ = 20, B = 2 − 20 + int $\langle\frac{20}{4}\rangle$ = −13
hence: JD = int <365.25 (2013 + 4716)> + int <30.6001 · 12>
+ 19.782638888 + 0.000416666 −
− 13 − 1524.5 = 2456616.283 + 0.000416666
after a manual intervention
JD = 2456616.283416666.

(ii) Compute T employing 3
First we find that: JD − JD_0 = 5071.283416666
and hence: T = $\frac{5071.283416}{36525}$ = 0.138844172
T = 1.38844172 10^{-1}.

At this point, the reader may ask whether or not it is necessary to perform those tedious arithmetic operations each time a position calculation—"fix"—by astronomical navigation is to be made? The answer is, of course, NO! Firstly, if the JD has been once computed and entered in the log, it is only necessary to count the days from there on. Secondly, you can also avoid calculating T more than once per day since most of the necessary corrections resulting form the daily variations in time such as hours, minutes and seconds can be added to the result obtained for a different UTC value.

For example, in order to calculate the Greenwich Hour Angle of Aries for a given instant UTC, it is only necessary to calculate the GHA ♈ once a day for UTC 0^h:00 and then add the product in hours and 15.04106865 to it. Similarly, the values of M(T), ℓ(T), and e(T) can be determined at any given instant T if the values are know for UTC 0^h:00.

6.3 The Elements of Our Orbit as a Function of the Time T Expressed by Polynomials

Thanks to all the astronomers who have contributed to the development of explicit formulae in terms of polynomials in T for the functions: e (T), M (T), ℓ (T) and ε_0 (T), we can now benefit from these results. The polynomial expressions for the aforementioned orbital elements can now be stated in accordance with the results provided by the International Astronomical Union, and are as follows: [60], [60], [62], [66]

Mean anomaly of the Sun—

1. $M(T) = 357°.52772 + 35999°.050340\ T - 0°.0001603\ T^2 - \frac{T^3}{300000}$

   Mean longitude of the Sun—

2. $\ell(T) = 280°.46646 + 36000°.76983\ T + 0°.0003032\ T^2$

   Eccentricity of the Earth's orbit.—

3. $e(T) = 0.016708634 - 0.000042037 \cdot T - 0.0000001267\ T^2$

   Mean obliquity of the ecliptic—

4. $\varepsilon_0(T) = 23°\ 26'\ 21''.448 - 46''.8150\ T - 0''.00059\ T^2 + 0''.001813\ T^3$

   $\varepsilon(T) = \varepsilon_0(T) + \Delta\varepsilon$, where $\Delta\varepsilon$ is given by expression (6), Sect. 3.4.

   In addition to these expressions, we will also need a polynomial expression for the hour angle of the first point of Aries that is given by:

5. $GHA\ ♈(T) = 100°.46061837 + 36000°.770053608\ T + 0.000387933\ T^2 - \frac{T^3}{38710000} + 15.04106865\ UTC$,

   and also a polynomial expression for the longitude of the ascending node of the moon, namely:

6. $\Omega(T) = 125°.04452 - 1934°.136261\ T + 0°.0020708\ T^2 + \frac{T^3}{450000}$.

   The later enters the equation of the equinox and, in particular, the formulae for the changes of longitude and obliquity of the ecliptic due to the effects of nutation. Of course, these formulae will only be used in cases where higher accuracy is required.

   For the sake of completeness, with reverence to polynomial expressions, the following formulae for the center C and the Greenwich Hour Angle are also included:

7. C (T) = (1°.914602 − 0°.004817 T − 0°.000014 T²) sin M + (0°.019993 − − 0°.000101 T) sin 2M + 0.000289 sin 3M
8. GHA ♈ (T) = 100°.46061837 + 0°.985647365 (JD⁰ − 2451545) + 15.04106864 UTC
—here JD⁰ denotes the JD at 0ʰ:00 UTC [2, 18], [2, 17]

As we proceed with the presentations and derivations for the formulae needed for a relatively simple ephemeris for the Sun and stars, we ought not to loose the necessary orientation that should enable us to reach our goal. Therefore, let us again state where we are going with our investigations.

It should be very clear by now that we are trying to derive formulae that will enable us to find the celestial positions of the Sun and the navigational stars at any instant $t > t_0$, if their positions are known at a fixed point of time t_0. This obviously entails two things, namely the calculations involving the effects of the motion of our coordinate system on the RA and declination of these celestial objects, and secondly, the calculations of the relative position of these bodies due to their own motion on the celestial sphere. Furthermore, it also entails calculating the effects of the motion of the Earth have on the coordinates to be observed, known as aberration and annual parallax.

In this chapter we shall once more review qualitatively the causes for the changes of the coordinate system as defined by the equator and the equinox, known as precession and nutation. We shall also explain the phenomena known as aberration and annual parallax and proper motion. Then we will also define the astronomical positions or places that will enable us to easily move from one epoch to another, i.e., to calculate the apparent position of a CO at any instant, provided its mean position at a given epoch is known. However, the quantitative aspects, i.e., the effects of these phenomena have on the RA and the declination of a CO will be covered in the next chapter.

6.4 Qualitative Aspects of Precession and Nutation

As has already been mentioned, the Earth is a giant, fast moving Gyro at 18.5 miles/s whose figure axis rotates about the normal to the ecliptic describing a cone with an angle ε_0—the obliquity of the ecliptic and completing one revolution in about 26,000 years (see Fig. 6.4.1, and period of precession). This effect is due to the gravitational forces exerted on the Earth by the Sun, Moon and planets. In addition, the periodic changes in the relative positions of the Sun, Moon and planets create additional periodic changes in the obliquity and longitude of the equinox causing the trajectory of the Earth axis on the celestial sphere to wobble. These periodic changes in the position of the axis of rotation are due to what is known as Nutation (see Fig. 6.4.2 and Sect. 8.6).

6.4 Qualitative Aspects of Precession and Nutation 267

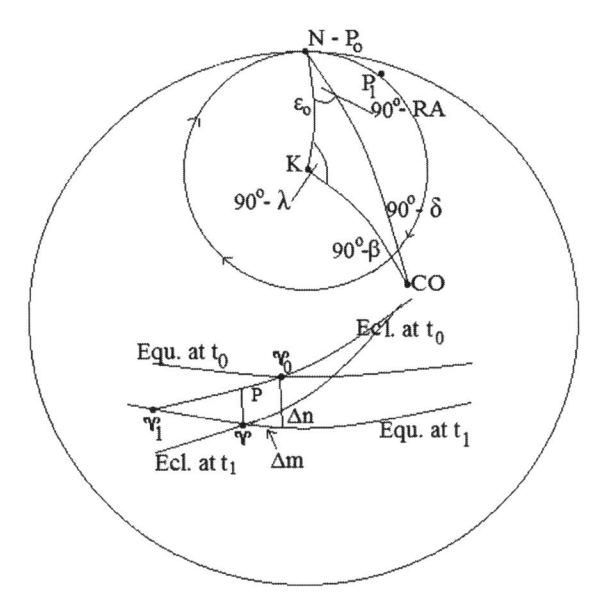

Fig. 6.4.1

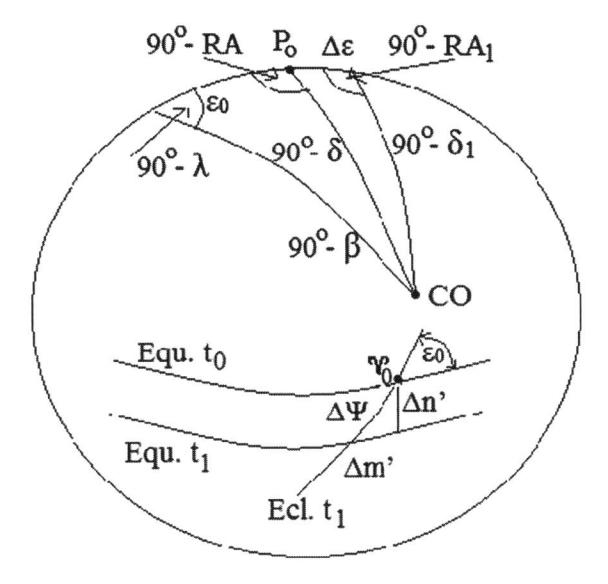

Fig. 6.4.2

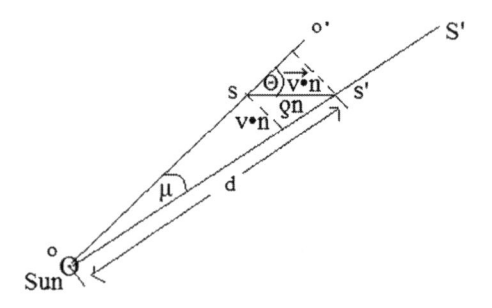

Fig. 6.5.1

6.5 The Concept of Proper Motion for Stars

For centuries, astronomers have known that some, if not all, stars are moving relative to the position of the Sun. Therefore, it is necessary to take the effects these motions have on the RA and declination into account in calculating their exact position on the celestial sphere. In order to understand the concept of said motions to the extent that explicit formulae can be derived, we first define the concept of proper motion with reference to Fig. 6.5.1.

In Fig. 6.5.1, a star moves within one year from s to s' along the straight line as defined by Fig. 6.5.1. The position it occupies on the celestial sphere is no longer S, but S'. The angle μ, subtended by the two rays OS and OS' is then referred to as the proper motion of μ of this star within one year.

Next let us represent the vector $n\vec{v}$ by its two components: $v \cdot n$, called the tangential component and ϱn, referred to as the radial component. Then, with reference to Fig. 6.5.1, we have:

(1) $v = |\vec{v}| \sin \Theta$, $\varrho = |\vec{v}| \cos \Theta$.

The tangental velocity v can be determined once the proper motion and the parallax Π are known. From Fig. 6.5.1 we deduce that:
$\sin \mu = \frac{n \cdot v}{d}$, and hence $\mu = \frac{n \cdot v}{d} \cdot \text{cosec } 1''$, but according to 11 in Sect. 6.7, $\Pi = \frac{a}{d} \cdot \text{cosec } 1''$, and therefore $v = \frac{\mu \cdot a}{n \Pi}$. Bu substituting the numerical values for the semi-major axis, namely $a = 149.6 \cdot 10^6$ km, and the numerical value for the seconds in a year, namely, $n = 3.15576 \, 10^7$ km into the last expression, we find that:

(2) $v = 4.74 \frac{\mu}{\Pi}$ km/s.

On the other hand, the component of the radial velocity can only be determined by means of spectroscoping methods and is given by:

(3) $\varrho = \frac{c \cdot \Delta \lambda}{\lambda}$ km/s, where c denotes the velocity of light, λ the wave length of light to be used, and Δλ the measured shift in wavelength due to the radial movement of the star (see also Doppler effect of electromagnetic waves).

(Note that we have made use of the fact that μ is a very small angle $\leq 10''$, and therefore, the triangle $S\hat{O}S'$ can be considered to be a small, plain rectangular triangle).

6.5 The Concept of Proper Motion for Stars

It follows then, that once v and ϱ are known, $|\vec{v}|$ and Θ, i.e., $\vec{v}$, can readily be calculated by employing the two equations (1).

Next let us consider two examples:
Examples:

1. **ALDEBARAN**: $\mu = 0".205$, $\Pi = 0".055$, v = 17.7 km/s.
   $\varrho = +54$ km/s., $|\vec{v}| = 56.8$ km/s., $\Theta = 18°.2$.
2. **CAPELLA**: $\mu = 0".439$, $\Pi = 0".075$, v = 27.7 km/s.
   $\varrho = +30.2$ km/s., $|\vec{v}| = 41.0$ km/s., $\Theta = 42°.5$.

6.6 Aberration

Since the Earth is moving with a tangential velocity of about 29.784585 km/s (about thirty times faster that the fastest bullet) around the Sun, and the velocity of light is about $3 \cdot 10^5$ km/s, we may conclude that the speed of our giant gyro-earth is small in comparison to the speed of light, but we may not conclude that it is negligible when it comes to astronomical observations. Therefore, we must treat such observations with reference to a moving and not to a stationary frame of reference. This also means that we cannot neglect the additional displacements of the observed positions of celestial objects.

In order to understand this phenomenon, known as ABERRATION, to the extent that we can calculate its effect on the RA and declination of those objects, let's first look at Fig. 6.6.1.

In Fig. 6.6.1, "$\overline{EF}$ denotes the direction of the moving Earth, i.e., the tangential velocity v. $\overline{Es}$ shows the direction for the observer E to the stellar object S if the Earth is assumed to be stationary, and $\overline{ES'}$ is the direction in which the observer sees the object when the Earth is moving with linear velocity $\vec{v}$. The distance Es and Ee

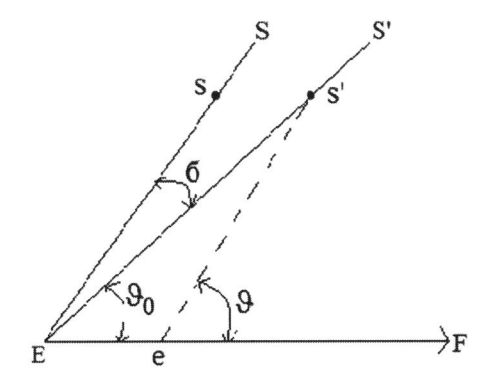

Fig. 6.6.1

are then related so that $\frac{Ee}{Es} = \frac{v}{c}$.[1] Here "c" denotes the velocity of light and $v = |\vec{v}|$ the tangential velocity of the Earth. It follows then, that by applying the SIN-TH of plane trigonometry we find:

$\sin 6/\sin \vartheta_0 = \frac{\overline{Ee}}{\overline{Es}} = \frac{v}{c}$ or $\sin 6 = \frac{v}{c} \cdot \sin \vartheta_0$. But since angle 6 is, in reality, very small, we have: $\sin 6 = \sin \ell" \cdot 6$, hence $6 = \frac{v}{c} \cdot \csc \ell" \sin \vartheta_0$. We can now define the constant of aberration κ as:

(1) $\kappa = \frac{v}{c} \cdot \csc \ell" = 20".5$, we also have:
(2) $6 = \kappa \cdot \sin \vartheta_0 = 20".5 \cdot \sin \vartheta_0$

Diurnal Aberration

So far, we have only considered the most important component of aberration, namely, the annual aberration. However, since the Earth not only moves on its orbit around the Sun, but also revolves around its axis of rotation completing one full revolution within twenty-four hours of Sidereal Mean Time (SMT), we must also consider this particular tangential component of the rotation velocity of the Earth in our calculus.

Expressed in seconds of SMT, this amounts to 86164 s. SMT = 24^h SMT. Therefore, the tangential velocity is: $v = \dfrac{2\pi r \cdot \cos \varphi}{86164 \text{ km/s}} = 0.465 \cdot \cos \varphi$ km/s, where φ denotes the latitude of the observer. Substituting this value into the formula (1) for the constant of aberration, we find that this constant, as it now refers to the DIURNAL aberration, is given by:

(3) $\kappa_D = \frac{0.465}{c} \csc \ell" \cos \varphi = 0".32 \cdot \cos \varphi \leq 0".32$.

Compared to the value of the annual aberration of about 20".5, we can pretend that this component does not have to enter our calculations for the effects of aberration on the RA and declination of COs. Furthermore, it also depends on the position of the observer and therefore only complicates matters.

Although the phenomena of aberration can and should be treated by applying the Theory of Special Relativity, we should refraining from doing so for the sake of simplicity. Similarly, we should also not consider any corrections for the light-time as it applies to the planets and the Moon for obvious reasons [58], [65].

Appendix

A less abstract and more physical explanation of the phenomenon ABBERATION can be conceived if one looks through a large telescope at a distant star S at the instant when the eye-piece of the telescope is focused on the star and its inclination is ϑ—see Fig. 6.6.1. The light from the star has reached, at this instant, the

---

[1] We deliberately have not used the addition of $\vec{c}$ and $\vec{v}$ so as to avoid invoking the Theory of Special Relativitiy.

objective at point S. Let us denote the focus-length of the telescope at E by ℓ. Then, the light will reach the eye-piece some $\tilde{t} = \frac{1}{c}$ seconds later, at which time the cross hairs of the eye-piece has moved to the point e, i.e., $\overline{Ee} = v \cdot \frac{1}{c}$ meters to the right of E. However, since the position of the telescope has not been altered, the focus is now on the point S' and hence, it appears as if the light is coming from S'.

Note that the apparent ray $\overline{Es'}$ forms an angle σ with the inclination ϑ_0 of the telescope, i.e., the ray originating from S with the inclination ϑ_0 has been moved forward in the direction of the moving Earth.

6.7 Annual Stellar Parallax, Definitions of Mean, True and Apparent Place of a Celestial Object

For the navigator who merely wishes to calculate stellar positions, the concept of stellar parallax is of no great importance since its numerical value never exceeds 0".8 and therefore may be neglected in most navigational applications. Nevertheless, I shall give a brief description of it together with the most relevant formulae because it is important to know how we can reach out to farther away places of the universe by being able to determine the distance of a star away from the Earth.

First, let's review the concept of a star's parallax defined as the angle Π under which the semi-major axis of the orbit of the Earth (a) appears when viewed from the star relative to a perspective perpendicular to the plane of the ecliptic. Then, per definition, Π is given by—Π: $\frac{\sin \Pi}{a} = \frac{\sin 90°}{d} = \frac{1}{d}$, and hence $\sin \Pi = \frac{a}{d}$, where "d" denotes the stars distance from the Sun. Since Π is very small, i.e., $\leq$ 0".8, we deduce from the last expression that:

(1) $\Pi = \frac{a}{d} \operatorname{cosec} \ell''$.

The annual stellar parallax is then defined as the angle Π_r subtended by the rays from the star towards the Sun and Earth as depicted by Fig. 6.7.1.

In Fig. 6.7.1, Earth's orbit about the Sun has been depicted as a circle. It follows then from the geometry of Fig. 6.7.1 that: $\Pi = \vartheta - \vartheta'$, and hence:

$\frac{\sin(\vartheta - \vartheta')}{r} = \frac{\sin \vartheta'}{d}$, which implies that:

(2) $\Pi_r = \vartheta - \vartheta' = \Pi \cdot \frac{r}{a} \cdot \sin \vartheta' \cong \Pi \cdot \sin \vartheta'$

As has already been mentioned, the derivations of the formulae for the effects on the RA and declination will be presented in the next chapter. However, before we can study them, we still need to define the relevant places of celestial objects as used in astronomy. Those places are defined as follows:

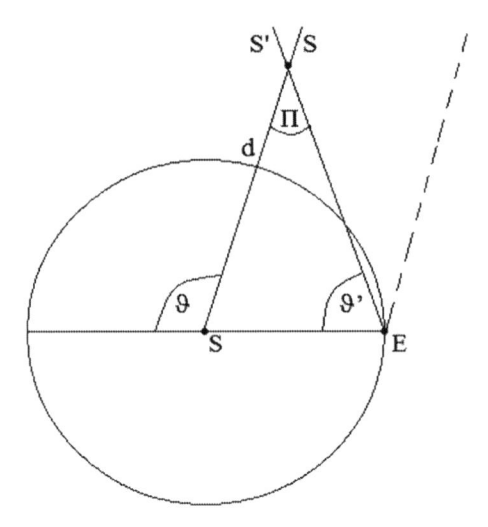

Fig. 6.7.1

Mean Place At Epoch

The mean place at epoch of a CO is its position on the celestial sphere centered at the Sun and referred to the mean equator and equinox at a specific time of the epoch—for instance, to the beginning of the year 2000AC.

Mean Equator and Equinox

The mean equator and equinox are defined with reference to the fixed planes of the ecliptic and equator, respective at the fixed time which coincides with the beginning of a specific year and referred to as Epoch.

Mean Place

Mean place at an epoch plus the effect of precession plus the proper motion computed for the interval between the epoch and the instant in question.

True Place

The true place of a CO is defined at any given instant as its position on the celestial sphere centered at the Sun and referred to the true equator and true equinox at that instant. Therefore, the equator and the equinox are now the actual equator and equinox at that particular instant and are referred to a fixed ecliptic at a fixed time being dependent on the precession and nutation computed for the interval between the epoch and the instant in question. Thus we have the following equation:

True place = Mean Place + Effect of Nutation at the date in question.

Apparent Place

The apparent place of a CO at any instant is defined to be its position on the celestial sphere centered at the Earth with reference to the true equator and equinox at that particular instant. Thus we have the equation:

Apparent Place = True Place + Effect of Aberration + Effect of Annual Parallax (if applicable).

6.7 Annual Stellar Parallax, Definitions of Mean, True and Apparent Place …

It follows then that what navigators actually need for calculating their positions are the GHA and declination of the apparent place of the CO in question. However, since navigators are located on the surface of the Earth and not at its center, and are also surrounded by an atmosphere that causes the light from a CO to be refracted, and since they may not be observing the center of the CO, they are obviously not measuring the altitude of the apparent place of the CO. Only after applying all the necessary corrections, such as instrument error, dip, refraction, parallax and semi-diameter to the observed altitude, may navigators conclude that they have found a valid approximation to the altitude of the apparent place of the CO.

Next we will derive all the necessary formulae for calculating the quantitative effects that the phenomena of precession, nutation, aberration, proper motion, and annual parallax have on the coordinates of the CO and thereby derive a suitable algorithm for calculated an ephemeris for the Sun and stars.

Chapter 7
Quantitative Treatise of Those Phenomena

7.1 Effects of Precession on the RA and the Approximate Method of Declination

In order to derive explicit formulae for the effects of precession, nutation, proper motion, aberration and annual parallax on the right ascension (a-RA) and declination δ of the COs, it is convenient to refer to their changing positions on the celestial sphere. By applying the corresponding formulae of spherical trigonometry, we obtain formulae that express their static positions and then by applying elementary calculus and or coordinate transformations, we shall be able to find the explicit formulae that give us the changes of a-RA and δ in terms of the variable time t.

Approximate Formulae

By applying the COS-TH. of spherical trigonometry to the triangle $S\widehat{K}P$ of Fig. 6.4.6 (Sect. 6.4), we deduce first that:

(i) $\sin \delta = \cos \varepsilon \cdot \sin \beta + \sin \varepsilon \cdot \cos \beta \sin \lambda$. Next we apply SIN-TH to find:
(ii) $\cos a \cdot \cos \delta = \cos \beta \cdot \cos \lambda$. And then by applying the ANALOG-TH., we obtain:
(iii) $\cos \beta \sin \lambda = \sin \delta \cdot \sin \varepsilon + \cos \delta \cdot \cos \varepsilon \sin a$.

Then by differentiating (i) with respect to λ, i.e., evaluating $\frac{d\delta}{d\lambda}$, and then by differentiating (ii) with respect to λ, and finally by employing (iii), we deduce that:

(1) $\left(\frac{da}{dt}\right)_L = \frac{d\lambda}{dt} \cdot (\cos \varepsilon + \sin \varepsilon \cdot \sin a \cdot \tan \delta); \left(\frac{d\delta}{dt}\right)_L = \frac{d\lambda}{dt} \cdot \sin \varepsilon \cdot \cos a$.

Here we have abbreviated the RA by a, i.e., a = RA, and the subscript "L" indicates that the rate of change so obtained corresponds to only one of the two components of change due to precession, namely to the "luni-solar" precession, resulting from the mutual gravitational effects of the Sun and Moon on the Earth.

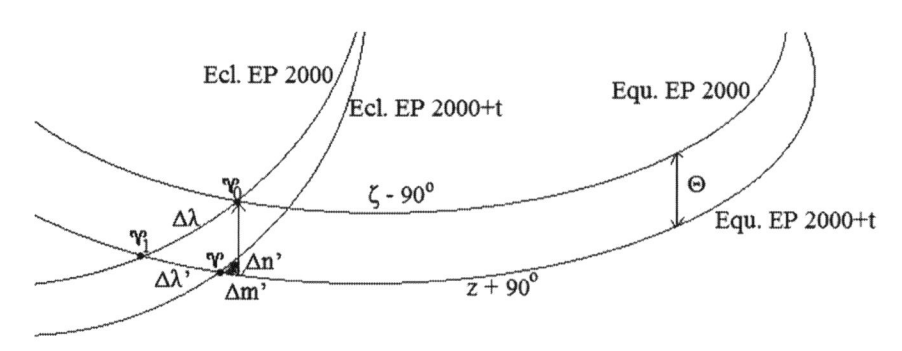

Up to this point in our arguments, formulae (1) are exact. Next, we must take into account the other component of precession, one that is due to the gravitational effects of the planets and is referred to as "planetary" precession. First, we apply the approximations suggested by Fig. 6.4.6 (Sect. 6.4), and Fig. 7.1.1.

Accordingly we have: $\Delta\lambda' = \Delta\lambda \cdot \cos\varepsilon - \Delta m'$, and hence:

(2) $\left(\dfrac{da}{dt}\right)_P = -\dfrac{d\lambda'}{dt} = \dfrac{dm'}{dt} - \dfrac{d\lambda}{dt} \cdot \cos\varepsilon$.

In this formula, the subscript "P" indicates that the rates of change refer to planetary precession. Similarly, we deduce from Fig. 6.4.6 (Sect. 6.4), and from Fig. 7.1.1, (above), that: $\dfrac{dn'}{dt} = \dfrac{d\lambda}{dt} \cdot \sin\varepsilon$ and when substituted into the second relation of (1), yields:

(3) $\left(\dfrac{d\delta}{dt}\right)_P = \dfrac{dn'}{dt} \cdot \cos a_0$, where a_0 and δ_0 denote the RA and declination respectively at the initial of epoch E_0.

Since the quantities $\dfrac{dm'}{dt}$ and $\dfrac{dn'}{dt}$ are almost constant, we shall denote them by "m" and "n", respectively, i.e., $m^S = \dfrac{dm'}{dt}$ and $n^S = \dfrac{dn'}{dt}$. By employing these expressions together with the Eqs. (1) (3), we deduce that the total rate of change for the RA is:

$\dfrac{da}{dt} = \left(\dfrac{da}{dt}\right)_L + \left(\dfrac{da}{dt}\right)_P = m^S + n^S \cdot \sin a_0 \cdot \tan\delta_0; \quad \dfrac{d\delta}{dt} = n'' \cdot \cos a_0 \cdot t,$ and

therefore, the final result is given by:

(4) $a - a_0 = (m^S + n^S \cdot \sin a_0 \cdot \tan\delta_0) \cdot t; \quad \delta - \delta_0 = n'' \cdot \cos a_0$. Note that "t" denotes the time expressed in years between epoch E_0 and E.

It has been found that the quantities of m^S and n^S depend on T as follows:

(5) $m^s = 3^s.07496 + 0^s.00186\,T, \quad n^s = 1^s.33621 - 0^s.00057\,T$
$m'' = 46''.1244 + 0''.0279\,T, \quad n'' = 20''.04315 - 0.00855\,T,$

where "T" is defined by Eq. (3), Sect. 6.2, and the superscript "s" indicates that the value is expressed in seconds of time.

7.2 Rotational Transformations and Rigorous Formulae for Precession

Before we can actually deal with the rigorous approach, we need to recall some results of elementary analytic geometry. First, let's consider the case where n = 2, i.e., of plane trigonometry. Let "$\vec{e_1}$" and "$\vec{e_2}$" denote the unit vectors which are orthogonal, i.e. $\vec{e_1} \cdot \vec{e_2} = 0$, and let "$\vec{x}$" denote a vector with components x, y in said frame of reference (See Fig. 7.2.2).

Then if we rotate the coordinate system spanned by $\vec{e_1}$ and $\vec{e_2}$ by the plane of angle α, measured counterclockwise, we generate a new orthogonal coordinate system denoted by $\vec{e_1'}$ and $\vec{e_2'}$, and the vector $\vec{x}$ stationary now has the new coordinates x', y'. By applying simple geometry, we deduce that:

$x' = \cos \alpha \cdot x + \sin \alpha \cdot y$

$y' = -\sin \alpha \cdot x + \cos \alpha \cdot y$, written in vector notations: $\vec{x'} = {'R}(\alpha) \cdot \vec{x}$, where

${'R}(\alpha) = \begin{pmatrix} \cos \alpha & \sin \alpha \\ -\sin \alpha & \cos \alpha \end{pmatrix}$.

The matrix R(α) is referred to as the rotational matrix in two dimensions. If we now apply the same mathematics to the three dimensional case, n = 3, we find that by turning the coordinate system $\vec{e_1}, \vec{e_2}, \vec{e_3}$ about $\vec{e_3}$, i.e., about the z-axis, the coordinates transformation is as follows:

$\vec{x'} = {'R_z}(\alpha) \cdot \vec{x}$, with ${'R_z}(\alpha) = \begin{pmatrix} \cos \alpha & \sin \alpha & 0 \\ -\sin \alpha & \cos \alpha & 0 \\ 0 & 0 & 1 \end{pmatrix}$. See Fig. 7.2.3.

Similarly, we now rotate the coordinate system $\vec{e_1'}, \vec{e_2'}, \vec{e_3'}$ so obtained about the y'-axis, by the angle β, we find that:

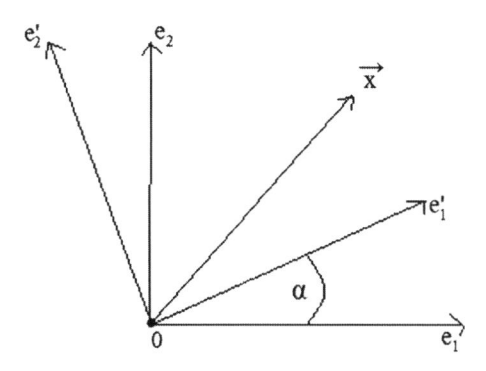

Fig. 7.2.2

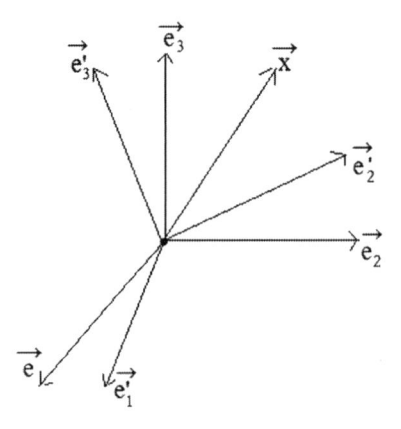

Fig. 7.2.3

$$\vec{x''} = {}'R_y(\beta) \cdot \vec{x'} = \begin{pmatrix} \cos\beta & 0 & -\sin\beta \\ 0 & 1 & 0 \\ \sin\beta & 0 & \cos\beta \end{pmatrix} \cdot \vec{x'}. \text{ But since } \vec{x'} = {}'R_z(\alpha) \cdot \vec{x},\text{ we}$$

find that: $\vec{x''} = {}'R_y(\beta) \cdot {}'R_z(\alpha) \vec{x}$. Finally, if we once more rotate the coordinate system $\vec{e_1''}, \vec{e_2''}, \vec{e_3''}$ by an angle γ—always counterclockwise, about $\vec{e_3''}$, i.e., about the new z''-axis, we then obtain:

$$X''' = R_{z''}(\gamma) \cdot \vec{x''} = R_{z''}(\gamma) \cdot R_y(\beta) \cdot R_z(\alpha) \cdot \vec{x},\text{ with } R_{z''}(\gamma) = \begin{pmatrix} \cos\gamma & \sin\gamma & 0 \\ \sin\gamma & \cos\gamma & 0 \\ 0 & 0 & 1 \end{pmatrix}.$$

Therefore, the rotation matrix $R(\alpha, \beta, \gamma)$ that comprises all three rotations is equal to the product of all single rotation matrices, i.e.:

(1) $R(\alpha, \beta, \gamma) = R_{z''}(\gamma) \cdot R_{y'}(\beta) \cdot R_z(\alpha)$, with

$$R_z(\alpha) = \begin{pmatrix} \cos\alpha & \sin\alpha & 0 \\ -\sin\alpha & \cos\alpha & 0 \\ 0 & 0 & 1 \end{pmatrix},\ R_{y'}(\beta) = \begin{pmatrix} \cos\beta & 0 & -\sin\beta \\ 0 & 1 & 0 \\ \sin\beta & 0 & \cos\beta \end{pmatrix},\text{ and}$$

$$R_{z''}(\gamma) = \begin{pmatrix} \cos\gamma & \sin\gamma & 0 \\ -\sin\gamma & \cos\gamma & 0 \\ 0 & 0 & 1 \end{pmatrix}.$$

Formula (1) will enable us to derive the rigorous solution once we have identified the angles α, β, and γ. From Fig. 6.4.6 (Sect. 6.4), we readily deduce that the first rotation takes place in the plane of the equator of epoch E_0 (here the year 2000AC) and must be negative. Therefore by denoting it by ζ, we conclude that: $\alpha = -\zeta$.

7.2 Rotational Transformations and Rigorous Formulae for Precession

The next rotation brings the equator of epoch E_0 into the new equator of epoch $E_0 + t$. The rotation is about the axis CM, where "C" denotes the center of the celestial sphere and "M" denotes the node of the equators. (See Fig. 7.1.1, Sect. 7.1.) In terms of our usual notations, this axis is the new y'-axis that is the result of the first transformation. Let us denote the angle of the second rotation by Θ which must be positive. Therefore, we find that $\beta = \Theta$. The result of the second transformation is that the axis of rotation of the Earth, denoted by "z" has been moved from P_0 to P_1 (See Fig. 6.4.6, Sect. 6.4).

This axis, now denoted by "z'" lies in the plane defined by CP_0P_1. We still have to rotate this plane about the new z'-axis in order for it to pass through the new equinox. (See Fig. 7.1.1, Sect. 7.1.) This time, the angle or rotation is negative and we denote it by "$-z$", i.e., we have $\gamma = -z$. Summing up these results and referring them to Fig. 6.4.6 (Sect. 6.4), and to Fig. 7.1.1 (Sect. 7.1), we have: $\alpha = -\zeta, \beta = \Theta, \gamma = -z$ and the rotation matrix in

(1) becomes:

(2) $\mathbb{R}(-\zeta, \Theta, -z) =$
$$\begin{Bmatrix} \cos z \cdot \cos \Theta \cdot \cos \zeta - \sin \zeta \cdot \sin z & -\cos z \cdot \cos \Theta \cdot \sin \zeta - \sin z \cdot \cos \zeta & -\cos z \cdot \sin \zeta \\ \sin z \cdot \cos \Theta \cdot \cos \zeta + \cos z \cdot \sin \zeta & -\sin z \cdot \cos \Theta \cdot \sin \zeta + \cos z \cdot \cos \zeta & -\sin z \cdot \sin \Theta \\ \sin \Theta \cdot \cos \zeta & -\sin \Theta \cdot \sin \zeta & \cos \Theta \end{Bmatrix}.$$

By expressing the vectors $\vec{x}_0$ and $\vec{x}$ of our transformation in polar coordinates, a, δ, i.e., by:

$$\vec{x}_0 = \begin{pmatrix} \cos \delta_0 \cdot \cos a_0 \\ \cos \delta_0 \cdot \sin a_0 \\ \sin \delta_0 \end{pmatrix}; \quad \vec{x} = \begin{pmatrix} \cos \delta \cdot \cos a \\ \cos \delta \cdot \sin a \\ \sin \delta \end{pmatrix},$$ where "a" and "δ" are the

RAs and declinations of the CO, we may then deduce the component equations from our transformation formula $\vec{x} = \mathbb{R}(-\zeta, \Theta, -z) \cdot \vec{x}_0$ as:

(3) $\cos \delta \cdot \cos(a - z) = \cos \Theta \cdot \cos \delta_0 \cdot \cos(a_0 + \zeta) - \sin \Theta \cdot \sin \delta_0$
$\cos \delta \cdot \sin(a - z) = \cos \delta_0 \cdot \sin(a_0 + \zeta)$
$\sin \delta = \sin \Theta \cdot \cos \delta_0 \cdot \cos(a_0 + \zeta) + \cos \Theta \cdot \sin \delta_0$

Finally, by defining the quantities A, B, and C by:

(4) $A = \cos \delta_0 \cdot \sin(a + \zeta)$
$B = \cos \Theta \cdot \cos \delta_0 \cdot \cos(a_0 + \zeta) - \sin \Theta \cdot \sin \delta_0$
$C = \sin \Theta \cdot \cos \delta_0 \cdot \cos(a_0 + \zeta) + \cos \Theta \cdot \sin \delta_0$

we find a and δ explicitly by evaluating:

(5) $a = z + \tan^{-1} \frac{A}{B}$, and $\delta = \sin^{-1} C$. [2, 19], [2, 16]

These formulae are relatively easy to evaluate numerically, and the problem of ambiguity in employing the multi-valued functions $\tan^{-1} x$ and $\sin^{-1} x$ can easily be resolved by referring the solutions to the corresponding quadrants.

In applying formula (2), we have assumed that the values of ζ, Θ, and z are known. These values are all time dependent and, therefore, can only be found by physical observation like the values of m and n in formulae (4) and (5), Sect. 7.1.

Thanks to the efforts of many distinguishes Astronomers, we can adopt the following corresponding polynomial expressions in terms of the powers of T as given by:

(6) $\zeta = (2306''.2181 + 1''.39656 T_0 - 0.000139 T_0^2)T + (0''.30188 - 0''.000344 T_0)T^2 + 0''.017998 T^3$

$\Theta = (2004''.3109 - 0''.85330 T_0 - 0''.000217 T_0^2)T - (0''.42665 + 0''.000217 T_0)T^2 + 0''.041833 T^3$

$z = (2036''.218 + 1''.39656 T_0 - 0''.000139 T_0^2)T + (1''.09468 + 0''.000066 T_0)T^2 + 0''.018203 T^3$

Where T_0 and T are defined by:

(7) $T_0 = \frac{(JD_0 - 2451545)}{36525}$, and $T = \frac{(JD - JD_0)}{36525}$.

Here T_0 is the time interval in Julian centuries between the year 2000AC and the starting epoch of JD_0 and T is the time interval expressed in the same units between the starting epoch JD_0 and the final epoch JD.

Note that in cases where the starting epoch coincides with the epoch of the year 2000AC, $T_0 = 0$, and the above expressions (6) simplify considerably.

7.3 Approximate Formulae for the RA Θ and Declination δ as the Result of Two Rotations Only

Now, with the aid of the exact formulae, we are able to cast some light on the validity of approximate formula (4), Sect. 7.1. Specifically, we can see that those formulae can only be deduced from the exact formulae in the small, i.e., in differential form. However, in the large, i.e., for finite angles of rotations, those formulae cannot be deduced from the exact equations. Therefore, higher derivatives of those erroneous formulae should not be used to construct Taylor-Series expansion for the functions a(t) and δ(t).

Go back to Fig. 6.4.6 (Sect. 6.4) and 7.1.1 (Sect. 7.1). You can see that the approximate solutions are based on the premise that the coordinate system attached to the Earth is only subjected to two rotations, not three, as it should be. Quantitatively speaking then, the authors of said method have rotated their coordinate system only about two very small angles, $\alpha = -\Delta \bar{m}$, and $\beta = -\Delta \bar{n}$. Therefore, the corresponding rotation matrices are:

$$R_z(-\Delta m') = \begin{pmatrix} 1 & -\Delta m' & 0 \\ \Delta m' & 1 & 0 \\ 0 & 0 & 1 \end{pmatrix}, \quad R_y(\Delta n') = \begin{pmatrix} 1 & 0 & -\Delta n' \\ 0 & 1 & 0 \\ \Delta n' & 0 & 1 \end{pmatrix},$$

hence $R(-\Delta m', \Delta n') = R_z(-\Delta m') \cdot R_y(\Delta n') = \begin{pmatrix} 1 & -\Delta m' & -\Delta n' \\ \Delta m' & 1 & 0 \\ \Delta n' & 0 & 1 \end{pmatrix}$. We

7.3 Approximate Formulae for the RA Θ and Declination δ ...

now deduce that: $\vec{x} = R(-\Delta m', \Delta n') \cdot \vec{x}_0 = \vec{x}_0 + \begin{pmatrix} -\Delta m' \cdot y_0 & -\Delta n' \cdot z_0 \\ \Delta m' \cdot x_0 \\ \Delta n' \cdot x_0 \end{pmatrix}$,

and therefore: $\left(\dfrac{d\vec{x}}{dt}\right)_{t_0} = \begin{pmatrix} -m \cdot y_0 & -n \cdot z_0 \\ m \cdot x_0 \\ n \cdot x_0 \end{pmatrix}$, from which we deduce the three

equations: $\left(\dfrac{dx}{dt}\right)_{t_0} = -m \cdot y_0 - n \cdot z_0 \quad \left(\dfrac{dy}{dt}\right)_{t_0} = m \cdot x_0 \quad \left(\dfrac{dz}{dt}\right)_{t_0} = n \cdot x_0$

By expressing the Cartesian coordinates by their corresponding polar coordinates (see Sect. 7.2) and by invoking the equations of differentials for both system of coordinates as given by:

$r \cdot d\delta = \sin\delta \cdot \cos a \cdot dx - \sin\delta \cdot \sin a \cdot dy + \cos\delta \cdot dz$
$r \cdot \cos\delta \cdot da = -\sin a \cdot dx + \cos a \cdot dy$
$dr = -\cos\delta \cdot \cos a \cdot dx + \cos\delta \cdot \sin a \cdot dy + \sin\delta \cdot dz$

we arrive at our approximate formulae (4), Sect. 7.1, namely:
$\left(\dfrac{da}{dt}\right)_{t_0} = m + n \cdot \sin a_0 \cdot \tan\delta_0$, and $\left(\dfrac{d\delta}{dt}\right)_{t_0} = n \cdot \cos a_0$, where we have identified $r = \ell, a_0 = a(t_0)$, and $\delta_0 = \delta(t_0)$. Therefore, we may conclude that the approximate solutions are the result of two and not three rotations, as has previously been pointed out.

On the other hand, if we apply the rotation matrix (1) to the three very small angles: $\alpha = -\Delta m'$, and $\beta = \Delta n'$, and $\gamma = \Delta z$, we deduce from expression (1) that:

$R(-\Delta m', \Delta n', \Delta z) = \begin{pmatrix} 1 & -\Delta z - \Delta m' & -\Delta n' \\ \Delta m' + \Delta z & 1 & 0 \\ \Delta n' & 0 & 1 \end{pmatrix}$

$\neq \begin{pmatrix} 1 & -\Delta m' & -\Delta n' \\ \Delta m' & 1 & 0 \\ \Delta n' & 0 & 1 \end{pmatrix} = R(-\Delta m', \Delta n')$

for all $\Delta z \neq 0$, and $R(-\Delta m', \Delta n', 0) \neq R(-\Delta m', \Delta n')$.

As we can now readily see, the approximate solution does not even in the infinitesimal sense constitute a sufficiently accurate approximation to the true solution. By sufficiently accurate, I mean, accurately enough to approximate higher derivatives of the true solution by the corresponding higher derivates of the approximate solution. Consequentially, it is not recommended to employ the so called "improved" solutions that are based on those erroneous derivates. For that very reason, the method of improved solution based on the secular variations had not been included in this book.

Having developed approximate and rigorous formulae for the effect of precession on the right ascension and the declination δ, we will now do the same for the effects of nutation on these coordinates.

7.4 Effects of Nutation on the RA and Declination

Nutation not only changes the equinox and therefore the longitude, but also changes the inclination of the Earth axis relative to the pole K of the ecliptic. (See Fig. 6.4.6, Sect. 6.4.) Because of this, we need to distinguish between the nutation in obliquity and the nutation in longitude. First, let's address the phenomenon of nutation in obliquity

Nutation in Obliquity

Examine Fig. 7 (Sect. 6.4). It clearly shows that nutation in obliquity $\Delta\varepsilon$ does not alter the latitude β and longitude λ of a CO. Because of this, we may differentiate (i), Sect. 7.1 with regards to ε and thereby find the differential $d\delta$:

$$d\delta = -d\varepsilon(\sin\beta \cdot \sin\varepsilon - \cos\beta \cdot \cos\varepsilon \cdot \sin\lambda) \cdot \sec\delta.$$

If we now employ the Analogue-Th., we may deduce that:
$\sin\beta \cdot \sin\varepsilon = \cos\beta \cdot \cos\varepsilon \cdot \sin\lambda - \cos\delta \cdot \sin a$. (See Fig. 6.4.6, Sect. 6.4) It follows, then, that from the above expression for $d\delta$ assumes:

(1) $\quad d\delta_N^0 = d\varepsilon \cdot \sin\tilde{a}$, or in differences: $\Delta\delta_N^0 = \Delta\varepsilon \cdot \sin\tilde{a}$. (For definition of $\tilde{a}$, see below.)

Next we then differentiate the equation (ii), Sect. 7.1 with regard to δ and find that:

$$da \cdot \sin a \cdot \cos\delta = -d\delta \cdot \cos a \cdot \sin\delta.$$

Substituting (1) into this expression yields:

$da_N = -d\varepsilon \cdot \cos\tilde{a} \cdot \tan\tilde{\delta}$. Interpreting this expression in terms of the differences Δa and $\Delta\varepsilon$, we obtain the finite expression:

(2) $\quad \Delta a_N^0 = -\Delta\varepsilon \cdot \cos\tilde{a} \cdot \tan\tilde{\delta}$, where $\tilde{a} = a_{1+\tau} = $ RA, referred to the equinox and equator of epoch $E_{1+\tau}$, i.e., of that particular instant of time τ and $\tilde{\delta} = \delta_{1+\tau}$, the declination referred to the same instant.

Now, let's derive the other relevant formula for the other component of nutation, namely, nutation in longitude.

Nutation in Longitude

Again referring to Fig. 6.4.6, Sect. 6.4, we can see that the effect of nutation in longitude is qualitatively the same as the effect of luni-solar precession and according to Eq. 1, Sect. 7.1, is given by:

(3) $\quad \Delta a_N^L = \Delta\Psi(\cos\varepsilon + \sin\varepsilon \cdot \sin\tilde{a} \cdot \tan\tilde{\delta})$, where we have merely changed the notations from $\Delta\lambda$ to $\Delta\Psi$, a to $\tilde{a}$, and δ to $\tilde{\delta}$. We also deduce from the second part of (1) that:

(4) $\quad \Delta\delta_N^L = \Delta\Psi \cdot \sin\varepsilon \cdot \cos\tilde{a}.$

7.4 Effects of Nutation on the RA and Declination

By combining the above results (1)–(4), we arrive at the final equations, namely:

(5) $\Delta a_N = \bar{a} - a_{1+\tau} = \Delta a_n^0 + \Delta a_n^L = \Delta\Psi(\cos\varepsilon + \sin a_{1+\tau} \cdot \sin\varepsilon \cdot \tan\delta_{1+\tau})$
$\quad - \Delta\varepsilon \cos a_{1+\tau} \cdot \tan\delta_{1+\tau}$

$\Delta\delta_n = \bar{\delta} - \delta_{1+\tau} = \Delta\delta_n^0 + \Delta\delta_n^L = \Delta\Psi \sin\varepsilon \cdot \cos a_{1+\tau} + \Delta\varepsilon \sin a_{1+\tau}$

By applying these formulae to any concrete situation, it is imperative to realize that in order to find the true place of a CO at any instant $E_{l+\tau}$, we must first know it's mean place at that instant. This implies that we must know its RA a_0 and declination δ_0 at the initial epoch E_0, and then we must calculate those values for the epoch $E_{l+\tau}$ and thereby determine its mean place at $E_{l+\tau}$. The notation used here for the final epoch $E_{l+\tau}$ signifies merely that for the sake of convenience we have employed the epoch E_1 at the beginning of the year preceding the actual instant $E_{l+\tau}$. The quantity τ in time is then equal the time elapsed since the beginning of the year and the instant under consideration and is expressed in fractions of a year. For example, let the instant for which we like to find the true place of a particular CO be J2013, June 10, $18^h\ 20^m\ 00^s$ UTC and E_0 = J2000. Then E_1 = J2013, τ = 0.442899077, and $E_{l+\tau}$ = J2013.442899077.

It may also help to memorize the following scheme for finding the true place of a CO:

$E_0 \Rightarrow E_1 \Rightarrow E_{1+\tau} = E$
$a_0 \Rightarrow a_1 \Rightarrow a_{1+\tau}| \Rightarrow \bar{a}$
$\delta_0 \Rightarrow \delta_1 \Rightarrow \delta_{1+\tau}| \Rightarrow \bar{\delta}$

By application of precession and proper motion	By application of nutation.

Note that the formulae (5) of this section loose validity if the COs are close to the celestial pole. In the latter case, the corresponding formulae in elliptical coordinates β and λ can be employed. In order to apply formulae (5), we still must provide expressions for the two important parameters, $\Delta\varepsilon$ and $\Delta\Psi$. However, since the derivation of those formulae is based on more advanced theoretical methods of Celestial Mechanics, and on the Theory of Perturbation of Systems of Differential Equations, and also on a large amount of observational data that has been compiled over more than two centuries, we can only adopt those expressions for the nutation in longitude $\Delta\Psi$, and nutation in obliquity $\Delta\varepsilon$ as stated below:

(6) $\Delta\Psi = -17''.200 \cdot \sin\Omega + 0''.206 \cdot \sin 2\Omega - 1''.319 \cdot \sin 2\ell + 0''.143 \cdot \sin M$
$\quad - 0''227 \cdot \sin 2\ell' + 0''.071 \cdot \sin m + \ldots\ldots\ldots\ldots\ldots\ldots$(many more terms)

(7) $\Delta\varepsilon = 9''.203 \cdot \cos\Omega - 0''.090 \cdot \cos 2\Omega + 0''.574 \cdot \cos 2\ell + 0''.022 \cdot \cos(2\ell + M)$
$\quad + 0''.098 \cdot \cos 2\ell' + 0''.020 \cdot \cos(2\ell' - \Omega) + \ldots\ldots\ldots\ldots\ldots$

[62, 70, 72]

The expressions in terms of polynomials of T for the node of the Moon $\Omega(T)$, the mean anomaly of the sun M (T), and the mean longitude of the Sun ℓ (t) are all given in Sect. 6.3 among the formulae (1)–(6). The expressions for the mean anomaly of m (T) and the mean longitude of the Moon $\ell'(T)$ are given below:

(8) $m(T) = 134°.96298 + 477198°.857398 \cdot T + 0°.0086972 \cdot T^2 + \dfrac{T^3}{56250°}$

$\ell'(T) = 218°.3165 + 481267°.8813 \cdot T$

In most applications to navigation, it will suffice to consider only the first terms in (6) and (7) resulting in the approximations below:

(9) $\Delta\Psi \cong -17''.2 \cdot \sin \Omega$, and $\Delta\varepsilon \cong 9''.203 \cdot \cos \Omega$

The explanation for encountering parameters pertaining to the orbit of the Moon in formulae (6) and (7) is that in analyzing nutation one must consider the three-body problem of mechanics, namely the analytic problem of accounting quantitatively for the mutual attraction of the Sun, Moon and Earth. Since, so far, no analytic solution for this problem which involves solving a non-linear system of second order differential equations, has been found, only approximate or numerical solutions are available.

One of those approximate methods for solving the three-body problem consists in first solving the prevailing two-body problem, say Sun-Earth, and then by modifying this solution, for instance by adding additional terms, and thereby finding an approximate solution to the real three-body problem (Sun-Moon-Earth) This method is generally known as the Method of Perturbation.

However, there is still another unsolved problem involved in solving the aforementioned problem analytically, namely, the problem of finding a realistic model for the density distribution inside and on the surface of the Earth which is required for calculating the moments of inertia of the gyro-Earth. (See Sect. 8.6.) Therefore, the use of other astronomical methods as for instance observations are necessary in order to find the numerical values of the coefficients in the series expansions of $\Delta\Psi$ and $\Delta\varepsilon$.

In concluding this analytic treatise of the effects of nutation on the RA and declination, it should be noted that in all applications to navigation, whenever a higher degree of accuracy is required, the use of an intermittent epoch E_1 and the use of the time parameter τ offer the advantage that the rigorous formulae (4) and (6), Sect. 7.2 have to be applied only once a year for any particular CO, and because of $\tau < 1$, the approximate formulae (4) can then be employed to go from E_1 to $E_{1+\tau}$ at any other instant of the year corresponding to τ.

Next we shall examine the problem of finding the quantitative effects of proper motion on the RA and declination δ. Actually, we should have already examined this before discussing nutation, since the components of proper motion are required in the computation of the coordinates $a_{1+\tau}$ and $\delta_{1+\tau}$ of the mean place of the epoch $E_{1+\tau}$.

7.5 Effects of Proper Motion on the RA and Declination δ

Similarly, as in the case of nutation, we are compelled to refer the phenomenon of proper motion to the celestial sphere for interpretation in order to be able to derive analytic expressions for the quantitative effects of proper motion on the RA a and the declination δ of a CO. Strictly from the theoretical point of view, we should invoke the Theory of Special Relativity since the moving objects under consideration attain velocities of the order 10^{-4} c, where c = 3 · 10^6 km/s. However, such consideration is outside the parameters and the objective of this book.

From Fig. 7.5.4 we readily deduce the components of proper motion with regards to the displacement of S to A along the great circle SC in direction of the RA denoted by μ_a and in the direction of δ denoted by μ_δ. Specifically we deduce that:

$$\overline{AB} = \mu \cdot \sin \phi \cdot \sin \ell'' = \mu_a \cdot \cos \delta \cdot \sin \ell'', \text{ and}$$
$$\overline{AS} = \mu \cdot \cos \phi \cdot \sin \ell'' = \mu_\delta \cdot \sin \ell'', \text{ and hence:}$$

(1) $\mu_a = \mu \cdot \sin \phi \cdot \sec \delta$, and $\mu_\delta = \mu \cdot \cos \phi$ (in arc sec.)

Note that the relations between these components and the actual changes in RA and declination are:

(2) $\mu_a = \left(\dfrac{da}{dt}\right)_{PM}$, and $\mu_\delta = \left(\dfrac{d\delta}{dt}\right)_{PM}$, where the subscript "PM" refers to the changes of a and δ, respectively, relative to the effects of proper motion.

In what follows, we shall assume that μ remains constant over a fraction of a century and also that the quantities μ, μ_a, μ_δ, and the corresponding changes are all expressed in circular measures, i.e., radians.

Although it is quite feasible to treat the cases where μ can be a time-dependent variable, but the resulting terms in the corresponding formulae would exhibit parameters, such as the tangential velocity v and the parallax Π of the CO, for which numerical values are not readily available to navigators unless they have a

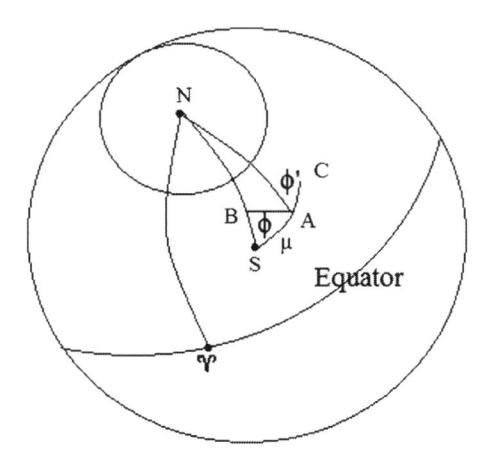

Fig. 7.5.4

recent star catalogue at their disposal. Furthermore, those additional terms in the more accurate formulae can be neglected in most of our practical applications.

In accordance with the aforementioned assumptions, $\mu_a = \mu_a(\phi, \delta)$ and $\mu_\delta = \mu_\delta(\phi, \delta)$ are continuous, differentiable functions of ϕ and δ only and their differentials are:

$d\mu_{a,\delta} = \frac{\partial \mu_{a,\delta}}{\partial \phi} d\phi + \frac{\partial \mu_{a,\delta}}{\partial \delta} d\delta$. Hence:

$\frac{d\mu_a}{dt} = \frac{\partial \mu_a}{\partial \phi} \frac{d\phi}{dt} + \frac{\partial \mu_a}{\partial \delta} \frac{d\delta}{dt}$, and $\frac{d\mu_\delta}{dt} = \frac{\partial \mu_\delta}{\partial \phi} \frac{d\phi}{dt} + \frac{\partial \mu_\delta}{\partial \delta} \frac{d\delta}{dt}$

Therefore, we require all the partial derivatives of μ_a and μ_δ, which can readily be deduced from (1), to be:

$\frac{\partial \mu_a}{\partial \phi} = \mu \cdot \cos \phi \cdot \sec \delta$, and

$\frac{\partial \mu_a}{\partial \delta} = \mu \cdot \sin \phi \cdot \sec \delta \cdot \tan \delta$, and

$\frac{\partial \mu_\delta}{\partial \phi} = -\mu \cdot \sin \phi$, and

$\frac{\partial \mu_\delta}{\partial \delta} = 0$

Since $\frac{d\delta}{dt}$ is known by virtue of (2), it remains only to derive a formula for $\frac{d\phi}{dt}$. Again, by referring to Fig. 7.5.4 and by applying the SIN-TH. of spherical trigonometry, we deduce that:

$\sin \phi \cdot \cos \delta = \sin \phi' \cdot \cos \delta' = $ const. Next, differentiating this expression with regards to δ yields:

$\frac{d}{d\delta}(\sin \phi \cdot \cos \delta) = -\sin \delta \cdot \sin \phi + \cos \delta \cdot \cos \phi \frac{d\phi}{d\delta} = 0$, i.e., $d\phi = \tan \phi \cdot \tan \delta \cdot d\delta$ from which it follows that:

$\frac{d\phi}{dt} = \tan \phi \cdot \tan \delta \cdot \mu_\delta$, where we have again made use of (2)

Substituting all the derivatives so obtained into the above equations for $\frac{d\mu_a}{dt}$ and $\frac{d\mu_\delta}{dt}$ yields:

(3) $\frac{d\mu_a}{dt} = 2\mu_a \cdot \mu_\delta \cdot \tan \delta \cdot \sin \ell''$, where μ_a and μ_δ are expressed in arc seconds.

(4) $\frac{d\mu_a}{dt} = 2\mu_a \cdot \mu_\delta \cdot 15 \cdot \tan \delta \cdot \sin \ell''$, where μ_a is expressed in seconds of time and μ_δ in arc sec.

and,

(5) $\frac{d\mu_\delta}{dt} = -\mu_a^2 \cdot \sin \delta \cdot \cos \delta \cdot \sin \ell''$, where μ_δ is expressed in arc seconds.

(6) $\frac{d\mu_\delta}{dt} = -\mu_a^2 \cdot 15^2 \cdot \sin \delta \cdot \cos \delta \cdot \sin \ell''$, where μ_δ is expressed in seconds of time.

In all those cases where a higher degree of accuracy is required, the following equations due to William Chauvenet can be employed:

7.5 Effects of Proper Motion on the RA and Declination δ

$\sin \gamma = \frac{\sin \Theta \sin (a+\zeta)}{\cos \delta}$, to calculate γ, and then:

(7) $\mu'_a = \sec \delta' \cdot (\mu_a \cos \delta \cdot \cos \gamma + \frac{\mu \cdot \delta}{15} \cdot \sin \gamma)$, and

$\mu'_\delta = -15\mu_a \cos \delta \cdot \sin \gamma + \mu_\delta \cdot \cos \gamma$, where the superscript "'" denotes the values corresponding to the new epoch E' and with ζ and Θ given by the formulae (6) Sect. 7.2. Note that the components of proper motion μ'_a and μ_a are expressed in seconds of time and that the components μ'_δ and μ_δ are expressed in arc seconds.

Finally, we are now in a position to state explicitly how the actual changes in Δa in RA and Δδ in declination are related to the components of proper motion and their derivatives. From the Taylor-Series expansion—approximation:

$\Delta a = \sum_{m=1}^{\infty} \frac{1}{n!} \cdot \left(\frac{d^n a}{da^n}\right)_0 \cdot t^n \cong \left(\left(\frac{da}{dt}\right)_0 + \frac{1}{2}\left(\frac{d^2 a}{dt^2}\right)_0 \cdot t\right) \cdot t$, we deduce that:

(8) $\Delta a = (\mu_a + \frac{1/2 d\mu_a}{dt} \cdot t) \cdot t$, and $\Delta \delta = (\mu_\delta + \frac{1/2 d\mu_\delta}{dt} \cdot t) \cdot t$, where "t" stands for the time between epoch E' and E.

Next, we shall derive explicit formulae for the quantitative effects of aberration on the RA and declination δ.

7.6 Effects of Aberration on the RA and Declination δ

Annual Aberration Only—See Sect. 6.6.

In treating said effects, we now assume that the observer is located at the center of the Earth, and therefore, the center of the celestial sphere now coincides with the center of the Earth-geocentric system—See Fig. 7.6.5.

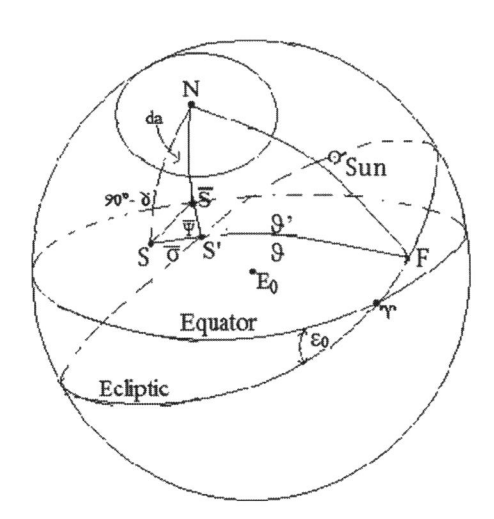

Fig. 7.6.5

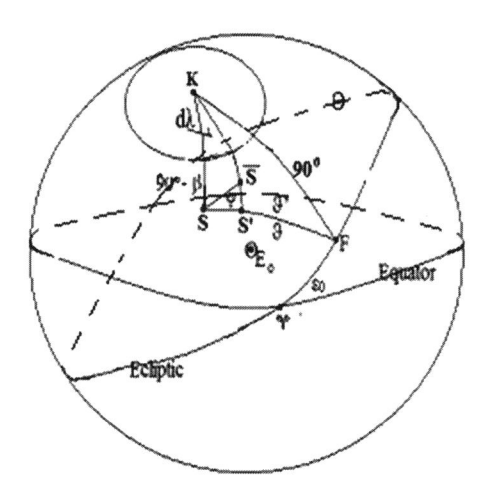

Fig. 7.6.6

Again, it is convenient to refer to the position S of the CO; its image S'; the position of the Sun ☉; and also the position of F that defines the direction of the moving Earth on the celestial sphere. Because we are now deriving the formulae for the finite displacements of $\Delta\lambda$ and $\Delta\beta$ of the CO and no longer rates of changes, it will simply suffice to employ the equations of spherical trigonometry.

First, lets look at the spherical triangle $S\widehat{K}F$ and apply the COS-TH. to find that:

(1) $\cos\vartheta = \sin\beta \cdot 0 + \cos\beta \cdot \sin 90° \cdot \cos(\lambda_\odot - \lambda - 90) = \cos\beta \cdot \sin(\lambda_\odot - \lambda)$,
where "$\lambda_\odot$" denotes the longitude of the true Sun—See Fig. 7.6.6.

Next we consider the small triangle $S\widehat{S'}\overline{S}$ to be a plane triangle and deduce that:

(2) $\overline{\overline{SS}} = \Delta\lambda \cdot \cos\beta = \sigma \cdot \sin\Psi$, and $\overline{\overline{SS'}} = -\Delta\beta = \sigma \cdot \cos\Psi$, where $\sigma \cong \kappa \cdot \sin\vartheta$—See (2), Sect. 2.6.

Again by applying the COS-TH. to the triangle $K\widehat{S}F$, we conclude that:
$$\cos 90° = \sin\beta \cdot \cos\vartheta + \cos\beta \cdot \sin\vartheta \cdot \cos(180° - \Psi)$$
$$= \sin\beta \cdot \cos\vartheta - \cos\beta \cdot \sin\vartheta \cdot \cos\Psi$$
and it follows that:

(3) $\cos\Psi = \tan\beta \cdot \cot\vartheta$.

Next we apply the SIN-TH. to the same triangle to obtain:
$$\frac{\sin 90°}{\sin(180° - \Psi)} = \frac{\sin\vartheta}{\sin(\lambda_\odot - \lambda - 90°)}, \text{ which yields:}$$

(4) $\sin\Psi = -\dfrac{\cos(\lambda_\odot - \lambda)}{\sin\vartheta}$. Combining the expressions (1)–(4) and using the expression (2), Sect. 6.6, again, we obtain:

7.6 Effects of Aberration on the RA and Declination δ

(5) $\Delta\lambda = -\kappa \cdot \sec\beta \cdot \cos(\lambda_\odot - \lambda)$

$\Delta\beta = -\kappa \cdot \sin\beta \cdot \sin(\lambda_\odot - \lambda)$

One important case that we will require in the derivations of algorithms for an ephemeris for the Sun is the special case of the Sun, itself, i.e., where $\lambda = \lambda_\odot$ and $\beta = 0$. In this case, (5) reduces to:

(6) $\Delta\lambda_\odot = -\kappa, \quad \Delta\beta = 0.$

By applying the same techniques to the triangles of Fig. 7.6.5, we obtain the corresponding formulae for the displacements in RA and declination as follows:

(7) $\Delta a = -\kappa \cdot \sec\delta \cdot (\cos a \cdot \cos\lambda_\odot \cdot \cos\varepsilon + \sin a \cdot \sin\lambda_\odot)$ and

$\Delta\delta = -\kappa \cdot \cos\lambda_\odot \cos\varepsilon \cdot (\tan\varepsilon + \cos\delta - \sin a \cdot \sin\delta) - \kappa \cdot \cos a \cdot \sin\delta \cdot \sin\lambda_\odot$

As an immediate consequence of Eq. (5), it follows that as the Earth moves around the Sun, the image S′ of S moves around S on an ellipse with the semi-major axis κ and the semi-minor axis κ · sin β. For, let us define:

$x = \Delta\lambda \cos\beta$ and $y = -\Delta\beta$.

then by employing the expressions (5), it follows that:

$\frac{x^2}{\kappa^2} + \frac{y^2}{\kappa^2 \cdot \sin^2\beta} = 1$. This is the equation of an ellipse about—(0, 0), i.e. about the position of S and it is referred to as the Aberrational Ellipse.

However, before we can write down an algorithm for the ephemeris of the Sun and the Stars we will have to derive explicit formulae for the quantitative effects of annual parallax on the coordinates of the COs.

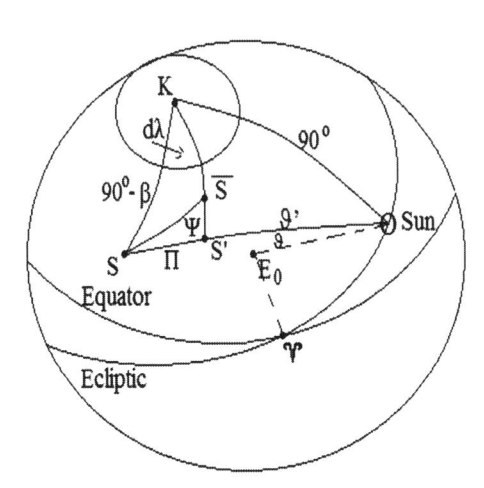

Fig. 7.7.7

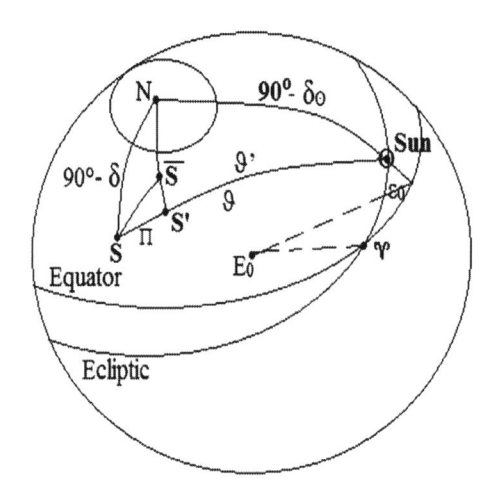

Fig. 7.7.8

7.7 Effects of Annual Parallax on the RA and Declination δ

The derivation of the required formulae for the computation of the effects of annual stellar parallax follows very closely the derivation of the corresponding formulae for the effects of aberration of the coordinates of the COs. Again, we need to refer back to the celestial sphere where we can identify the positions and angles shown in Fig. 6.7.1, Sect. 6.7. On the celestial sphere, things look, then, as depicted in Figs. 7.7.7 and 7.7.8 below.

First, let's consider the spherical triangle $K\widehat{S}\odot$ and apply the COS-TH. to its sides as follows:

$\cos 90° = \cos(90° - \beta) \cdot \cos \vartheta + \sin(90° - \beta) \cdot \sin \vartheta \cdot \cos(180° - \Psi)$, and therefore

$\sin \beta \cdot \cos \vartheta - \cos \beta \cdot \sin \vartheta \cdot \cos \Psi = 0$, resulting in:

(1) $\cos \Psi = \tan \beta \cdot \cot \vartheta$.

Next we apply SIN-TH. to the same triangles and obtain:

$\dfrac{\sin 90°}{\sin(180° - \Psi)} = \dfrac{\sin \vartheta}{\sin(\lambda_\odot - \lambda)}$, and resulting in:

(2) $\sin \Psi = \dfrac{\Pi}{\Pi_a} \cdot \sin(\lambda_\odot - \lambda)$, where $\Pi_a = \Pi \cdot \sin \vartheta$—(See (2) Sect. 6.7)

By looking at the very small triangle, $S\widehat{S'}\overline{S}$, we realize that it can be considered to be a plane triangle. Therefore, we deduce that:

7.7 Effects of Annual Parallax on the RA and Declination δ

(3) $\overline{SS} = \cos\beta \cdot \Delta\lambda = \Pi_a \cdot \sin\Psi$, and $\overline{SS} = -\Delta\beta = \Pi_a \cdot \cos\Psi$. Combining the expressions (1)–(3), we finally obtain:

(4) $\Delta\lambda = \Pi \cdot \sec\beta \cdot \sin(\lambda_\odot - \lambda)$ and
$\Delta\beta = -\Pi \cdot \sin\beta \cdot \cos(\lambda_\odot - \lambda)$, where $\lambda_\odot$ again denotes the longitude of the true Sun.

Exactly as in the case of aberration, we may also conclude that the image S′ of S prescribes an ellipse about S with semi-major axis Π and semi-minor axis $\Pi \cdot \sin\beta$. This is an immediate consequence of the expressions (4). For by defining:
$x = \Pi \cdot \sin(\lambda_\odot - \lambda)$, and $y = \Pi \cdot \sin\beta \cdot \cos(\lambda_\odot - \lambda)$, we deduce that:
$\frac{x^2}{\Pi'^2} + \frac{y^2}{\Pi'^2 \sin^2\beta}$, the equation of the ellipse of annual parallax.

Finally, let's also examine the formulae for the effects of annual parallax on the RA and declination δ. However, since the actual derivation of these formulae follow exactly the same patterns of some of the previous ones, I only need to state the explicit results. Look at Fig. 7.7.8.

From this we can readily deduce the final results:

(5) $\Delta a = \Pi \cdot \sec\delta \cdot (\cos a \cdot \cos\varepsilon \cdot \sin\lambda_\odot - \sin a \cdot \cos\lambda_\odot)$ and
$\Delta\delta = \Pi \cdot (\cos\delta \cdot \sin\varepsilon \cdot \sin\lambda_\odot - \cos a \cdot \sin\delta \cdot \cos\lambda_\odot - \sin a \cdot \sin\delta \cdot \cos\varepsilon \cdot \sin\lambda_\odot)$

where "a" denotes the RA and δ the declination [55, 63].

7.8 Calculating the Apparent RA and Declination δ, and the Equation of the Equinox

To sum up the results of this chapter. so far, we have compiled and examined all the necessary formulae for an ephemeris of the Sun[1] and Stars. What remains to be done is to show how to put all these pieces together in order to execute the algorithms with ease. One way of showing how to apply these formulae is to use the following scheme:

(I) Going from one epoch E_0^0 to an intermittent epoch E_1^0, and from there to the actual one $E_{1+\tau}^0$—Symbolically:
$E_0^0 \Rightarrow E_1^0 \Rightarrow E_{1+\tau}^0$.

(II) Going from the mean place of the initial epoch to the mean place of an intermittent epoch, and from there to the mean place of the final epoch, and from there to the true place, and finally to the apparent places—Symbolically:

$M_P(E_0^0) \Rightarrow M_P(E_1^0) \Rightarrow M_P(E_{1+\tau}) \Rightarrow T_N(E_{1+\tau}) \Rightarrow A_A(E_{1+\tau}) \Rightarrow A_{AP}(E_{1+\tau})$

---

[1] Less perturbation.

(III) Calculating the RA, a, and declination δ by applying at each step the indicated correction:

$a_0 \Rightarrow$ by applying precession + proper motion	$a_1 \Rightarrow$ by applying precession + proper motion	$a_{1+\tau} \Rightarrow$ by applying nutation	$a_n \Rightarrow$ by applying aberration	$a_A \Rightarrow a_{AP} = a$ by applying annual parallax

(Note that the superscript "^{0}" in E^0 indicates an epoch that begins at the first day of the year at 0^h:00 UTC. Furthermore, M_P denotes a mean place, T_N the true place, and A_P and A_{PA} the apparent places.)

The total effects on the RA, a, and declination δ can then be computed as follows:

$$a - a_0 = (a_{AP} - a_A) + (a_A - a_N) + (a_N - a_{1+\tau}) + (a_{1+\tau} - a_1) + (a_1 - a_0)$$

which can be written as:

(1) $a = a_0 + \Delta_{PP}a_0 + \Delta_{PP}a_1 + \Delta_N a_{1+\tau} + \Delta_A a_N + \Delta_{AP}a_N$
$\delta = \delta_0 + \Delta_{PP}\delta_0 + \Delta_{PP}\delta_1 + \Delta_N \delta_{1+\tau} + \Delta_A \delta_N + \Delta_{AP}\delta_N$, and

where Δ_{PP} refers to the change due to precession and proper motion, and Δ_N to the change due to nutation. Furthermore, Δ_A refers to the change due to aberration, and Δ_{AP} to the change due to annual stellar parallax.

In some applications where either the use of the exact formulae for precession is not required or where it is imperative to employ the exact equations throughout, you may go directly from the initial epoch E_0^0 to epoch $E_{1+\tau}$. Therefore, in those cases, the terms $\Delta_{PP}a_1$ and $\Delta_{PP}\delta_1$ will not appear in the expression (1). Furthermore, the difference operator δ_{PP} for the combined effect of precession and proper motion can be split into the operators Δ_P and Δ_{PM} where Δ_P denotes solely the effect of the precession, and Δ_{PM} signifies the effects of proper motion only so that:

(2) $\Delta_{PP}a_0 = \Delta_P a_0 + \Delta_{PM}a_0$, and $\Delta_{PP}\delta_0 = \Delta_P \delta_0 + \Delta_{PM}\delta_0$.

THE EQUATION OF THE EQUINOX

So far we have used the coordinate system of the celestial sphere, i.e., the coordinates RA and declination only. In order to apply this system of coordinates to navigation, we must relate it to our fixed terrestrial coordinate system defined in terms of latitude and longitude with reference to the actual equator of the Earth and the Greenwich meridian. This is readily accomplished by employing the concept of the Greenwich Hour Angle (GHA) of any CO. The equation that we have been using in Part I of this book and in Sect. 5.1 that links the RA to the GHA of any CO is:

(3) GHA* = GHA ♈ + SHA* = GHA ♈—RA*. Here "*" denotes the CO.

This formula clearly shows that we must also increase the accuracy of the GHA ♈ in order to obtain more accurate values for the GHA of the CO. We also

7.8 Calculating the Apparent RA and Declination δ ...

know that the formula (5) Sect. 6.3, which gives the GHA ♈ in terms of a polynomial in powers of T, cannot account for the effects of nutation on said coordinates. Therefore, we must find an improved formula for the GHA ♈ that takes nutation into account.

This can be accomplished by adopting the concept of the "fictitious star"—First point of Aries denoted by GHA ♈. Since this star moves along the equator, its declination is zero and since it is a fictitious star, the effects of proper motion, aberration, and annual parallax are all zero. Hence according to (3) we have:

GHA ♈ = GHA ♈$_0$ − RA ♈, where ♈$_0$ denotes the position of the First Point of Aries at a fixed epoch E$_0$, i.e., RA ♈$_0$ = 0. According to formula (1), we then conclude that: ΔRA ♈$_0$ = Δ_PRA ♈$_0$ + Δ_NRA ♈$_0$. Taking also into account that RA ♈$_0$ = 360° − ΔRA ♈$_0$, we conclude that:

(4) GHA ♈∗ = GHA ♈$_0$ + ΔRA ♈$_0$ = (GHA ♈$_0$ + Δ_PRA ♈$_0$) + Δ_NRA ♈$_0$.

Since δ = 0, we can deduce from formula (5), Sect. 7.4 that Δ_NRA ♈$_0$ = $\Delta\Psi$ · cos ε, and by applying formula (9), Sect. 7.4, we arrive at:

(5) Δ_NRA ♈$_0$ = $\Delta\Psi$ · cos ε = −17″.2 · sin Ω · cos ε = −0°.004777777 · sin Ω · cos ε.

Substituting (5) into (4) and noting that the sum of the first two terms in (4) is equal to the expression (5) Sect. 6.3, we conclude that:

(6) GHA ♈ ∗ (T) = GHA ♈(T) + E, where E, the Equation of the Equinox, is given by:
(7) E = −0°.0047777777 · sin Ω · cos ε.

By multiplying expression (6) by $\frac{1}{15}$, we find the Sidereal Time.

In concluding of this chapter, we still need to examine one other technical problem related to the evaluation of the formulae provided herein. The starting point of all the numerical calculations is calculating the value for the time T of the instant, in terms of Julian centuries, i.e., of T as defined by (3), Sect. 6.2. For PC and sophisticated calculators, this does not pose a problem. However, for the user who only has a simple scientific calculator with a ten digit arithmetic at his disposal, there arises a difficult situation once seconds of time have to be expressed in terms of T since one second of UTC corresponds to 3.168808777 · 10^{-10} in T time.

For instance, if one tries to add this number to the number one, the calculator will give the result as one which is obviously incorrect. Therefore the question arises: How can one account for seconds of time in all those formulae that depend on the variable T if one only has a simple calculator to hand?

One method for overcoming this difficulty is to define first a value of T that does not account for the hours, minutes, and seconds, denoted by T$_0$, and given by:

(8) $T_0 = \dfrac{JD^0 - JD_O}{36525}$, where JD^0 is the Julian Date of the instant at 0^h.00 UTC.

Then if UTC stands for the hours, minutes, and seconds since 0^h.00, expressed in hours, we have:

(9) UTC = h + m · 0.016666666 + s · 0.000277777, where "h" denotes the hours, "m" the minutes, and "s" the seconds.

By the definition of T, we deduce that:

(10) $T = \dfrac{JD^0 + \frac{UTC}{24} - JD_O}{36525} = \dfrac{JD^0 - JD_O}{36525} + \dfrac{UTC}{24 \cdot 36525} = T_0 + \Delta T$, with

$\Delta T = 1.140771161 \cdot 10^{-6}$ UTC. It follows then that ΔT satisfies: $3.168808777 \cdot 10^{-10} \leq \Delta T \leq 2.737850787 \cdot 10^{-5}$.

On the other hand, we know that $T_0 < \ell$ and that it will fall in the range of $(10^{-2}, 10^{-1})$. It follows then that the operation $T_0 + \Delta T$ cannot be executed on a simple calculator. However, since in all of our applications T appears only in conjunction with a coefficient denoted by A, so that actually only terms containing $A \cdot T = A \cdot T_0 + A \cdot \Delta T$ are involved in our formulae, we can be assured that the addition of $A \cdot \Delta T$ can be executed on a simple calculator because the magnitude of A is always sufficiently large. Also note that because of the range of ΔT it follows that all terms containing ΔT^2 and higher powers of ΔT can be neglected in our calculations so that the approximation $T^2 \cong T_0^2 + 2T_0 \cdot \Delta T$ can be employed. (See numerical examples in Sect. 6.4.)

Chapter 8
Ephemerides

8.1 Low Accuracy Ephemeris for the Sun, a Numerical Example

This low accuracy ephemeris is based on an algorithm for computing the GHA and declination of the Sun that utilizes only the linear parts of the expressions provided in Chap. 7. Therefore, the approximations so obtained are the result of a linearization of the exact equations and are derived from said expressions by neglecting—truncating—all terms that contain powers of two or higher in T. Furthermore, this algorithm also does not take into consideration terms in (l), Sect. 7.8 that account for the effects of nutation and aberration.[1]

It also strictly adheres to the initial epoch of the year 2000AC, since only for this epoch is all relevant data provided for in Appendix A.

With regards to accuracy, it can be said, that in many applications to navigation, the accuracy attained by using this ephemeris comes sometimes even close to the results given in the NA and, in general, the error should not exceed $\pm 1'$.

---

[1] It also does not account for the perturbation caused by the Moon and the bulge of the Earth

Algorithm I

Step 1: Input	Year Y	Month M	Day d	Hour h	Minute $\bar{m}$	Second s
Step 2: Calculate	$UTC = h + \bar{m} \cdot 0.016666666 + s \cdot 0.000277777$ =					
Step 3: Calculate or use recorded value	$y = \begin{pmatrix} Y-1, \; m = M+12, \; \text{if } M \leq 2; \; A = \text{int} <y/100>\;;\; B = 2 - A + \text{int} <A/4> \\ Y, \quad m = M, \text{ if } M > 2. \quad \text{if } Y \geq 1582, \; D = d + UTC/24 \end{pmatrix}$ = $JD = \text{int} <365.25\,(y+4716)> \; + \text{int} <30.6001\,(m+1)> \; + D + B - 1524.5$ =					
Step 4: Calculate	$T_0 = \dfrac{JD^0 - 2451545}{36525}, \quad \Delta T = 1.140771161\,10^{-6} \cdot UTC$ $T = T_0 + \Delta T$ $JD^0 = JD - UTC/24$ =					
Step 5: Calculate	$GHA = 100°.46061837 + 36000.77005 \cdot T_0 + 15.04106865 \cdot UTC$ =					
Step 6: Calculate	$M(T) = 357°.52772 + 35999.05034 \cdot T_0 + 0.04106667845 \cdot UTC$ =					
Step 7: Calculate	$\ell(T) = 280°.46646 + 36000.76983 \cdot T_0 + .04106864 \cdot UTC$ =					
Step 8: Calculate	$e(T) = 0.016708634 - 0.000042037 \cdot T_0$ =					
Step 9: Calculate	$C(T) = \dfrac{180 \cdot e}{\pi} \cdot \left[2 \sin M + e \cdot 5/4 \cdot \sin 2M + e^2 \cdot (13/12 \sin 3M - 1/4 \sin M)\right]$ =					
Step 10: Calculate	$\varepsilon_0(T) = 23°.439291111 - 0.013004166 \cdot T_0$ =					
Step 11: Calculate	$\lambda = C + 1, \quad \nu = C + M$ =					
Step 12: Calculate	$RA \odot = \tan^{-1}(\cos \varepsilon_0 \cdot \tan \lambda)^*, \; SHA \odot = 360° - RA \odot$ =					
Step 13: Calculate	$GHA\odot = GHA \; \Upsilon + SHA\odot$					
Step 14: Calculate	$\delta = \sin^{-1}(\sin \varepsilon_0 \cdot \sin \lambda)$ =					
Step 15 Calculate	$r = \dfrac{1 - e^2}{1 + e \cos \nu} \cdot AU.$ = $\Pi = \dfrac{8''.794}{r} = HP\odot$ = $SD = \dfrac{959''.63}{r}$ =					

*Note that $\tan^{-1} x$ is a multi-valued function with branches that are 180° apart, i.e., $\tan^{-1} x = \tan_0^{-1} x \pm k \cdot 180°, k = 0, 1, 2\ldots \tan_0^{-1}$ denotes the value of zero-branch as used in most calculators. To remove the inherent ambiguity it is simply necessary to recall that the $RA\odot$ of the true Sun can only differ by no more than $\pm 4°2'$ from the $RA\odot = 1(T)$ of the fictitious Sun

8.1 Low Accuracy Ephemeris for the Sun, a Numerical Example

The reader should also recall that Π in the last formula of this algorithm denotes the horizontal parallax as needed in sight reductions.

Next, let's look at a practical example in order to illustrate how this algorithm actually works:

Example Find the GHA and the declination δ of the true Sun $\odot$ on December 24, 2013 at $14^h\ 25^m\ 36^s$ UTC.

Solution:

1. Input: Y = 2013, M = 12, d = 24, h = 14, $\bar{m}$ = 25, s = 36.
2. Calculate: UTC = 14.39333329
3. Calculate: JD^0 = 2456650.5
4. Calculate: T_0 = 0.139780971 and ΔT = 1.641949909 $\cdot\ 10^{-5}$
5. Calculate: GHA Υ = 309°.1743264 = 309° 10' 27".6
6. Calculate: M (T) = 350°.1010164
7. Calculate: ℓ (T) = 273°.2801346
8. Calculate: e (T) = 1.6702758 $\cdot\ 10^{-2}$
9. Calculate: C (T) = −0.335937345
10. Calculate: ε_0 (T) = 23°.43747338
11. Calculate: λ = 272°.9441973, ν = 349°.7650791
12. Calculate: RA $\odot$ = 273°.2084226, SHA $\odot$ = 86°.79157738
13. Calculate: GHA $\odot$ = 35°.96590378 = 35° 58 (NA = 35° 58' 36")
14. Calculate: δ = −23°.40469131 = −23° 24' 16".8 (NA = −23° 24'.1)
15. Calculate: r = 0.983530723 AU. Π = 8".94125602 = 0'.149020933
    SD = 16'.26126018, AU = 1.49597870 $\cdot\ 10^8$ km.

In all cases where higher accuracy is required, we need to include terms of T^2 in our approximations of the "exact" formulae and also include the effects of nutation and aberration on the coordinates. The resulting algorithm will produce results that can be classified as approximation of intermediate accuracy that should suffice for the purpose of navigation. It should also be noted that the Julian Date JD does not have to computed every time a "fix" is required since it suffices to calculate it, for example, once a month when it is recorded in the ship's Log and then the days are added as time goes on.

In the above example, all the numerical calculations were carried out with a simple, inexpensive, scientific calculator.

8.2 Intermediate Accuracy Ephemeris for the Sun

As a direct application of some of the explicit expressions provided in the previous chapter, we can now write down an algorithm by taking terms of up to order two in T and of up to order four in e into account in our formulae and also account for the corrections for nutation and aberration. In order to provide the user with the option of employing the results of the low accuracy algorithm, we shall use the superscript " ' " for all the quantities of our improved algorithm #II.

298   8 Ephemerides

In case the user opts for the latter, all, except for three improved quantities, can be calculated by simply adding one or two additional corrective terms to the results of the low accurate algorithm. We then have:

Algorithm II

Step 1: Input	Year Y	Month M	Day d	Hour h	Minute $\bar{m}$	Second s
Step 2: Calculate	UTC = h + $\bar{m}$ · 0.016666666 + s · 0.000277777 =					
Step 3: Calculate or use recorded value	$y = \begin{pmatrix} Y-1, & m = M+12, & \text{if } M \leq 2; \\ Y, & m = M, & \text{if } M > 2. \end{pmatrix}$ A = int < y / 100 > ; B = 2 − A + int < A/4 > if Y ≥ 1582, D = d + UTC/24 JD = int < 365.25 (y + 4716) > + int < 30.6001 (m + 1) > + D + B − 1524.5 =					
Step 4: Calculate	$T_0 = \dfrac{JD^0 - 2451545}{36525}$, $\Delta T = 10.140771161 \cdot 10^{-6} \cdot$ UTC $JD^0 =$ JD − UTC / 24 $T = T_0 + \Delta T$, $T^2 \cong T_0^2 + 2 \cdot T_0 \cdot \Delta T$ =					
Step 5: Calculate	$M'(T) = 357°.52772 + 35999.05034 \cdot T_0 + 0.04106667845 \cdot$ UTC $- 1.603 \cdot 10^{-4} \cdot (T_0 + 2.281542322 \cdot 10^{-6} \cdot$ UTC$) \cdot T_0$ $= M(T) - 1.603 \cdot 10^{-4} \cdot (T_0 + 2.281542322 \cdot 10^{-6} \cdot$ UTC$) \cdot T_0$ =					
Step 6: Calculate	$\ell'(T) = 280°.46646 + 36000.76983 \cdot T_0 + 0.04106864 \cdot$ UTC $+ 3.032 \cdot 10^{-4} \cdot (T_0 + 2.281542322 \cdot 10^{-6} \cdot$ UTC$) \cdot T_0$ $= \ell(T) + 3.032 \cdot 10^{-4} \cdot (T_0 + 2.281542322 \cdot 10^{-6} \cdot$ UTC$) \cdot T_0$ =					
Step 7: Calculate	$e'(T) = e(T) = 0.016708634 − 0.000042037 \cdot T_0$ =					
Step 8: Calculate	$C'(T) = \dfrac{180 \cdot e}{\pi} \cdot [2 \sin M' + e' \cdot 5/4 \cdot \sin 2M' + e'^2 \cdot (13/12 \sin 3M' - 1/4 \sin M')]$ $+ e'^4 \cdot (103/96 \cdot \sin 4M' - 11/24 \cdot \sin 2M')]$ $= C(T) + e'^4 \cdot (103/96 \cdot \sin 4M' - 11/24 \cdot \sin 2M') \cdot \dfrac{180}{\pi}$ =					
Step 9: Calculate	$\Omega'(T) = 125°.04452 - 1934.136261 \cdot T_0 - 2.206406868 \cdot 10^{-3} \cdot$ UTC $+ 2.0708 \cdot 10^{-3} \cdot (T_0 + 2.281542322 \cdot 10^{-6} \cdot$ UTC$) \cdot T_0$ =					
Step 10: Calculate	$\varepsilon'_0(T) = \varepsilon_0(T) + \Delta\varepsilon = 23°.439291111 − 0.013004166 \cdot T_0 + \Delta\varepsilon$ $= 23°.439291111 − 0.013004166 \cdot T_0 + 0.002556388 \cdot \cos \Omega'$ =					
Step 11: Calculate	GHA' ♈ $= 100°.46061837 + 36000.77005 \cdot T_0 + 15.04106865 \cdot$ UTC $+ 3.8793 \cdot 10^{-4} \cdot (T_0 + 2.281542322 \cdot 10^{-6} \cdot$ UTC$) \cdot T_0 − 0.0047777777 \cdot$ $\cdot \cos \varepsilon' \cdot \sin \Omega'$ $=$ GHA ♈ $+ 3.87933 \cdot 10^{-4} \cdot (T_0 + 2.281542322 \cdot 10^{-6} \cdot$ UTC$) \cdot T_0$ $- 0.0047777777 \cdot \cos \varepsilon \cdot \sin \Omega'$ =					

(continued)

8.2 Intermediate Accuracy Ephemeris for the Sun

(continued)

Step 12: Calculate	$\lambda' = C' + \ell' + \Delta\lambda_\ominus + \Delta\psi = C' + \ell' - \kappa - 0.004777777 \cdot \sin\Omega'$ $= C' + \ell' - 0°.005694444 - 0°.004777777 \cdot \sin\Omega',$ $=$ and $v' = C' + M' - 0.004777777 \cdot \sin\Omega'$ $=$
Step 13: Calculate	$RA\,\odot' = \tan^{-1}(\cos\varepsilon' \cdot \tan\lambda'),\ SHA\,\odot = 360° - RA\,\odot'$ $=$
Step 14: Calculate	$GHA\,\odot' = GHA'\,\Upsilon + SHA\,\odot'$ $=$
Step 15: Calculate	$\delta' = \sin^{-1}(\sin\varepsilon' \cdot \sin\lambda')$ $=$
Step 16: Calculate	$r' = \dfrac{1 - e'^2}{1 + e'\cos v'} \cdot AU.$ $=$ $\Pi' = \dfrac{8''.794}{r'}$ $=$ $SD = \dfrac{959''.63}{r'}$ $=$

Next, let's consider an example where the input is the same as in the previous example in order to have a tangible comparison.

Example Find the GHA and the declination δ of the true Sun $\odot$ on December 24, 2013 at $14^h\ 25^m\ 36^s$ UTC.

Solution:

1. Input: Y = 2013, M = 12, d = 24, h = 14, $\bar{m}$ = 25, s = 36.
2. Calculate: UTC = 14.39333329
3. Calculate: JD^0 = 2456650.5
4. Calculate: T_0 = 0.139780971 and ΔT = 1.641949909 · 10^{-5}
5. Calculate: M' (T) = 350°.1010133
6. Calculate: ℓ' (T) = 273°.2801405
7. Calculate: e' (T) = 1.6702758 · 10^{-2}
8*. Calculate: C' (T) = 0−0.335937386
9. Calculate: Ω' (T) = 214°.65573583
10. Calculate: ε' (T) = 23°.43537054
11. Calculate: GHA' Υ = 309°.177444923
12*. Calculate: λ' = 272°.9412255, v' = 349°.7677928
13*. Calculate: RA $\odot'$ = 273°.2051343, SHA $\odot'$ = 86°.79486571
14*. Calculate: GHA' $\odot$ = 35°.972318633 = 35° 58' 20''.4

15*. Calculate: $\delta' = -23°.40265784 = -23°\ 24'\ 10''$
16*. Calculate: $r' = 0.983530723$ AU. $\Pi' = 0'.149020933$
    $SD' = 16'.26126018$, AU $= 1.49597870\ 10^8$ km

Again, all the numerical evaluations of this section have been performed with a simple calculator. However, for obvious reasons, it is highly recommended that the potential user have a more sophisticated programmable calculator in addition to the simple one.

In closing this section, the reader is reminded that besides the two algorithms already presented in this section, a third has been previously outlined in Sect. 8.6 and has been designated as an emergency ephemeris.

In the next section, I will construct an algorithm for an ephemeris for the stars… in particular, for the 57 most important stars to navigation for which all the necessary data has been provided in Appendix A, making this book a practical manual for navigation and not just a collection of navigational formulae.

8.3 Low Accuracy Ephemeris for the Stars

Because of the database available in Appendix A in the back of this book, it is convenient to choose 2000AC for the epoch E_0 as the initial epoch for this ephemeris. In the case of a low accuracy ephemeris it would suffice to go directly from E_0 to the final epoch E without invoking the exact formulae (4), (5), and (6), Sect. 7.2. Furthermore, you may also neglect the effects of aberration and annual parallax as well as the equation of the equinox. Then, according to the formulae of Sect. 7.8 the true place of a star, taken as a good approximation to the apparent place, can be found by executing the following algorithm:

Algorithm III

Step 1: Input	Year Y	Month M	Day d	Hour h	Minute $\bar{m}$	Second s
	RA*	Declination	RA component of PM		δ component of PM	
	a_0	δ_0	μ_a		μ_δ	
Step 2: Calculate	\multicolumn{6}{l}{UTC $= h + \bar{m} \cdot 0.016666666 + s \cdot 0.000277777$}
	=					
Step 3: Calculate	\multicolumn{6}{l}{$y = \begin{pmatrix} Y-1,\ m = M+12,\ \text{if } M \leq 2; \\ Y,\ m = M,\ \text{if } M > 2. \end{pmatrix}$ $A = \text{int} <y/100>$; $B = 2 - A + \text{int} <A/4>$ $D = d + UTC/24$, if $Y \geq 1582$,}
	=					
	\multicolumn{6}{l}{$JD = \text{int} <365.25\ (y+4716)> + \text{int} <30.6001\ (m+1)> + D + B - 1524.5$}
	=					
Step 4: Calculate	\multicolumn{6}{l}{$T_0 = \dfrac{JD^0 - 2451545}{36525}$, JD^0 constant from $0^h:00$, $JD^0 = UTC/24$}
	=					
	\multicolumn{6}{l}{$T = T_0 + \Delta T$, and $\Delta T = 1.140771161 \cdot 10^{-6} \cdot UTC$}
	=					

(continued)

8.3 Low Accuracy Ephemeris for the Stars

(continued)

Step 5: Calculate:	GHA Υ = $100°.46061837 + 36000.77005 \cdot T_0 + 15.04106865 \cdot$ UTC =
Step 6: Calculate	$m^s = 3^s.07496 + 0^s.00180 \cdot T_0$ = $n^s = 1^s.33621 - 0^s.00057 \cdot T_0$ = $m'' = 15 \cdot m^s$ = $n'' = 15 \cdot n^s$ =
Step 7: Calculate:	$\Delta_P a_0^s = (m^s + n^s \cdot \sin a_0 \cdot \tan \delta_0) \cdot 10^2 \cdot T$ = $\Delta_P \delta_0 = n'' \cdot \cos a_0 \cdot 10^2 \cdot T, \quad T = T_0 + \Delta T$ =
Step 8: Calculate	$\Delta_{PM} a_0^s = (\mu_a^s + \mu_a^s \cdot \mu_\delta'' 15 \cdot \tan \delta_0 \cdot \sin 1'' \cdot 10^2 \cdot T) 10^2 T$, $\sin 1'' = 4.84813681 \cdot 10^{-6}$ = $\Delta_{PM} \delta_0 = (\mu_\delta'' - \frac{1}{2}(\mu_a^s \cdot 15)^2 \cdot \sin \delta_0 \cdot \cos \delta_0 \cdot \sin 1'' \cdot 10^2 \cdot T) \cdot 10^2 T$ =
Step 9: Calculate	$\Omega(T) = 125°.04452 - 1934.136261 \cdot T_0 - 2.206406868 \cdot 10^{-3} \cdot$ UTC =
Step 10: Calculate:	$\Delta\Psi = -17''.200 \cdot \sin \Omega$, and $\Delta\varepsilon = 9''.203 \cdot \cos \Omega = 0°.002556388 \cdot \cos \Omega$ =
Step 11: Calculate	$\varepsilon(T) = \varepsilon_0(T) + \Delta\varepsilon = 23°.439291111 - 0.013004166 \cdot T_0 + \cdot 0.002556388$ $\cdot \cos \Omega$ =
Step 12: Calculate	$a_\tau = a_0 + \Delta_P a_0 + \Delta_{PM} a_0$ and $\delta_\tau = \delta_0 + \Delta_P \delta_0 + \Delta_{PM} \delta_0$, in degress = $\Delta_N a_\tau'' = \Delta\psi \cdot (\cos \varepsilon + \sin \varepsilon \cdot \sin a_\tau \cdot \tan \delta_\tau) - \Delta\varepsilon \cdot \cos a_\tau \tan \delta_\tau$ = $\Delta_N \delta_\tau'' = \Delta\psi \cdot \sin \varepsilon \cdot \cos a_\tau + \Delta\varepsilon \cdot \sin a_\tau$ =
Step 13: Calculate	$a = a_\tau + \Delta_N a_\tau$ = and $\delta = \delta_\tau + \Delta_N \delta_\tau$ =
Step 14: Calculate	GHA* = GHA Υ + SHA* = and SHA* = 360° − a =

Next, let's consider an example to determine the true place (approximation for apparent place) of the navigational star **ALPHERATZ** for which we have the required data in Appendix A. For the time of observation we will use the time and date used in the previous examples. (Just a suggestion... but if you record the computed JD from the previous examples, you will avoid having to use the subroutine for finding it in this one.)

Example: Star—ALPHERATZ

Step 1: Input	Year	Month	Day	Hour	Minute	Second
	2013	12	24	14	23	36
	RA* $0^h.8^m\,23^s.3$	Declination $29°.5'\,26''$	RA Component of PM $0^s.136$		δ Component of PM $-0''.163$	
	a_0 $2°.097083328$	δ_0 $29°.0905555$	μ_a $0°.000566666$		μ_δ $-0°.000045277$	
Step 2: Calculate	UTC = 14.39333329					
Step 3: Calculate	JD = 2456650.5 $T_0 = 0.139780971$ $\Delta T = 1.641949909 \cdot 10^{-5}$ T = 0.1397973914					
Step 4: Calculate	$10^2 \cdot T = 13.97973914$					
Step 5: Calculate	GHA Υ = 309°.1743264 = 309° 10' 27".6					
Step 6: Calculate	$m^s = 3^s.075219903$ $n^s = 1^s.336130325$ $n'' = 20''.04195487$					
Step 7: Calculate	$\Delta_P a_0 = 43^s.3710675 = 0°.180712781$ $\Delta_P \delta_0 = 0°.077776014$					
Step 8: Calculate	$\Delta_{PM} a_0 = 1^s.901098268 = 0°.007921242$ $\Delta_{PM} \delta_0 = -2''.279535221 = -0°.000633204$					
Step 9: Calculate:	$\Omega(T) = 214°.65573583$					
Step 10: Calculate	$\Delta\Psi = 9''.780680262$ $\Delta\varepsilon = -7''.570236834$					
Step 11: Calculate	$\varepsilon(T) = 23°.43537054$					
Step 12: Calculate	$a_\tau = 2°.285717351$ $\delta_\tau = 29°.16770831$ $\Delta_N a_\tau = 13''.28235845 = 0°.003689544$ $\Delta_N \delta_\tau = 3''.584900608 = 0°.000995805$					
Step 13: Calculate	a = 2°.289406901 $\delta = 29°.16870412 = 29 \circ .10'\,7''.3$ (NA = 29°.10' 18")					
Step 14: Calculate	SHA* = 357°.7105931 = 357° 43' (NA = 357° 43'.1) GHA* = 306°.8849195 = 306° 53'					

8.4 Intermediate Accuracy Ephemeris for the Stars

In all cases where higher accuracy is required, you need to invoke the exact equations for the effect of precession, the equation of the equinox and also take into account the displacement due to aberration and, if necessary, the displacement due to annual parallax. You should also choose the starting epoch to coincide with the JD corresponding to the year 2000AC resulting in $T'_0 = 0$ and hence in a considerable simplification of the formulae (6) in Sect. 7.2 for the parameters ζ, Θ, and z of the rotational transformation (3), (4), and (5), Sect. 7.2.

Algorithm IV

Step 1: Input	Year	Month	Day	Hours	Minutes	Seconds
	Y	M	d	h	$\bar{m}$	s
	RA*	Declination	RA—Component of PM		δ—Component of PM	
	a_0	δ_0	μ_a		μ_δ	
Step 2: Calculate	$\text{UTC} = h + \bar{m} \cdot 0.016666666 + s \cdot 0.000277777$					
	=					
Step 3: Calculate or Count	$y = \begin{pmatrix} Y-1, \ m = M+12, \ \text{if } M=2; \ A = \text{int} <y/100>; \ B = 2 - A + \text{int} <A/4> \\ Y, \ m = M, \ \text{if } M>2, \ D = d + \text{UTC}/24, \ \text{and for all } Y = 1583AC \end{pmatrix}$					
	=					
	$JD = \text{int} <365.25 \, (y+4716)> + \text{int} <30.6001 \, (m+1)> + D + B - 1524.5$					
	=					
Step 4: Calculate	$T_0 = \dfrac{JD^0 - 2451545}{36525}$, JD^0 counts from $0^h : 00$, $JD^0 = JD - \text{UTC}/24$					
	=					
	$T = T_0 + \Delta T$, and $\Delta T = 1.140771161 \cdot 10^{-6} \cdot \text{UTC}$					
	=					
Step 5: Calculate	$\Omega(T) = 125°.04452 - 1934.136261 \cdot T_0 - 2.206406868 \cdot 10^{-3} \cdot \text{UTC}$					
	=					
Step 6: Calculate	$\Delta\Psi = -17''.200 \cdot \sin\Omega$, and $\Delta\varepsilon = 9''.203 \cdot \cos\Omega$					
	=					
Step 7: Calculate	$\varepsilon(T) = \varepsilon_0(T) + \Delta\varepsilon = 23°.439291111 - 0.013004166 \cdot T_0 + \cdot 0.002556388 \cdot \cos\Omega$					
	=					
Step 8: Calculate	$\text{GHA } \Upsilon = 100°.46061837 + 36000.77005 \cdot T_0 + 15.04106865 \cdot \text{UTC}$ $+ 0.000387933 \cdot T_0^2 - 0.004777777 \cdot \cos\varepsilon \cdot \sin\Omega$					
	=					
Step 9: Calculate	$\xi = 2306''.2182 \cdot T + 0''.30188 \cdot T^2$					
	=					
	$\Theta = 2004''.3109 \cdot T - 0''.42665 \cdot T^2$					
	=					
	$z = 2306''.2181 \cdot T + 1''.09468 \cdot T^2$					
	=					

(continued)

(continued)

Step 10: Calculate	$A = \cos \delta_0 \cdot \sin(a_0 + \xi)$ = $a_P = z + \tan^{-1} \dfrac{A}{B}$ = $\delta_P = \sin^{-1} C$ = $B = \cos \Theta \cdot \cos \delta_0 \cdot \cos(a_0 + \xi) - \sin \Theta \cdot \sin \delta_0$ = $C = \sin \Theta \cdot \cos \delta_0 \cdot \cos(a_0 + \xi) + \cos \Theta \cdot \sin \delta_0$ =
Step 11: Calculate	$\Delta_{PM} a_0 = (\mu_a^s + \mu_a^s \cdot 15 \mu_\delta'' \tan \delta_0 \cdot \sin 1'' \cdot 10^2 \cdot T) \cdot 10^2 \cdot T, \quad \sin 1'' = 4.8484368 \cdot 10^{-6}$ = $\Delta_{PM} \delta_0 = (\mu_\delta'' - 1/2 (\mu_a^s \cdot 15)^2 \cdot \sin \delta_0 \cdot \cos \delta_0 \cdot \sin 1'' \cdot 10^2 \cdot T) \cdot 10^2 \cdot T$ =
Step 12: Calculate	$a_\tau = a_p + \Delta_{PM} a_0$ = $\delta_\tau = \delta_p + \Delta_{PM} \delta_0$ = $\Delta_N a_\tau = \Delta\psi (\cos \varepsilon + \sin \varepsilon \cdot \sin a_\tau \cdot \tan \delta_\tau) - \Delta\varepsilon \cdot \cos a_\tau \cdot \tan \delta_\tau$ = $\Delta_N \delta_\tau = \Delta\psi \cdot \sin \varepsilon \cdot \cos a_\tau + \Delta\varepsilon \cdot \sin a_\tau$ =
Step 13: Calculate	$a_T = a_\tau + \Delta_N a_\tau$ = $\delta_T = \delta_\tau + \Delta_N \delta_\tau$ = $\Delta_A a_T = -\kappa \cdot \sec \delta_T (\cos a_T \cos \lambda_\odot \cos \varepsilon + \sin a_T \cdot \sin \lambda_\odot)$ = $\Delta_A \delta_T = -\kappa \cdot \cos \lambda_\odot \cdot \cos \varepsilon (\tan \varepsilon \cdot \cos \delta_T - \sin a_T \cdot \sin \delta_T) - \kappa \cdot \cos a_T \cdot \sin \delta_T \sin \lambda_\odot$ = $\kappa \cong 20''.5$ $\lambda_\odot$: longitude of true Sun to be found by using the subroutine for the ephemeris for the Sun or using approximation*
Step 14: Calculate	$\Delta_{AP} a_T = \Pi \cdot \sec \delta_T (\cos a_T \cos \varepsilon \cdot \sin \lambda_\odot - \sin a_T \cdot \cos \lambda_\odot)$ = $\Delta_{AP} \delta_T = \Pi \cdot (\cos \delta_f \cdot \sin \varepsilon \cdot \sin \lambda_\odot - \cos a_T \cdot \sin \delta_T \cos \lambda_\odot - \sin a_T \cdot \sin \delta_T \sin \lambda_\odot)$ = Π: star's geometirc parallax; $\lambda_\odot$: Sun's longitude [see note *]
Step 15: Calculate	$a = a_T + \Delta_A a_T + \Delta_{AP} a_T$ = $\delta = \delta_T + \Delta_A \delta_T + \Delta_{AP} \delta_T$ =
Step 16: Calculate	GHA* = GHA ♈ + SHA* = SHA = 360° − a =

*For the purpose of calculating Δ_A and Δ_{AP}, one may resort to approximating $\lambda_\odot$ by the mean longitude l(T)—see formula (2), Sect. 2.3

8.4 Intermediate Accuracy Ephemeris for the Stars

For the same of having a tangible comparison for the achieved accuracy at hand, consider again the data of the previous example:

*Example: Star * ALPHERATZ*
Calculate the GHA* and the declination.

Step 1: Input	Year 2013	Month 12	Day 24	Hours 14	Minutes 23	Seconds 36
	RA* $0^h.8^m 23^s.3$	Declination 29°5' 26"	RA— Component of PM $0^s.136$		δ—Component of PM $-0^s.136$	
	a_0 2°.097083328	δ_0 29°.0905555	μ_a 0°.000566666		μ_δ $-0°.000045277$	
Step 2: Calculate	UTC = 14.39333329					
Step 3: Calculate	JD = 2456650.5					
Step 4: Calculate	$T_0 = 0.139780971$, $\Delta T = 1.641949909 \cdot 10^{-5}$, $T = 0.13979739$					
Step 5: Calculate	$\Omega(T) = 214°.65573583$					
Step 6: Calculate	$\Delta\psi = 9''.780680262$, $\Delta\varepsilon = -7''.570236834$					
Step 7: Calculate	$\varepsilon(T) = 23°.43537054$					
Step 8: Calculate	GHA Υ = 309°.1768267 = 309°10'36".6					
Step 9: Calculate	$\xi = 0°.089558103$, $\Theta = 0°.077830304$, $z = 0°.089562407$					
Step 10: Calculate	A = 0.033341687, B = 0.872554826, C = 0.487377067					
Step 11: Calculate	$a_P = z + \tan^{-1} A / B = 2°.27785922$					
Step 12: Calculate:	$\delta_P = \sin^{-1} C = 29°.1683291$					
Step 13: Calculate	$\Delta_{PM} a_0 = 0°.007921242$, $\Delta_{PM}\delta_0 = -0°.000633204$					
Step 14: Calculate	$a_\tau = a_0 + \Delta_p a_0 + \Delta_{PM} a_0 = 2°.285780462$					
Step 15: Calculate	$\delta_\tau = \delta_0 + \Delta_p \delta_0 + \Delta_{PM} \delta_0 = 29°.1676959$					
Step 16: Calculate	$\Delta_N a_\tau = 0°.003689544$					
Step 17: Calculate	$\Delta_N \delta_\tau = 0°.000995803$					
Step 18: Calculate	$a_T = a_\tau + \Delta_N a_\tau = 2°.283470006$					
Step 19: Calculate	$\delta_T = \delta_\tau + \Delta_N \delta_\tau = 29°.1686917$					
Step 20: Calculate	$\Delta_A a_T = -0''.16776099 = -0°.0000466$					
Step 21: Calculate	$\Delta_A \delta_T = 9''.623708627 = 0°.002673252$					
Step 22: Calculate	$\Delta_{AP} a_T = -0''.033628231 = -0°.00000934$					
Step 23: Calculate:	$\Delta_{AP}\delta_T = -0''.011275716 = -0°.000003132$					
Step 24: Calculate	a = 2°.289414066 = 2°17'21".9					

(continued)

(continued)

Step 25: Calculate	$\delta = 29°.17136182 = 29°10'17''$
Step 26: Calculate	$SHA^* = 357°.7105859 = 357°42'38''$
Step 27: Calculate	$GHA^* = 306°.887685 = 306°53'14''.89$

Concerning the accuracy of the results thus obtained it should be understood that it is no longer meaningful to compare them relative to data that exhibits a lesser degree of accuracy. By taking only the linear terms of the expressions which depend on T (see Eqs. (1) to (8), Sect. 6.3) into account and by using the first two terms in the Equation of the Center, the introduction of a simplified time parameter τ yields the algorithm presented in Sect. 8.5. Again, the errors should not exceed $\pm 1'$ in all relevant cases.

8.5 Compressed Low Accuracy Ephemeris for the Sun and Stars for the Years 2014±

Algorithm V

Input	Year Y	Month M	Day d	Hour h	Minute $\bar{m}$	Second s
	RA*	Declination	R component of PM		δ component of PM	
	a_0	δ_0	μ_a		μ_δ	
(1)	D: Count the full days from day 01/01/2014 to the day of the instant in question at 0^h:00. =					
(2)	$UTC = h + \bar{m} \cdot 0.016666666 + s \cdot 0.000277777$ =					
(3)	$\tau = 14 + 2.73785078 \cdot 10^{-3} \cdot D$ =					
(4)	$GHA\ \Upsilon = 100°.46061837 + 360°.77005 \cdot \tau + 15.04106865 \cdot UTC$ =					
SUN						
(5S)	$M(\tau) = 357°.52772 + 359°.990503 \cdot \tau + 0.04106667845 \cdot UTC$ =					
(6S)	$\ell(\tau) = 280°.46646 + 360°.007698 \cdot \tau + 0.04106864 \cdot UTC$ =					
(7S)	$e(\tau) = 0.016708634 - 0.00000000420 \cdot \tau$ =					
(8S)	$\varepsilon_0(\tau) = 23°.439291111 - 0.000130042 \cdot \tau$ =					

(continued)

8.5 Compressed Low Accuracy Ephemeris …

(continued)

(9S)	$C(\tau) = \dfrac{180}{\pi} \cdot e \cdot (2 \sin M + e \cdot \dfrac{5}{4} \cdot \sin 2M)$ $\lambda_\odot(\tau) = C(\tau) + \ell(\tau)$
(10S)	$RA\ \odot = \tan^{-1}(\cos \varepsilon_0 \cdot \tan \lambda_\odot)^*$ $SHA\ \odot = 360° - RA\ \odot$
(11S)	$GHA\ \odot = GHA\ \Upsilon + SHA\ \odot$ $E.T. = 4 \cdot (\ell(\tau) - RA\ \odot)$
(12S)	$\delta = \sin^{-1}(\sin \varepsilon_0 \sin \lambda_\odot)$ $v = C + M$ $SD = \dfrac{959''.63}{r}$ $r = \dfrac{1-e^2}{1+e\cos v} \cdot AU$ $\Pi = \dfrac{8''.794}{r}$

STARS

(5*)	$m'' = 46''.1244 + 0.000279 \cdot \tau$ $n'' = 20''.0431 - 0.000085 \cdot \tau$
(6*)	$a = a_0 + 2.777777 \cdot 10^{-4} \cdot (m'' + n'' \cdot \sin a_0 \cdot \tan \delta_0 + \mu_a'') \cdot \tau$ $\delta = \delta_0 + 2.777777 \cdot 10^{-4} \cdot (n'' \cdot \cos a_0 + \mu_\delta'') \cdot \tau$
(7*)	$GHA^* = GHA\ \Upsilon + SHA^*$ $SHA^* = 360° - a$

*Select the correct branch of $\tan^{-1} x$ by assuring that the RA is no more than about $\pm 4°$ away from $l(\tau)$

Example 1: SUN

Input: 2013 12 24 14 23 36

(1) D = −8
(2) UTC = 14.39809719
(3) τ = 13.97809719
(4) M = 350°.1010462
(5) ℓ = 273°.2801664
(6) e = 0.016702763
(7) ε₀ = 23°.43747337
(8) GHA ♈ = 309°.1743661
(9) C = −0.335804634, λ☉ = 272°.9443614
(10) RA ☉ = 273°.2086018, SHA ☉ = 86°.79139818
(11) GHA ☉ = 35°.96576428 ≅ 35° 58', ET = 17^s.2
(12) δ = −23°.40468763 = −23° 24' 16".9

Example 2: STAR* ALPHERATZ

Input: 2013 12 24 14 23 36
2°.097083328 29°.0905555 0".000566666 − 0".000045277

(1) D = −8
(2) UTC = 14.39333329
(3) τ = 13.97809719
(4) GHA ♈ = 309°.1743661
(5) m" = 46".12829989, n" = 20".04191189
(6) a = 2°.277777057, δ = 29°.16832204 = 29° 10' 6"
(7) SHA* = 357°.7222229 = 357° 43' 20"
    GHA* = 306°.896589 = 306° 53' 47".7

8.6 The Earth Viewed as a Gyro

In the first part of this book, our working assumption was that the Earth was a SPHEROID. Now, it's time to imagine that the Earth is simply a sphere with bulges near the equator. By employing this model, we can readily see that there are unequal forces of attraction $\overrightarrow{F_+}$ and $\overrightarrow{F_-}$ on these bulges. (See Fig. 8.6.1.)

Since the net gravitational torque on a spherical planet is zero, we only have to deal with the resulting torque exerted by the forces $\overrightarrow{F_+}$ and $\overrightarrow{F_-}$. These torques are: $\overrightarrow{\Gamma_+} = \vec{a} \times \overrightarrow{F_+}$, and $\overrightarrow{\Gamma_-} = -\vec{a} \times \overrightarrow{F_-}$. Hence, the resulting torque is: $\overrightarrow{\Gamma} = -\vec{a} \times (\overrightarrow{F_+} - \overrightarrow{F_-})$.

A similar situation exists with regards to the attraction by the Moon and the Planets. By adding up these additional torques and averaging with the torque of the

8.6 The Earth Viewed as a Gyro

Fig. 8.6.1

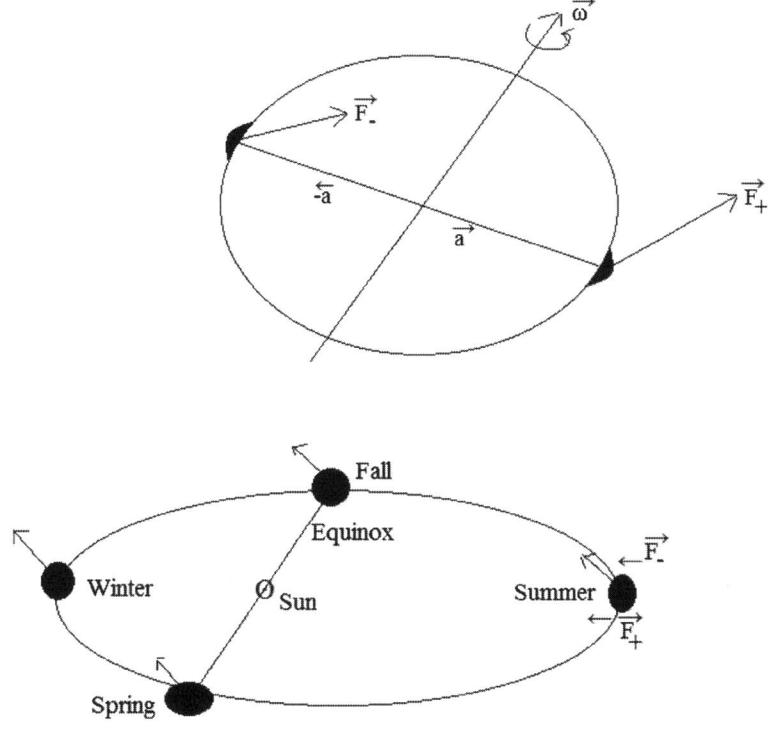

Fig. 8.6.2

Sun, we may conclude that there exists a torque $\vec{\Gamma}$ on the Earth that is perpendicular to the figure axis of the Earth and the normal to the ecliptic. (See Fig. 8.6.2.)

By taking into account the motion of the Moon and Planets relative to the Earth and Sun, we can see that there exists additional forces exerted by the mutual gravitational attraction of said bodies on the Earth resulting in additional torques that try to pull the equator of the Earth into the plane of the ecliptic. Therefore, the angle ε between the Earth and the ecliptic is no longer constant but a periodic function of time.

It is also important to realize that it is a gross oversimplification to assume that the Earth is a SPHERICAL GYRO, i.e., a spheroid with three equal major movements of inertia. However, it is sufficiently realistic to assume that the Earth "represents" a SYMMETRIC GYRO, i.e. a spheroid with two equal major moments of inertia. It follows then that the figure axis of the Earth $\vec{k}$, the axis of angular momentum $\vec{L}$, and the axis of rotation $\vec{\omega}$ lie in a common plane. (See Fig. 8.6.3.)

In actuality, the angles between the vectors $\vec{k'}$, $\vec{L}$, and $\vec{\omega}$ are very small indeed, as we will see later.

310   8 Ephemerides

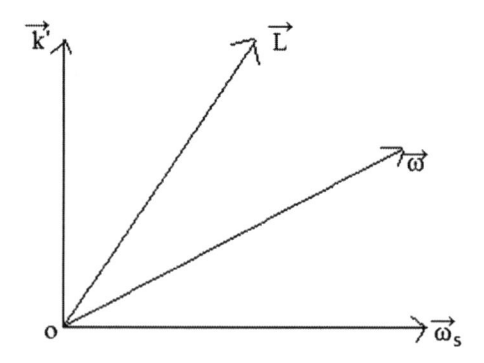

Fig. 8.6.3

Another very important consideration is to distinguish between secular and periodic forces and torques. Accordingly, their net effects on the motion of the axis of the Earth are divided in secular and periodic changes of the axis of the Earth and therefore, on the coordinates of all celestial bodies (CB).

Summarizing the aforementioned phenomena, we can state that the axis of the rotation $\vec{\omega}$, as well as the figure axis $\vec{k'}$ prescribe a circular cone about the axis of the angular momentum which, in turn, prescribes a circular cone about the normal $\vec{k}$ to the ecliptic, resulting in a conical motion which "wobbles" about the normal to the ecliptic. (See Fig. 8.6.4.)

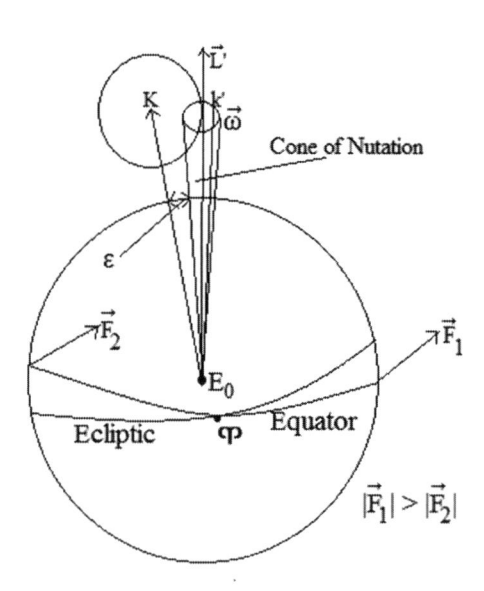

Fig. 8.6.4

8.6 The Earth Viewed as a Gyro

Accordingly, the secular components of the motion of the Earth-axis are referred to as PRECESSION, and the periodic components are referred to as NUTATION. Since our objective has been to calculate the effects of precession and nutation on the equatorial coordinates of the CBs, it is appropriate to give a brief outline of the underlying analytic treatment of these two phenomena without going deep into the theory of Celestial Mechanics, but rather to explain where the limitations of said theory are.

As the result of Newton's Law of Inertia, we can write down the equation of the motion of the "rigid spheroid" Earth as:

(1) $\frac{d}{dt}\vec{L} = \vec{\Gamma}$, where $\vec{L}$ denotes the angular momentum and $\vec{\Gamma}$ torque.

It can also be readily shown that:

(2) $\vec{L} = \Pi(I)\vec{\omega}$, where $\Pi(I)$ denotes the tensor of inertia – think of it as a special matrix in three dimensions, namely:

(3) $\Pi(I) = \begin{pmatrix} I_{xx} & -I_{xy} & -I_{xz} \\ -I_{xy} & I_{yy} & -I_{yz} \\ -I_{xz} & -I_{yz} & I_{zz} \end{pmatrix}$, where I_{xx} denotes the moment of inertia about the x-axis; I_{yy} the moment of inertia about the y-axis; and I_{zz} the moment of inertia about the z-axis. The elements I_{xy}, I_{xz} and I_{yz} are referred to as the moments of derivation, for obvious reasons.

It follows then from (2) and (3) that, in general, and in particular in case of the Earth, the angular momentum and the axis of rotation do not have the same direction in space, as has been pointed out earlier. This is true even in the case of zero external forces, i.e., $\Gamma = 0$ [76, 77]

It can now be shown that the tensor of inertia (3) can be reduced to a much simpler form, namely:

(4) $\Pi(I) = \begin{pmatrix} I_1 & 0 & 0 \\ 0 & I_2 & 0 \\ 0 & 0 & I_3 \end{pmatrix}$, by transformation of the space-like coordinate system $\vec{i}, \vec{j}, \vec{k}$ to a fixed body coordinate system $\vec{i'}, \vec{j'}, \vec{k'}$. The scalars I_1, I_2, I_3 represent the major-principal moments of inertia about the three major-principal axes $\vec{i'}, \vec{j'}, \vec{k'}$, which are again mutually perpendicular to each other. The origin $\vec{0}$ of said coordinate system coincides with the center mass of the rigid body. In all cases where two of those three major-principal moment are equal, such a gyro should be referred to as a symmetric gyro and arranged so that $I_1 = I_2 = I_s$; $I_3 = I$

If we choose as axis of symmetry the figure axis denoted by the vector $\vec{k'}$, and as the other corresponding two major-principal axes the vectors $\vec{s'}$ and $\vec{s''}$ that are mutually perpendicular and also perpendicular to $\vec{k'}$, i.e., $\vec{k'} \cdot \vec{s'} = 0$, $\vec{k'} \cdot \vec{s''} = 0$ and $\vec{s'} \cdot \vec{s''} = 0$, then I becomes the moment of inertia about $\vec{k'}$

and I_s the moment of inertia about $\vec{s'}$ and $\vec{s''}$.
Next, let us define the scalar Υ by:

(5) $\Upsilon = \frac{I}{I_s}$, then it can be shown that $1 < \Upsilon < 2$. [77]

If we now write the vector of rotation $\vec{\omega}$ as the sum of a vector $\vec{\omega_s}$ that is perpendicular to $\vec{k'}$, i.e., $\vec{k'} \cdot \vec{\omega_s} = 0$, and the vector $\omega_z \cdot \vec{k'}$, we obtain:

(6) $\vec{\omega} = \vec{\omega_s} + \omega_z \cdot \vec{k'}$.

Substituting (6) into Eq. (2) and using (4) and (5), we conclude that

(7) $\vec{L} = \frac{I}{\gamma} \cdot \vec{\omega_s} + I \cdot \omega_z \vec{k'}$. Eliminating $\vec{\omega_s}$ from (6) yields:

(8) $\vec{\omega} = \frac{\gamma}{I} \cdot \vec{L} - (\gamma - 1) \cdot \omega_z \vec{k'}$. Differentiating (8) and employing Eq. (1) yields:

(9) $\vec{\Gamma} = \frac{I}{\gamma} \left[\frac{d\vec{\omega}}{dt} + (\gamma - 1) \cdot \left(\frac{d\omega_z}{dt} \cdot \vec{k'} + \omega_z \cdot \frac{d\vec{k'}}{dt} \right) \right]$, the equation of the gyro.

Recalling that any radian vector $\vec{r}$ that rotates with an angular velocity ω about a unit vector $\vec{u}$ has a linear or tangential velocity $\frac{d\vec{r}}{dt}$ that can be expressed by $\frac{d\vec{r}}{dt} = \vec{\omega} \times \vec{r}$, where $\vec{\omega} = \omega \cdot \vec{u}$. Therefore, we may use the expressions: $\frac{d\vec{k'}}{dt} = \vec{\omega_x} \times \vec{k'}$, $\frac{d\vec{\omega_s}}{dt} = \vec{k'} \times \vec{\omega_s}$, and $\frac{d\vec{L}}{dt} = \vec{k'} \times \vec{L}$ in our calculus.

By multiplying, i.e., by using the dot-product, Eq. (9) by $\vec{k'}$ and taking into account Eq. (6) as well as the fact that $\vec{k'} \cdot (\vec{r} \times \vec{k'}) = 0$ for any vector $\vec{r}$, we may conclude that: $\Gamma_{z'} = \vec{k'} \cdot \vec{\Gamma} = I \cdot \frac{d\omega_{z'}}{dt}$. Therefore, for all external forces that result in $\Gamma_z = 0$, we have: $\frac{d\omega_z}{dt} = 0$, and hence –

(10) $\omega_z = $ const., a very important result indeed.

On account of expressions (10) and (6), Equation (9) reduces to:

(11) $\vec{\Gamma} = \frac{I}{\gamma} \cdot \left(\frac{d\vec{\omega_s}}{dt} + \gamma \cdot \omega_z \frac{d\vec{k'}}{dt} \right)$.

A very important special case results if we neglect the first term in (11) the result of this approximation is the much simpler equation:

(12) $\vec{\Gamma} = I \cdot \omega_z \cdot \frac{d\vec{k'}}{dt}$, referred to as the equation of the precession. It should be noted that in most applications the quantity $\frac{d\vec{\omega_s}}{dt}$ is very small in comparison to the second term in Eq. (11). A further simplification in solving (12) results if we consider the cased where $\left| \vec{\Gamma} \right| = $ const. Then, together with $\omega_z = $ const., we obtain a constant rate of precession, i.e., $\left| \frac{d\vec{k'}}{dt} \right| = $ const.

Next, let's apply these results to a simple model of the gyro-Earth and also illustrate of the most important properties of a gyro.

8.6 The Earth Viewed as a Gyro

Examples: Gyro-Earth—A Simplified Approach

Here we assume that $\vec{\Gamma} = R \cdot \vec{k'} \times (-m_B g_S \vec{k}) = m_B \cdot g_S \cdot R \cdot \vec{k} \times \vec{k'}$, where "R" denotes the radius of the Earth, "m_B" denotes the mass of the Earth's bulges, and "g_S" denotes the gravitational constant related to the mutual attraction of the Sun and Earth.

It follows from the above equation that $\left|\vec{\Gamma}\right| = R \cdot m_B \cdot g_S \sin \phi$ where "ϕ" denotes the angle between $\vec{k}$ and $\vec{k'}$, i.e., the angle between the figure axis of the Earth and the normal to the ecliptic. Our simplified model eliminates the effects of nutation since $\left|\vec{\Gamma}\right| = $ const. implies that $\omega_z = $ const. On the other hand, we can deduce from Eq. (12) that:

$$\vec{\Gamma} = I \cdot \omega_z \cdot \frac{d\vec{k'}}{dt} = I \cdot \omega_z \cdot \vec{\omega_p} \times \vec{k'} = I \cdot \omega_z \cdot \omega_p \vec{k} \times \vec{k'} = m_B g_S R \vec{k} \times \vec{k'},$$

where "$\vec{\omega_p}$" denotes the angular velocity of precession:

(13) $\quad \omega_p = \dfrac{m_B \cdot g_S \cdot R}{I \cdot \omega_z}.$

Of course, the constants $I, m_B,$ and g_S are only known approximately and therefore, numerical values for the precession can only be obtained experimentally. From astronomical observations, it has been determined that:

$$\omega_p = \frac{360°}{25800 \text{ y}} = \frac{50''.23}{\text{year}} = \frac{1°.5918 \cdot 10^{-6}}{\text{hour}}, \text{ as compared to } \omega_z = \frac{15°}{\text{hour}}. \text{ Hence}$$

$\omega_p \ll \omega_z$, indeed.

Gyro Compass—a simplified approach

In order to illustrate the most important aspect of a gyro, let's consider a solid wheel mounted on an axel that is supported by movable bearing on two ends. (See Fig. 8.6.5 below.)

The wheel is now spun around increasing the angular velocity until one of its supports can be removed. Instead of dropping to the floor, its axis, that is now perpendicular to the force of gravity, wanders around the second support that rotates freely in the plane that is parallel to the floor. The mathematical treatment is exactly the same as in the Gyro-Earth example and the rate of precession is again give by Eq. (13). However, in this case, the parameters are known.

In conclusion, we can say that although we know the fundamental Eqs. (9) and (11) of the gyro, we still will need the necessary initial values as wells the knowledge of the resulting torque and the principal moments of inertia in order to solve them.

Of course, finding these parameters is a highly theoretical task reserved for the discipline of Celestial Dynamics. But it should be obvious by now that we will have to employ experimental facts to arrive at viable numerical results involving many

Fig. 8.6.5

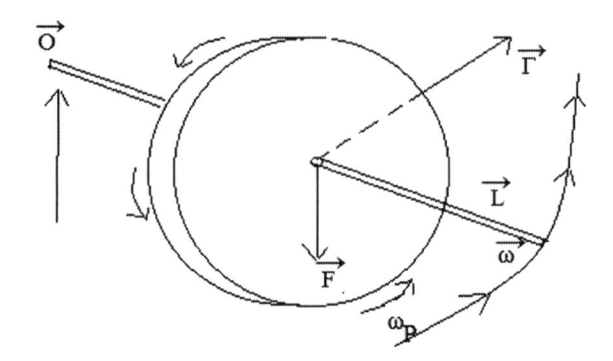

other disciplines, as for instance, geology, oceanography, experimental astronomy and numerical analysis to mention just a few.

So far as we know, at this point of history, exact values for the major-principal moments of inertia are not known and the multi-body of dynamics can only be solved approximately by numerical methods or by series expansions that are based on the theory of perturbation resulting in hundreds of terms necessary to come even close to support the experimental results (see also related literature [74–80]).

Appendix A
Condensed Catalogue for the 57 Navigational Stars and Polaris

EPOCH 2000/12^h:00

Name	Constellation	SAO#	Mag	RA h/m/s	δ_0″/″	μ_α''	μ_δ''	Π''
Acamar	Eridanus	216113	3.24	02/53/15.7	−40/18/17	−0.045	0.018	0.035
Achernar	Eridanus	232481	0.46	01/37/42.9	−57/14/12	0.095	−0.035	0.026
Acrux	Crux	257964	1.33	12/26/35.9	−63/05/57	−0.036	−0.036	0.008
Adhara	Canis Major	172676	1.5	06/58/37.5	−28/58/20	0.004	0.003	0.001
Aldebaran	Taurus	94027	0.85	04/36/55.2	16/30/33	0.063	−0.19	0.048
Alioth	Ursa Major	28553	1.77	12/54/01.7	55/57/35	0.112	−0.006	0.009
Alkaid	Ursa Major	44752	1.86	13/47/32.4	49/18/48	−0.122	−0.011	0.035
Al Nair	Grus	230992	1.74	22/08/14	−46/57/40	0.129	−0.151	0.057
Alnilam	Orion	132346	1.7	05/36/12.8	−01/12/7	0.001	−0.002	0.002
Alphard	Hydra	136871	1.98	09/27/35.2	−08/39/31	−0.014	0.033	0.022
Alphecca	Corona Borealis	83893	2.23	15/34/41.3	26/42/53	0.121	−0.089	0.049
Alpheratz	Andromeda	73765	2.06	00/08/23.3	29/05/26	0.136	−0.163	0.032
Altair	Aquila	125122	0.77	19/50/47	08/52/06	0.538	0.386	0.198
Ankaa	Phoenix	215093	2.39	00/26/17	−42/18/22	0.203	−0.396	0.035
Antares	Scorpius	184415	0.96	16/29/16.4	−26/25/55	−0.01	−0.02	0.024
Arcturus	Boötes	100944	−0.04	14/15/37.7	19/10/57	−1.093	−1.998	0.09
Atria	Triangulum Australe	253700	1.92	16/48/39.9	−69/01/40	0.014	−0.034	0.031
Avior	Carina	235932	1.86	08/22/30.8	−59/30/35	−0.026	0.014	
Bellatrix	Orion	112740	1.64	05/25/07.9	06/20/59	−0.009	−0.014	0.029
Betelgeuse	Orion	113271	0.5	05/55/10.3	07/24/25	0.026	0.009	0.005
Canopus	Carina	234480	−0.72	06/23/57.1	−52/41/45	0.022	0.021	0.028
Capella	Auriga	40186	0.08	05/16/41.4	45/59/59	0.076	−0.425	0.073
Castor	Gemini	60179	1.98	07/34/36	31/53/18	−0.171	−0.098	0.067
Deneb	Cygnus	49941	1.25	20/41/25.9	45/16/49	0.003	0.002	0.006
Denebola	Leo	99809	2.14	11/49/36	14/34/19	−0.407	−0.114	0.082
Diphda	Cetus	147420	2.04	00/43/35.4	−17/59/12	0.234	0.033	0.061
Dubhe	Ursa Major	15384	1.79	11/03/43.7	61/45/03	−0.119	−0.067	0.038
Elnath	Taurus	77168	1.65	05/26/17.5	28/36/27	0.022	−0.175	0.028

(continued)

(continued)

Name	Constellation	SAO#	Mag	RA h/m/s	δ_0/'/''	μ_α''	μ_δ''	Π''
Eltanin	Drago	30653	2.23	17/56/36	51/29/20	−0.008	−0.019	0.025
Enif	Pegasus	127029	2.39	21/44/11.2	09/52/30	0.031	−0.001	0.006
Fomalhaut	Piscis Austrinus	191524	1.16	22/57/39.1	−29/37/20	0.333	−0.165	0.149
Gacrux	Crux	240019	1.63	12/31/09.9	−57/06/48	0.023	−0.262	
Gienah	Corvus	157176	2.59	12/15/48	−17/32/31	−0.161	0.023	
Hadar	Centaurus	252582	0.61	14/03/49.4	−60/22/23	−0.032	−0.019	0.009
Hamal	Aries	75157	2.00	02/07/10.4	23/27/45	0.19	−0.148	0.048
Kaus Australis	Sagittarius	210091	1.85	18/24/10.3	−34/23/0.5	0.038	−0.124	0.023
Kochab	Ursa Minor	8102	2.08	14/50/42.3	74/09/20	−0.031	0.012	0.039
Markab	Pegasus	108378	2.49	23/04/45.7	15/12/19	0.063	−0.042	0.038
Menkar	Cetus	110920	2.53	09/02/16.8	04/05/23	−0.009	−0.078	0.009
Miaplacidus	Carina	250495	1.68	09/13/12	−69/43/02	−0.162	0.108	0.021
Mirfak	Perseus	38787	1.79	03/24/19.4	49/57/40	0.024	−0.025	0.016
Nunki	Sagittarius	187448	2.02	18/55/15.9	−26/17/48	−0.013	−0.054	
Peacock	Pavo	246574	1.94	20/25/38.9	−56/44/06	0.007	−0.089	
Pollux	Gemini	78666	1.14	07/45/18.9	28/01/34	−0.628	−0.046	0.091
Procyon	Canis Minor	115756	0.38	07/39/18.1	05/13/30	−0.71	−1.023	0.288
Raselhague	Ophiuchus	102932	2.08	17/34/56.1	12/33/38	0.12	−0.226	0.067
Regulus	Leo	98967	1.35	10/08/22.3	11/58/02	−0.248	0.006	0.045
Rigel	Orion	131907	0.12	05/14/32.3	−08/12/06	0.00	−0.001	0.013
Rigel Kentaur	Centaurus	252838	−0.01	14/39/35.9	−60/50/07	−3.642	0.699	0.751
Sabik	Ophiuchus	160332	2.43	17/10/22.7	−15/43/29	0.039	0.098	0.052
Schedar	Cassiopeia	21609	2.23	00/40/30.5	56/32/14	0.053	−0.032	0.015
Shaula	Scorpius	208954	1.63	17/33/36.5	−37/06/14	−0.001	−0.029	
Sirius	Canis Major	151881	−1.46	06/45/08	−16/42/58	−0.553	−1.205	0.375
Spica	Virgo	157923	0.98	13/25/11.6	−11/09/41	−0.041	−0.028	0.023
Suhail	Vela	219504	1.78	08/09/32	−47/20/12	−0.004	0.006	0.017
Vega	Lyra	67174	0.03	18/36/56.3	38/47/01	0.202	0.286	0.123
Zubenelgenubi	Libra	158840	2.75	14/50/52.1	−16/02/30	−0.106	−0.067	0.05
POLARIS	Ursa Minor	308	2.02	02/31/48.7	89/15/51	0.038	−0.015	0.00

Appendix B
Greek Alphabet

A α a	Alpha	N ν	Nu
B β б	Beta	Ξ ξ	Xi
Γ γ	Gamma	O o	Omicron
Δ δ	Delta	Π π ϖ	Pi
E ε	Epsilon	P ρ	Rho
Z ζ	Zeta	Σ σ ς	Sigma
H η	Eta	T τ	Tau
Θ θ ϑ	Theta	Υ υ	Upsilon
I ι	Iota	Φ φ φ	Phi
K κ	Kappa	X χ	Chi
Λ λ	Lambda	Ψ ψ	Psi
M μ	Mu	Ω ω	Omega

Appendix C
Star Charts

North Sky Chart

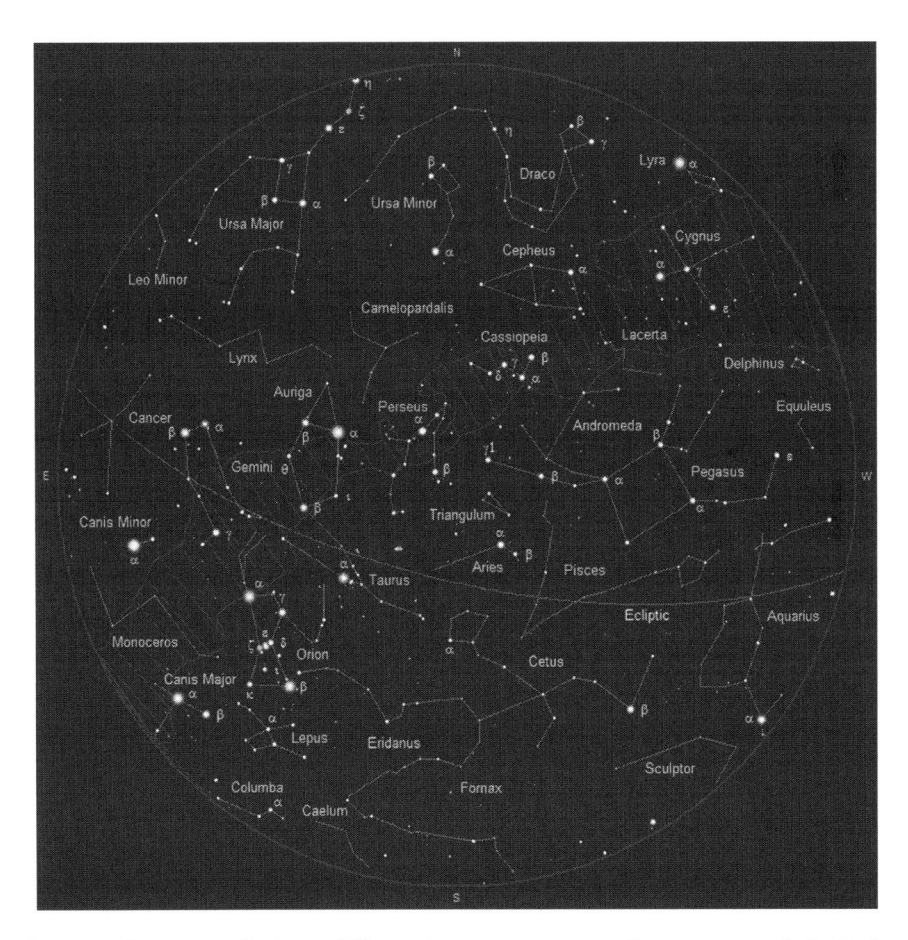

Image credit www.mapsharing.org/MS-maps/map-pages-space-map/2-space-map-north-sky.html

© Springer International Publishing AG 2018
K.A. Zischka, *Astronavigation*,
DOI 10.1007/978-3-319-47994-1

320   Appendix C: Star Charts

South Sky Chart

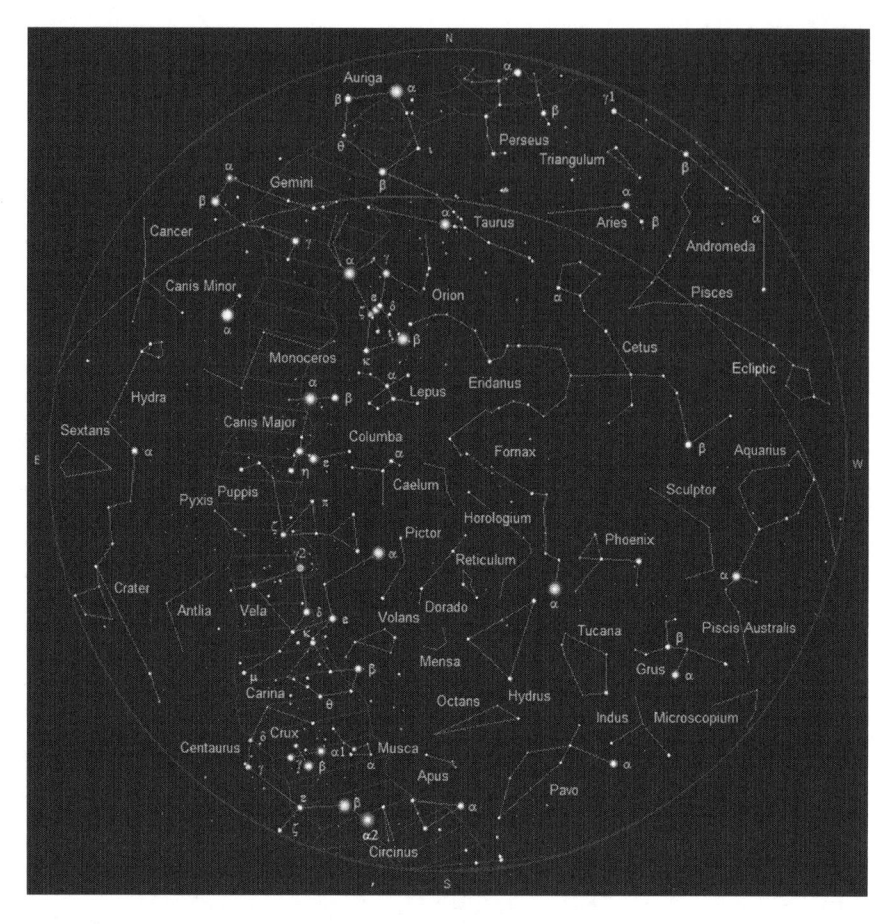

Image credit www.mapsharing.org/MS-maps/map-pages-space-map/3-space-map-south-sky.html

References

1. Admiralty Navigation Manual, vol. III, Her Majesty's Stationary Office, London
2. Air Navigation, AF Manual 51-40, vol. I, U.S. Government Printing Office, Washington, DC
3. Abalakin 1985-Refraction Tables of Pulkova Observatory, 5th edn. Nauka, Leningrad
4. Auer, L., Standish, E.H.: Astronomical refraction-computation method for all zenith angles. Astronomical J. **119**, 2472–2424
5. Baillard, M.-J.: Astronomical refractions for the HP-41, the museum of HP calculators (2006)
6. Ball, R.S.: A Treatise of Spherical Astronomy. Cambridge University Press (1915)
7. Becker, F.: Einführung in die Astronomie. Bibliographisches Institut, Mannheim (1960)
8. Bowditch, Nathaniel, American Practical Navigator, DMAHC, Published No. 9, GPO., vols. I and II, pp. 1977/75
9. Burlirsch, R., Rutishauser, H.: Mathematische Hilfsmittel des Ingenieur. Springer, Berlin (1965)
10. Burch, D.: Emergency Navigation. International Marine Publishing Company, Camden, Maine (1986)
11. Burch, D.: Celestial Navigation with the 2102-D Star Finder—A Users Guide. Paradise Cay Publications, San Francisco, CA (1986)
12. Casey, J.: A Treatise on Spherical Trigonometry and its Applications to Geodesy and Astronomy. Wexford College Press (2008)
13. Chauvenet, W.: A Manual of Spherical and Practical Astronomy, vol. I. Dover Publications Inc, New York (1906)
14. Cotter, C.H.: The Astronomical and Mathematical Foundation of Geography. American Elsevier Publishers Inc, New York (1966)
15. Conell, P.H.: Refraction Tables Arranged for use at the Royal Observatory Greenwich, Greenwich Observations for 1898 (1900). Appendix I, Her Majesty Stationary Office, London
16. Dutton's Navigation and Piloting. United States Naval Institute, Annapolis, Maryland (1956)
17. Hoskinson, A.J., Duerksen,A.: Manual of Geodetic Astronomy. United States Government Printing Office, Washington, DC (1952)
18. Hildebrand, F.B.: Introduction to Numerical Analysis, 2nd edn. McGraw-Hill Book Company, New York (1974)
19. Kennedy, J.B., Neville, A.M.: Basic Statistical Methods for Engineers and Scientists, 2nd edn. A Dun-Donnelley Publisher, New York
20. Kennedy, M.: Understanding Map Projections. ESRI Press Editor (2001)
21. Kurzynska, K.: On the accuracy of the refraction tables of Pulkova-observatory, 5th edn. Astron. Nachr. 309,213 (1988)

22. Kurzynska, K.: Precision in determination of astronomical refraction from aerological data. Astron. Nachr. **308**, 323 (1987)
23. Kurzynska, K.: Local effect in pure astronomical refraction. Astron. Nachr. **309**, 57 (1988)
24. Letcher, J.S. Jr.: Self-contained celestial navigation with H.O. 208. International Marine Publishing Company, Camden, Maine.
25. Lipcanu, M.: A direct method for the calculation of astronomical refraction. In: Proceedings for the Romanian Academy, Series A, vol. 6
26. McNally, D.: Positional Astronomy. Müller Educational, London (1974)
27. Moody A. B.: Celestial Navigation in the Computer Age. Van Nostrand Reinhold Book (1992)
28. Karl, J.K.: Celestial Navigation in the GPS Age. Van Nostrand Reinhold Book (1982)
29. Mueller, I.I.: Spherical and Practical Astronomy as Applied to Geodesy. Continuum International Publishing Group (1969)
30. Newcomb, Simon: A Compendium of Spherical Astronomy, vol. 1906. The Macmillan Company, New York (2008)
31. Nautical Almanac, Commercial Edition. Paradise Cay Publications, Inc.
32. Press, E.: Understanding Map Projections. ESRI Press Editor, AR ArcGIS9 (2004)
33. Davidson, N.: Sky Phenomena, SteinerBooks (2004)
34. Pulkova, : Refraction Tables of Pulkova Observatory, 4th edn. Academy of Science Press, Moscow, Leningrad (1956)
35. Ralston, A., Rabinowitz, P.: A First Course in Numerical Analysis. McGraw-Hill Book Company, New York (1978)
36. Richardson, M.: Plane and Spherical Trigonometry. The Macmillan Company, New York
37. Schroeder, W.: Practical Astronomy. Werner Laurie Ltd., London (1955)
38. Seidelmann, K.P.: Explanatory Supplement to the Astronomical Almanc. University Science Books, Sausalito, CA (2006)
39. Snyder, J.P.: Flattening the Earth. University of Chicago Press, (1997)
40. Snyder, J.P.: Map Projections—A Working Manual. University of Michigan Library (1987)
41. Smart, W.M.: Textbook on Spherical Astronomy. Cambridge University Press, London (1986)
42. Taff, L.G., Taff, R.: Computational Spherical Astronomy. John Wiley & Sons (1981)
43. The Astronomical Ephemeris and Nautical Almanac, Nautical Almanac Office. United States Naval Observatory, Washington, DC
44. The Almanac for Computers: Nautical Almanac Office. United States Naval Observatory, Washington, DC (1987)
45. Young, A.T.: Understanding astronomical refraction. Observatory **126**, 1191 (2006)

Amendments

46. Fischer, D.: Latitude Hooks and Azimuth Rings. International Marine Publisher, Camden Maine (1995)
47. Wough, A.E.: Sundials, Their Theory and Construction. Dover Publications, New York (1975)
48. Rohr, R.R.J.: Sundials History, Theory and Practice. Dover Publications, New York (1965)
49. Moffit, F.H., Bouchard, H.: Surveying. Intext Educational Publisher, New York (1975)
50. Attwood, Cl: Logarithms, Anti-logarithms, and Logarithmic Trigonometrical Functions, Also, Trigonometrical Tables and Formulae. Pergamon Press, New York (1958)
51. Watkins, R., Janiczek, P.M.: Sight reduction with matrices. Navigation 27
52. DeWil, C.: Optimal estimation of a multi-star fix. Navigation 21
53. DeWil, C.: Some remarks on sight reduction with matrices. Navigation 26
54. Boulet, D.: Methods of Orbit Determination for the Micro Computer. Willian-Bell (1991)

Astronomy

55. Becker, F.: Einführung in de Astronomie, B.R. Mannheim (1960)
56. Chauvenet, W.: Manual of Spherical and Practical Astronomy. Dover Pblication (1891)
57. Eichhorn, H.: Astronomy of Star Positions. Ungar, N.Y. (1974)
58. Green, R. M.: Spherical Astronomy. Cambridge University Press, (1985)
59. McNally, D.: Positional Astronomy. Muller Educational (1974)
60. Newcomb, S.A.: Compendium of Spherical Astronomy. The McMillan Co., (1906)
61. Nautical Almanac, Commercial Edition, Paradise Cay Publication (2013)
62. Seidelman, P. K.: Explanatory Supplement to the Astronomical Almanac (2006)
63. Smart, W.M.: Textbook on Spherical Astronomy. Cambridge University Press (1962)
64. Taff, L.G.: Computational Spherical Astronomy. John Wiley & Sons (1981)
65. Woolard, E.W., Clemence, G.M.: Spherical Astronomy. Academic Press, N.Y. (1966)

Astronomical Algorithms

66. Almanac for Computers; U.S. Naval Observatory, Washington, D.C. (1985)
67. Bretagnon, P., Simon, L.: Planetary Programs and Tables. Willmann-Bell (1986)
68. Fras, A.J.: Mathematical Astronomy with a Pocket Calculator. John Wiley & Sons (1978)
69. Kolbe, G.: Long Term Almanac 2000–2080. Star Path Publication (2008)
70. Montenbruck, O.: Practical Ephemeris Calculations. Springer (1987)
71. Montenbruck, O.: Astronomy on the Personal Computer. Springer (1989)
72. Meeus, J.: Astronomical Formulae for Calculators. Willmann-Bell (1982)
73. Meeus, J.: Astronomical Algorithms. Willmann-Bell (1991)
74. Smith, P. D.: Practical Astronomy with your Calculator. Cambridge University Press (1979)

Celestial Mechanics

75. Brouwer, D., Clemence, G.H.: Methods of Celestial Mechanics. Academic Press (1991)
76. Bucerius, H., Schneider, M.: Himmelsmechanic, vols. I and II, B. I. Mannheim (1955)
77. Grammel, R.: Der Kreisel, Berlin (1950)
78. Joos, G.: Lehrbuch der Theoretischen Physik, Akademische Verlagsgesellschaft (1959)
79. Plummer, H.C.: An Introductory Treatise on Dynamical Astronomy. Dover Publication (1918)
80. Szebehely, V.G.: Adventures in Celestial Mechanics. University of Texas Press (1989)
81. Boulet, D.L.: Methods of Orbit Determination for the Micro-Computer (1991)

Catalogues

82. The Bright Star Catalogue, 5th Revised Edition, Year 2000
83. Sky 2000 Astrophysical Observatory's Star Catalogue
84. Smithsonian Astrophysical Observatory's Star Catalogue
85. Combined General Catalogue of Variable Stars

Index

A

Aberration, 258, 266, 269, 270
Addition Theorem, 212
Air Force Manual 2, 130
Algorithm, 9, 10, 60, 68, 74, 76, 87, 94, 97, 102, 115, 159, 160, 170, 222, 227, 230, 260, 261, 273, 289, 291, 295, 297, 299, 300, 306
Altitude, 37, 38, 43, 44, 49, 52–55, 57, 60, 66, 69, 71, 77, 78, 80, 83, 87, 94, 103, 105, 108, 110, 112, 115, 116, 120, 121, 123–125, 131, 132, 135, 139, 142, 147, 150–152, 154, 157, 158, 164, 167, 170, 173, 177, 179–181, 183, 185, 191, 196, 201, 204–206, 208, 273
Altitude Measurements, 44, 57, 173
Analogue Formula, 38, 39, 111
Apparent altitude, 196, 200
Apparent position, 191, 194, 266
Apparent Solar Time (AST), 198
Approximate position, 43, 44, 60, 64, 69, 70, 93, 94, 103, 105, 106, 164, 197
Archimedes, 17, 32, 139
Aries, 36, 40, 62, 125, 148, 154, 156, 197, 199, 241–244, 246, 254, 256, 259, 260, 265, 293
Astrolabe, 144, 232, 233, 238
Atomic Time Signal, 150
Azimuth, 29, 37, 38, 42–44, 46, 50, 60, 61, 71, 72, 79, 81, 83, 84, 90, 94, 103, 105–107, 110, 111, 113–116, 120, 123, 125, 135–137, 139, 142, 148, 158, 159, 164, 167, 171, 195, 205–207, 231

B

Bernoulli's Law, 233
Blunder, 55, 56, 67, 72, 76, 78, 91

Boyle-Mariotte's Law, 186, 187
Bubble sextant, 131, 132, 147, 189, 208, 231, 235

C

Capt. Cook, 17
Capt. Vancouver, 17
Cartesian Coordinates, 63–66, 68, 281
Cassini's Model, 178
Celestial equator, 36, 142, 241, 242, 259
Celestial Object (CO), 27, 34, 36, 37, 43–45, 52, 55, 60, 64–66, 68, 77, 78, 91, 116, 125, 133–135, 140–143, 146, 197, 231, 242, 259, 260, 266, 269, 271, 272
54 Celestial Poles, 36, 283
Celestial sphere, 34, 36, 39, 44, 87, 139, 144, 243, 246, 259, 266, 268, 272, 275, 279, 285, 287, 288, 290, 292
Centroidal point, 200, 203
Chebychev approximation, 182, 183
Civil Calendar, 199
Compressed low accuracy ephemeris, 306
Computed Time, 149, 168, 197, 199
Convergent approximations, 77
Coordinated Universal Time (UTC), 198, 252
Coriolis Correction, 191
Coriolis Force, 189, 208
Cosine Theorem (COS-TH), 18

D

Dead-Reckoning (DR), 29, 131
Dead Reckoning Position (DRP), 80, 128, 150
Declination, 36, 38, 42, 43, 61, 62, 65, 77, 112, 113, 115–118, 125, 134, 136, 137, 142, 143, 150–152, 155, 156, 158–160, 164, 167, 172, 231, 241–243, 249–252, 258,

260, 266, 267, 269–271, 273, 275, 276, 279–285, 287, 289, 291–293, 295, 297, 299, 305
Differential Equation of Refraction, 176
Dip, 73, 138, 147, 173, 183, 185–188, 190, 191, 206, 232, 273

E
Eccentricity, 198, 246, 252–254, 257, 265
Ecliptic System of Coordinates, 34, 39
Einstein, 17, 33
Elliptic Geometry, 32
Equation of Calculated Time (ECT), 149, 157, 159–161, 170
Equation of the center, 241, 246, 248, 252, 257, 258, 306
Equation of the Equinoxes (EE), 197
Equation of the Great-Circle Route, 21, 22
Equation of the Vertical, 192, 194
Equation of Time, 119, 120, 136, 198, 241, 245, 248–250, 256, 257
Equatorial System of Coordinates, 37
Equidistant Azimuthal Projection, 144
Error analysis, 44, 55, 64, 67, 68, 70, 73, 77, 94, 97, 114, 200
Euclidean Geometry, 17, 18, 30–33, 145
Exact position, 44, 71, 77, 79, 87, 95, 103, 116, 148, 190, 268

F
First Branch, 41
Fix, 11, 29, 55, 94, 130, 143, 264, 297
Flattening of the Earth, 17
Formula for Noon, 152
Formula for the Magnitude, 140
Four-Parts Formula, 213, 215, 251
Fundamental Equations of Navigation (FEN), v, xiii

G
Gauss Curve, 201
Gnomonic Navigation, 20
GPS, 34, 74, 80, 131, 135, 147, 169, 234
Great-Circle Navigation, 20, 23, 44
Greenwich Hour Angle (GHA), 36, 45, 46, 48, 61, 62, 65, 67, 71, 77, 90, 100, 105, 116, 142, 154–156, 158, 198, 199, 241–245, 251, 252, 258, 265, 273, 292, 295, 299
Greenwich Mean Time (GMT), 135–137, 148, 149, 151, 155, 198, 252

H
Horizon System of Coordinates, 34, 142

I
Ill Conditioned, 54, 66, 77
Illumination, 139, 140
Incompatible Parameters (ICP), 66, 76
Inner Product, 64, 66, 92
Instrument Error, 60, 191, 273
Integral Representation of Refraction, 176
Intercept, 82, 83
Intermediate Accuracy Ephemeris, 297, 303
Intermediate Value Theorem, 226

J
Julian Calendar, 257, 260

K
Kamal, 147, 180, 181
Kepler's Equation, 246, 248
King-Pin Formula, 219

L
Lagrange's Formula, 178
Lagrangen Polynomials, 220, 229
Lambert Conformal Conic Projection Charts, 21
Laplace Formula, 177, 187
Latitude, 3, 4, 6, 7, 10–13, 17, 18, 20, 21, 24, 34, 36, 38, 40, 42, 43, 51, 52, 54–56, 60, 65, 67, 71–73, 75, 76, 79, 86, 88, 90, 94–96, 104, 111–118, 120, 122–126, 136, 137, 142, 148, 151–153, 157, 159, 162, 164, 167, 168, 176, 190, 206, 207, 231, 270, 292
Least-Square Approximations, 67, 91
Linear Regression, 201
Lines of Position (LOP), 11, 27, 29, 30, 53, 54, 60, 64, 67, 73, 77, 79, 81–85, 90, 94, 95, 100, 103, 128–132, 148
Lobachevsky, 17, 18
Local Hour Angle, 38, 42, 43, 56, 84, 104, 116, 118, 125, 142, 151, 152, 154, 197, 198, 231, 252
Local Mean Time (LMT), 134, 136–138, 149, 154–156, 198
Longitude, 3, 7, 11–13, 20, 22, 24, 34–36, 40, 47, 51, 52, 57, 60, 65, 68, 71, 74, 94, 112, 115, 116, 118, 119, 122–127, 133, 135–138, 142, 148, 150, 151, 153, 155–157, 160–162, 164, 166, 168, 242, 243, 246, 249, 254, 265, 266, 282, 283, 288, 291, 292, 304
Longitude of Epoch, 245, 246, 255
Longitude of Perigee, 245, 246, 253
Lost Time, 150, 153

Index

Low Accuracy Ephemeris, 295, 300
Loxodrome Navigation, 7, 9
Lumens, 139
Luminous Flux, 139
Luminous Intensity, 139, 140

M

Magellan, 17
Main Branch, 41, 58, 73, 217
Mariotte-Gay-Lussac's Law, 177, 178, 187
Matrices, 64, 67, 70, 76, 91, 93, 278, 280
Matrix Algebra, 92
Mean anomaly, 241, 245, 246, 249, 252, 255, 265, 283
Mean longitude, 198, 241, 245, 247, 252, 255, 265, 283
Mean Solar Time (MST), 198, 252, 253
Mean Sun, 119, 136, 198, 199, 242, 245, 246, 252
Mercator Plotting Sheets, 3
Meridian passage, 116, 119–123, 151, 154
Most Probable Position (MPP), 60, 68, 128–132, 200

N

Nautical Almanac (NA), 61, 62, 115, 134, 166, 167, 169, 185, 195, 197, 253
Navigational triangle, 29, 30, 34, 36–38, 42, 44, 46, 52, 78, 81, 84, 96, 97, 103, 105, 110, 111, 117, 125, 133, 138, 150, 159
Newton-Cotes Formulae, 228
Newton-Raphson, 161, 179, 226, 227
Non-Convergent Approximations, 77
Non-Mercator Plotting Sheets, 3
Normal Distribution Curve, 201
Normal Equation, 92, 93
Nutation, 197, 258, 260, 265, 266, 272, 273, 275, 281–285, 292, 293, 295, 297, 310, 312

O

Obliquity of Ecliptic, 40, 246, 252, 253, 265, 266

P

Padé approximations, 219, 220, 222, 223, 225
Parallactic angle, 46, 51, 79, 80, 103, 108, 109
Parallax of the Moon, 159, 167
Parallax of the Sun, 138, 170, 194
Planetary Parallax, 191
Plotting sheets, 3, 10–14, 17, 21, 26, 29, 80, 81, 87, 94, 211

Precession, 197, 198, 258, 260, 266, 272, 273, 275, 277, 281–283, 292, 303, 310, 312, 313
Probable Error (E), 132, 185, 200, 201, 205
Probable Error of the Mean, 200
Pythagoras Theorem (Py.-TH), 209

Q

Quadrature Formulae, 228, 229

R

Refraction, 68, 135, 138, 147, 170, 173–188, 191, 207, 208, 228, 230, 232, 273
Regression, 201, 203, 204
Regula Falsi, 60, 89, 90
Rhumb-line, 3, 5–7, 9, 12–14, 17, 20, 21, 23, 44, 124
Riemann, 17, 33
Right Ascension (RA), 36, 155, 160, 242, 245, 249, 258, 275, 281
Robinson Crusoe, 158, 166

S

Scientific calculator, 21, 102, 217, 233, 241, 262, 293, 297
Secant Method, 89, 95, 159, 160
Second Branch, 41
Semi-diameter, 133, 135, 138, 153, 170, 171, 173, 183, 191, 194, 208, 248, 273
Sextant, 43, 52, 123, 131, 142, 147, 148, 152, 166, 167, 171, 173, 184, 189–191, 193, 194, 196, 197, 199, 200, 205, 206, 208, 231, 232
Sextant error, 173, 183, 200
Sidereal Apparent Time (SAT), 197
Sidereal Clock, 128, 148, 198
Sidereal Hour Angle (SHA), 36, 57, 61–63, 71, 76, 113, 125–128, 142, 143, 149, 160, 162, 168, 242, 249, 251, 292, 297, 299, 307
Sidereal Mean Time (SMT), 149, 197, 199, 252, 253, 270
Sidereal Year (SY), 197
Simpson's Rule, 229
Sine Theorem (SIN-TH), 18
Snell's Law, 174, 175, 178, 187
Solar Time, 196
Spherical Geometry, 17, 33
Standard Error, 200, 201, 204, 205
Standard Error of the Mean, 200
Star of Destination, 148, 149
Steradian, 139
Stereographic Projection, 144, 146

Substellar Points (SP), 27, 32, 35, 36, 66, 91, 95
Sumner's Method, 44, 60, 63, 94–96

T
Taylor-Series expansion, 18, 280, 287
Terrestrial Navigation, 8, 14, 27, 34
Theory of Probability, 80, 129, 201–203
Time Completely Lost, 150, 164, 199
Time Lost At Sea, 164
Time of Maximum Altitude (LMA), 116, 117, 119–124, 151, 152
Transverse Mercator Projection, 14, 17, 18
Trapezoidal Rule, 230
Tropical Year (TY), 197, 199, 246
True Longitude of the Sun, 248

U
Universal Time (UT), 197, 198, 200, 242, 243, 252

V
Vector, 63–65, 69, 73, 77, 91–93, 130, 131, 145, 189, 210–214, 257, 268, 277, 279, 309, 311
Vector Representations, 63, 64, 77, 145
Visual Position, 191

W
Way Points, 15, 16, 21, 23, 25

Z
Zenith Stars, 148, 149

Druck

Canon Deutschland Business Services GmbH
Ferdinand-Jühlke-Str. 7
99095 Erfurt